Applications of Nature-Inspired Computing in Renewable Energy Systems

Mohamed Arezki Mellal
M'Hamed Bougara University, Algeria

A volume in the Advances in Environmental Engineering and Green Technologies (AEEGT) Book Series

IGI Global
PUBLISHER of TIMELY KNOWLEDGE

Published in the United States of America by
IGI Global
Engineering Science Reference (an imprint of IGI Global)
701 E. Chocolate Avenue
Hershey PA, USA 17033
Tel: 717-533-8845
Fax: 717-533-8661
E-mail: cust@igi-global.com
Web site: http://www.igi-global.com

Library of Congress Cataloging-in-Publication Data

Names: Mellal, Mohamed Arezki, editor.
Title: Applications of nature-inspired computing in renewable energy systems / Mohamed Arezki Mellal, editor.
Description: Hershey PA : Engineering Science Reference, [2022] | Includes bibliographical references and index. | Summary: "This book discusses the latest research on nature-inspired computing approaches applied to the design and development of renewable energy systems and provides new solutions to the renewable energy domain such as microgrids, wind power, and artificial neural networks"-- Provided by publisher.
Identifiers: LCCN 2021055372 (print) | LCCN 2021055373 (ebook) | ISBN 9781799885610 (hardcover) | ISBN 9781799885627 (paperback) | ISBN 9781799885634 (ebook)
Subjects: LCSH: Renewable energy sources--Data processing. | Natural computation--Industrial applications.
Classification: LCC TJ808 .A66 2022 (print) | LCC TJ808 (ebook) | DDC 621.042--dc23/eng/20211202
LC record available at https://lccn.loc.gov/2021055372
LC ebook record available at https://lccn.loc.gov/2021055373

This book is published in the IGI Global book series Advances in Environmental Engineering and Green Technologies (AEEGT) (ISSN: 2326-9162; eISSN: 2326-9170)

British Cataloguing in Publication Data
A Cataloguing in Publication record for this book is available from the British Library.

For electronic access to this publication, please contact: eresources@igi-global.com.

Advances in Environmental Engineering and Green Technologies (AEEGT) Book Series

Sang-Bing Tsai
Zhongshan Institute, University of Electronic Science and Technology of China, China & Wuyi University, China
Ming-Lang Tseng
Lunghwa University of Science and Technology, Taiwan
Yuchi Wang
University of Electronic Science and Technology of China Zhongshan Institute, China

ISSN:2326-9162
EISSN:2326-9170

Mission

Growing awareness and an increased focus on environmental issues such as climate change, energy use, and loss of non-renewable resources have brought about a greater need for research that provides potential solutions to these problems. Research in environmental science and engineering continues to play a vital role in uncovering new opportunities for a "green" future.

The **Advances in Environmental Engineering and Green Technologies (AEEGT)** book series is a mouthpiece for research in all aspects of environmental science, earth science, and green initiatives. This series supports the ongoing research in this field through publishing books that discuss topics within environmental engineering or that deal with the interdisciplinary field of green technologies.

Coverage

- Pollution Management
- Water Supply and Treatment
- Renewable Energy
- Alternative Power Sources
- Policies Involving Green Technologies and Environmental Engineering
- Electric Vehicles
- Green Technology
- Contaminated Site Remediation
- Waste Management
- Radioactive Waste Treatment

IGI Global is currently accepting manuscripts for publication within this series. To submit a proposal for a volume in this series, please contact our Acquisition Editors at Acquisitions@igi-global.com or visit: http://www.igi-global.com/publish/.

The Advances in Environmental Engineering and Green Technologies (AEEGT) Book Series (ISSN 2326-9162) is published by IGI Global, 701 E. Chocolate Avenue, Hershey, PA 17033-1240, USA, www.igi-global.com. This series is composed of titles available for purchase individually; each title is edited to be contextually exclusive from any other title within the series. For pricing and ordering information please visit http://www.igi-global.com/book-series/advances-environmental-engineering-green-technologies/73679. Postmaster: Send all address changes to above address.

Titles in this Series

For a list of additional titles in this series, please visit: www.igi-global.com/book-series

Modeling and Control of Static Converters for Hybrid Storage Systems
Arezki Fekik (Akli Mohand Oulhadj University, Bouira, Algeria) and Nacereddine Benamrouche (University of Tizi Ouzou, Aleria)
Engineering Science Reference • © 2022 • 355pp • H/C (ISBN: 9781799874478) • US $225.00

Disruptive Technologies and Eco-Innovation for Sustainable Development
Ulas Akkucuk (Boğaziçi University, Trkey)
Engineering Science Reference • © 2022 • 340pp • H/C (ISBN: 9781799889007) • US $195.00

Handbook of Research on Microbial Remediation and Microbial Biotechnology for Sustainable Soil
Junaid Ahmad Malik (Government Degree College, Bijbehara, India)
Engineering Science Reference • © 2021 • 806pp • H/C (ISBN: 9781799870623) • US $295.00

Handbook of Research on Strategic Management for Current Energy Investments
Serhat Yüksel (Istanbul Medipol University, Turkey) and Hasan Dinçer (Istanbul Medipol University, Turkey)
Engineering Science Reference • © 2021 • 422pp • H/C (ISBN: 9781799883357) • US $295.00

Forest Fire Danger Prediction Using Deterministic-Probabilistic Approach
Nikolay Viktorovich Baranovskiy (National Research Tomsk Polytechnic University, Russia)
Engineering Science Reference • © 2021 • 297pp • H/C (ISBN: 9781799872504) • US $195.00

Role of IoT in Green Energy Systems
Vasaki Ponnusamy (Universiti Tunku Abdul Rahman, Malaysia) Noor Zaman (Taylor's University, Malaysia) Low Tang Jung (Universiti Teknologi PETRONAS, Malaysia) and Anang Hudaya Muhamad Amin (Higher Colleges of Technology, UAE)
Engineering Science Reference • © 2021 • 405pp • H/C (ISBN: 9781799867098) • US $195.00

Combating and Controlling Nagana and Tick-Borne Diseases in Livestock
Caleb Oburu Orenge (Department of Veterinary Anatomy and Physiology, Faculty of Veterinary Medicine and Surgery, Egerton University, Kenya)
Engineering Science Reference • © 2021 • 440pp • H/C (ISBN: 9781799864332) • US $185.00

Artificial Intelligence and IoT-Based Technologies for Sustainable Farming and Smart Agriculture
Pradeep Tomar (Gautam Buddha University, India) and Gurjit Kaur (Delhi Technological University, India)
Engineering Science Reference • © 2021 • 400pp • H/C (ISBN: 9781799817222) • US $215.00

IGI Global
PUBLISHER of TIMELY KNOWLEDGE

701 East Chocolate Avenue, Hershey, PA 17033, USA
Tel: 717-533-8845 x100 • Fax: 717-533-8661
E-Mail: cust@igi-global.com • www.igi-global.com

List of Reviewers

Tahir Cetin Akinci, *Istanbul Teknik Universitesi, Turkey*
Sameer Al-Dahidi, *German Jordanian University, Jordan*
R. Ganesh Babu, *SRM TRP Engineering College, India*
K. Balachander, *Karpagam University, India*
Paula Bastida-Molina, *Universitat Politècnica de València, Spain*
Aaqib Bulla, *National Institute of Technology, India*
Harpreet Kaur Channi, *Chandigarh University, India*
Madeleine Combrinck, *Northumbria University, UK*
Bishwajit Dey, *IIT ISM Dhanbad, India*
Nitesh Funde, *Soongsil University, India*
Wai Shin Ho, *Universiti Teknologi Malaysia, Malaysia*
Kittisak Jermsittiparsert, *Chulalongkorn University, Thailand*
Md Mustafa Kamal, *Aligarh Muslim University, India*
Rajeev Kumar, *Krishna Institute of Engineering and Technology, India*
Siva Mangipudi Kumar, *Gudlavalleru Engineering College, India*
Praveen Laws, *NYU Abu Dhabi, UAE*
Mojtaba Nedaei, *Islamic Azad University, Iran*
Provas Kumar Roy, *Kalyani Government Engineering College, India*
Yendaluru Raja Sekhar, *Vellore Institute of Technology, India*
M. Selvi, *Vellore Institute of Technology, India*
Jan Shair, *Tsinghua University, China*
Bashar Shboul, *University of Sheffield, UK*
K. Mohana Sundaram, *KPR Institute of Engineering and Technology, India*
Naser Taheri, *Technical and Vocational University, Iran*
Edward J. Williams, *University of Michigan, USA*

Table of Contents

Preface xv

Chapter 1
Some Words About Nature-Inspired Computing 1
Mohamed Arezki Mellal, M'Hamed Bougara University, Algeria

Chapter 2
Nature-Inspired Approach Using Seasonal Comparison of Wind Speed With Spectral and Statistical Analysis 10
Tahir Cetin Akinci, Istanbul Technical University, Turkey & University of California, Riverside, USA
Ramazan Caglar, Istanbul Technical University, Turkey
Gokhan Erdemir, Istanbul Sabahhattin Zaim University, Turkey
Aydin Tarik Zengin, Istanbul Sabahattin Zaim University, Turkey
Serhat Seker, Istanbul Technical University, Turkey

Chapter 3
Nature-Inspired Algorithm Applied to a Renewable Energy-Integrating Hydro-Thermal Power Plant 21
Sunanda Hazra, Central Institute of Petrochemicals Engineering and Technology, India
Provas Kumar Roy, Kalyani Government Engineering College, India

Chapter 4
Ant Colony Optimization Algorithm for Electrical Power Systems Applications: A Literature Review 37
Ragab A. El-Sehiemy, Kafrelsheikh University, Egypt
Almoataz Y. Abdelaziz, Ain Shams University, Egypt

Chapter 5
Application of Optimization to Sizing Renewable Energy Systems and Energy Management in Microgrids: State of the Art and Trends 60
Khaled Dassa, University of Boumerdes, Algeria
Abdelmadjid Recioui, University of Boumerdes, Algeria

Chapter 6
Artificial Neural Network: A New Tool for the Prediction of Hydrate Formation Conditions 95
Anupama Kumari, Indian Institute of Technology, Roorkee, India
Mukund Madhaw, DHC Trading India Private Limited, Delhi, India
C. B. Majumder, Indian Institute of Technology, Roorkee, India
Amit Arora, Shaheed Bhagat Singh State University, Ferozepur, India

Chapter 7
DNA Computing: Future of Renewable Smart Computation Systems 116
Mandrita Mondal, Indian Statistical Institute, Kolkata, India

Chapter 8
Hybrid Optimization Methods Application on Sizing and Solving the Economic Dispatch Problems of Hybrid Renewable Power Systems 136
Muhammet Tahir Guneser, Karabuk University, Turkey
Abdurazaq Elbaz, Karabuk University, Turkey
Cihat Seker, Karabuk University, Turkey

Chapter 9
Multi-Objective Optimal Performance of a Hybrid CPSD-SE/HWT System for Microgrid Power Generation 166
Bashar Shboul, The University of Sheffield, UK & Al al-Bayt University, Jordan
Ismail Al-Arfi, The University of Sheffield, UK
Stavros Michailos, The University of Sheffield, UK
Derek Ingham, The University of Sheffield, UK
Godfrey T. Udeh, The University of Sheffield, UK & University of Port Harcourt, Nigeria
Lin Ma, The University of Sheffield, UK
Kevin Hughes, The University of Sheffield, UK
Mohamed Pourkashanian, The University of Sheffield, UK

Chapter 10
Numerical and Experimental Investigations on a Bio-Inspired Design of Darrieus Vertical Axis Wind Turbine Blades With Leading Edge Tubercles 211
Nishant Mishra, Shiv Nadar University, India
Punit Prakash, Shiv Nadar University, India
Anand Sagar Gupta, Shiv Nadar University, India
Jishnav Dawar, Shiv Nadar University, India
Alok Kumar, Shiv Nadar University, India
Santanu Mitra, Shiv Nadar University, India

Chapter 11
Computational Analysis of Symmetric and Cambered Blade Darrieus Vertical Axis Wind Turbine With Bio-Mimicked Blade Design 225
Punit Prakash, Shiv Nadar University, India
Praveen Laws, New York University, Abu Dhabi, UAE
Nishant Mishra, Shiv Nadar University, India
Santanu Mitra, Shiv Nadar University, India

Chapter 12
Study on Reliability Design of the Domestic Compressor Subjected to Repetitive Internal Stresses by Parametric Accelerated Life Testing 241
Seongwoo Woo, Ethiopian Technical University, Ethiopia
Dennis L. O'Neal, Baylor University, USA
Yimer Mohammed Hassen, Ethiopian Technical University, Ethiopia

Chapter 13
Energy Resources and Their Consumption 267
Harpreet Kaur Channi, Chandigarh University, India

Compilation of References 277

About the Contributors 319

Index 325

Detailed Table of Contents

Preface xv

Chapter 1
Some Words About Nature-Inspired Computing 1
Mohamed Arezki Mellal, M'Hamed Bougara University, Algeria

The use of artificial intelligence (AI) in various domains has drastically increased during the last decade. Nature-inspired computing is a strong computing approach that belongs to AI and covers a wide range of techniques. It has successfully tackled many complex problems and outperformed several classical techniques. This chapter provides the original ideas behind some nature-inspired computing techniques and their applications, such as the genetic algorithms, particle swarm optimization, grey wolf optimizer, ant colony optimization, plant propagation algorithm, cuckoo optimization algorithm, and artificial neural networks.

Chapter 2
Nature-Inspired Approach Using Seasonal Comparison of Wind Speed With Spectral and Statistical Analysis 10
Tahir Cetin Akinci, Istanbul Technical University, Turkey & University of California, Riverside, USA
Ramazan Caglar, Istanbul Technical University, Turkey
Gokhan Erdemir, Istanbul Sabahhattin Zaim University, Turkey
Aydin Tarik Zengin, Istanbul Sabahattin Zaim University, Turkey
Serhat Seker, Istanbul Technical University, Turkey

Seasonal analysis of wind speed includes elements of its evaluation and analysis for wind energy production in complex geographical areas. These analyses require wind energy systems to be set up, integrated, operated, and designed according to seasonal differences. Istanbul wind speed data were collected hourly and analyzed seasonally. When the results of the analysis are examined, no significant increase in seasonal transitions was observed, while certain changes were observed between summer and winter. Here, statistical analysis, Weibull distribution function, and signal processing-based PSD analysis for wind speed is performed. In addition, correlation analysis was made between the seasons. Although significant results were obtained in signal-based analyses, results were obtained for seasonal transitions in correlation analyses. Seasonal spectral densities were calculated in the spectral analysis of wind speed data. This study has important implications in terms of extraction of seasonal characteristics of wind speed, resource assessment, operation, investment, and feasibility.

Chapter 3
Nature-Inspired Algorithm Applied to a Renewable Energy-Integrating Hydro-Thermal Power Plant 21
Sunanda Hazra, Central Institute of Petrochemicals Engineering and Technology, India
Provas Kumar Roy, Kalyani Government Engineering College, India

Due to the rising requirement on energy sources and the global doubts for using fossil fuel because of its consequences on the climate changes and the global warming caused by hazardous gases, the scientific research has shifted to the renewable energy. To minimize the usage of thermal power generation plants and to meet the rising load demand, a thermal-integrated wind-hydro-system is taking an important role in renewable power systems. A proficient nature-inspired optimization is proposed for solving economic and emission dispatch for the hydro-thermal-wind (HTW) scheduling problem. Further, the opposition-based learning have been incorporated with the chemical reaction optimization for improving the performance of the algorithm. To investigate the performance of oppositional chemical reaction optimization algorithm, the algorithm is tested on two different cases. Along with this, some statistical tests have also been performed. The results obtained by the OCRO algorithm are compared with other recently proposed methods to establish its robustness.

Chapter 4
Ant Colony Optimization Algorithm for Electrical Power Systems Applications: A Literature Review 37
Ragab A. El-Sehiemy, Kafrelsheikh University, Egypt
Almoataz Y. Abdelaziz, Ain Shams University, Egypt

Optimization has been an active area of research for several decades. As many real-world optimization problems become increasingly complex, better optimization algorithms are always needed. Recently, meta-heuristic global optimization algorithms have become a popular choice for solving complex and intricate problems, which are otherwise difficult to solve by traditional methods. This chapter reviews the recent applications of ant colony optimization (ACO) algorithm in the field of electrical power systems. Also, the progress of the ACO algorithm and its recent developments are discussed. This chapter covers the aspects like (1) basics of ACO algorithm, (2) progress of ACO algorithm, (3) classification of electrical power system applications, and (4) future of ACO for modern power systems application.

Chapter 5
Application of Optimization to Sizing Renewable Energy Systems and Energy Management in Microgrids: State of the Art and Trends 60
Khaled Dassa, University of Boumerdes, Algeria
Abdelmadjid Recioui, University of Boumerdes, Algeria

The smart grid is the aggregation of emerging technologies in both hardware and software along with practices to make the existing power grid more reliable and ultimately more beneficial to consumers. The smart grid concept is associated with the production of electricity from renewable energy sources (RES). For the distant isolated regions, microgrids (MG) with RES are offering a suitable solution for remote and isolated region electrification. The improper sizing would lead to huge investment cost which could have been avoided. The objective of this chapter is to review the state-of-the-art studies on the use of optimization techniques to renewable energy design and sizing. The chapter reviews the optimization techniques employed at different components of the microgrid including the energy sources, storage elements, and converters/inverters with their control systems.

Chapter 6
Artificial Neural Network: A New Tool for the Prediction of Hydrate Formation Conditions 95
Anupama Kumari, Indian Institute of Technology, Roorkee, India
Mukund Madhaw, DHC Trading India Private Limited, Delhi, India
C. B. Majumder, Indian Institute of Technology, Roorkee, India
Amit Arora, Shaheed Bhagat Singh State University, Ferozepur, India

The analysis and collection of data is an integral part of all research fields of the modern world. There is a need to perform forward mathematical modeling to improve the operations and calculations with modern technologies. Artificial neural network signifies the structure of the human brain. They can provide reasonable solutions quickly for the problems that classical programming cannot solve. An in-depth systematic study is presented in this chapter related to artificial neural network applications (ANN) for predicting the equilibrium conditions for gas hydrate formation, which can assist in designing future dissociation technology for gas hydrate so that this white gold can make world energy free for the future generation. This chapter can also help to develop a novel inhibitor for gas hydrate formation and save millions of dollars for the oil and gas industry.

Chapter 7
DNA Computing: Future of Renewable Smart Computation Systems.. 116
Mandrita Mondal, Indian Statistical Institute, Kolkata, India

The modern era of classical silicon-based computing is at the edge of a number of technological challenges which include huge energy consumption, requirement of massive memory space, and generation of e-waste. The proposed alternative to this pitfall is nanocomputing, which was first exemplified in the form of DNA computing. Recently, DNA computing is gaining acceptance in the field of eco-friendly, unconventional, nature-inspired computation. The future of computing depends on making it renewable, as this can cause a drastic improvement in energy consumption. Thus, to save the natural resources and to stop the growing toxicity of the planet, reversibility is being imposed on DNA computing so that it can replace the traditional form of computation. This chapter reflects the foundation of DNA computing and renewability of this multidisciplinary domain that can be produced optimally and run from available natural resources.

Chapter 8
Hybrid Optimization Methods Application on Sizing and Solving the Economic Dispatch Problems of Hybrid Renewable Power Systems.. 136
Muhammet Tahir Guneser, Karabuk University, Turkey
Abdurazaq Elbaz, Karabuk University, Turkey
Cihat Seker, Karabuk University, Turkey

Renewable energy systems are spread all over the world due to the security problems encountered in accessing fossil fuels, the desire to reduce the environmental damage and to respond to the rapid increase in energy demand. However, the problems are experienced in renewable energy technologies in sustainable supply and reduction of production costs. Obtaining the optimum power distribution planning between photovoltaic, wind, biomass, and other systems depending on the relevant parameters and optimizing the distribution of energy supply-demand planning among the same sources can be applied as an effective solution by using several single optimization methods or new updated hybrid versions of them. In this chapter, common methods were evaluated and an application of crow and particle swarm as a hybrid method was examined in a certain region of Libya for a PV/wind hybrid renewable power system.

Chapter 9
Multi-Objective Optimal Performance of a Hybrid CPSD-SE/HWT System for Microgrid Power Generation 166
Bashar Shboul, The University of Sheffield, UK & Al al-Bayt University, Jordan
Ismail Al-Arfi, The University of Sheffield, UK
Stavros Michailos, The University of Sheffield, UK
Derek Ingham, The University of Sheffield, UK
Godfrey T. Udeh, The University of Sheffield, UK & University of Port Harcourt, Nigeria
Lin Ma, The University of Sheffield, UK
Kevin Hughes, The University of Sheffield, UK
Mohamed Pourkashanian, The University of Sheffield, UK

A new integrated hybrid solar thermal and wind-based microgrid power system is proposed. It consists of a concentrated parabolic solar dish Stirling engine, a wind turbine, and a battery bank. The electrical power curtailment is diminished, and the levelised cost of energy is significantly reduced. To achieve these goals, the present study conducts a dynamic performance analysis over one year of operation. Further, a multi-objective optimisation model based on a genetic algorithm is implemented to optimise the techno-economic performance. The MATLAB/Simulink® software was used to model the system, study the performance under various operating conditions, and optimise the proposed hybrid system. Finally, the model has been implemented for a specific case study in Mafraq, Jordan. The system satisfies a net power output of 1500 kWe. The developed model has been validated using published results. In conclusion, the obtained results reveal that the optimised model of the microgrid can substantially improve the overall efficiency and reduce the levelised cost of electricity.

Chapter 10
Numerical and Experimental Investigations on a Bio-Inspired Design of Darrieus Vertical Axis Wind Turbine Blades With Leading Edge Tubercles 211
Nishant Mishra, Shiv Nadar University, India
Punit Prakash, Shiv Nadar University, India
Anand Sagar Gupta, Shiv Nadar University, India
Jishnav Dawar, Shiv Nadar University, India
Alok Kumar, Shiv Nadar University, India
Santanu Mitra, Shiv Nadar University, India

Various improvements can be made to Darrieus vertical axis wind turbines (VAWT) for maximum performance in an urban environment. One such improvement is the inclusion of bio-inspired leading-edge tubercles to increase the aerodynamic performance. These structures, found on the flippers of humpback whales, are believed to aid the mammal in quick maneuvering. The objective of the chapter is to investigate and compare the performance of a Darrieus type VAWT with the inclusion of leading edge tubercles. The performance of the turbine with leading-edge tubercles on the blades is compared with the turbine with normal blade, computationally (with computational fluid dynamics using transition SST turbulence model) and experimentally. The focus lies on building an experimental setup to compare the performance of leading-edge tubercles with the baseline turbine.

Chapter 11
Computational Analysis of Symmetric and Cambered Blade Darrieus Vertical Axis Wind Turbine With Bio-Mimicked Blade Design 225
Punit Prakash, Shiv Nadar University, India
Praveen Laws, New York University, Abu Dhabi, UAE
Nishant Mishra, Shiv Nadar University, India
Santanu Mitra, Shiv Nadar University, India

Vertical axis wind turbine suffers from low performance, and the need for improvement is a challenge. This work addresses this problem by using computational fluid dynamics. This chapter aims to analyze and compare symmetric and cambered Darrieus turbine. These analyses are usually carried for straight leading-edge blades, and cambered resembles more the natural shape of the wing of birds and other aquatic mammals, which helps them generate extra lift during movement. Moreover, recent studies suggest better performance was observed for NACA0018 symmetric aerofoil blades, and a similar trend has been observed for NACA2412 cambered aerofoil profiles. Turbine models having symmetric NACA0018 and cambered NACA2412 profiles have been studied. By comparing the symmetric model with cambered blade models, differences in coefficient of torque have been presented. OpenFOAM is used for performing the 2D simulation with dynamicOverset-FvMesh for motion solver with overset mesh method. Meshed geometry was constructed with GMSH codes and the simulation uses overPimpleDyMFoam algorithm as a solver.

Chapter 12
Study on Reliability Design of the Domestic Compressor Subjected to Repetitive Internal Stresses by Parametric Accelerated Life Testing 241
Seongwoo Woo, Ethiopian Technical University, Ethiopia
Dennis L. O'Neal, Baylor University, USA
Yimer Mohammed Hassen, Ethiopian Technical University, Ethiopia

This chapter explains the parametric accelerated life testing (ALT) to recognize design defects in mechanical products. A life-stress model and a sample size formulation are suggested. A compressor is used to demonstrate this method. Compressors were failing in the field. At the first ALT, the compressor failed due to a fractured suction reed valve. The failure modes were similar to those valves returned from the field. The fatigue of the suction reed valves came from an overlap between the suction reed valve and the valve plate. The problematic design was modified by the trespan dimensions, tumbling process, a ball peening, and brushing process for the valve plate. At the second ALT, the compressor locked due to the intrusion between the crankshaft and thrust washer. The corrective action plan performed the heat treatment to the exterior of the crankshaft made of cast iron. After the design modifications, there were no troubles during the third ALT. The lifetime of compressor was secured to have a B1 life 10 years.

Chapter 13
Energy Resources and Their Consumption.. 267
Harpreet Kaur Channi, Chandigarh University, India

Power is a significant cause of economic growth and crucial to the sustainability of the economy. Energy consumption is an indicator of a nation's economic growth. Economic growth is focused, among other aspects, on the long-term acquisition of affordable, existing resources, and their use does not pollute the environment. Industrialization serves economic growth and consumes energy. In 2018, 68% of total capital power was consumed by largest energy-intensive areas. When fossil fuel is the primary source of energy, energy consumption is positively correlated with ecosystem cleanliness. Fossil fuels account for more than 70% of the decent energy expectations of India and other economies. In this chapter, problems related to non-renewable energy sources are discussed, and emphasis is given to use more renewable sources.

Compilation of References 277

About the Contributors 319

Index... 325

Preface

Renewable energy is crucial to preserve the environment. This energy involves various systems that must be optimized and assessed to provide better performance; however, the design and development of renewable energy systems remains a challenge. It is crucial to implement the latest innovative research in the field in order to develop and improve renewable energy systems.

Applications of Nature-Inspired Computing in Renewable Energy Systems discusses the latest research on nature-inspired computing approaches applied to the design and development of renewable energy systems and provides new solutions to the renewable energy domain. Covering topics such as microgrids, wind power, and artificial neural networks, it is ideal for engineers, industry professionals, researchers, academicians, practitioners, teachers, and students.

Chapter 1
Some Words About Nature-Inspired Computing

Mohamed Arezki Mellal
https://orcid.org/0000-0003-0667-8851
M'Hamed Bougara University, Algeria

ABSTRACT

The use of artificial intelligence (AI) in various domains has drastically increased during the last decade. Nature-inspired computing is a strong computing approach that belongs to AI and covers a wide range of techniques. It has successfully tackled many complex problems and outperformed several classical techniques. This chapter provides the original ideas behind some nature-inspired computing techniques and their applications, such as the genetic algorithms, particle swarm optimization, grey wolf optimizer, ant colony optimization, plant propagation algorithm, cuckoo optimization algorithm, and artificial neural networks.

1. INTRODUCTION

Nowadays, artificial intelligence (AI) is largely used in almost all domains. It includes various approaches, notably nature-inspired algorithms. These algorithms involve mathematical models inspired by the principles of biology, such as human anatomy and evolution, animals and insects, and by the principles of natural phenomena, such as fractals, water cycle, and galaxy gravity. Stochastic models are used in this kind of algorithms to explore and exploit the search space for finding the optimal solutions in a reasonable number of iterations and time. Many mathematical and engineering benchmarks have been investigated to prove the effectiveness of these algorithms. Despite these algorithms having some limitations, the advantages are much more numerous and real-world applications are reaffirming this matter, notably in renewable energy systems, such as fuel cells, photovoltaic cells, wind turbines, biomass, geothermal, and hybrid systems.

A comprehensive illustration of nature-inspired algorithms may take the whole book; therefore, this chapter presents the basic principles and applications of some of them. It is organized as follows. Sec-

DOI: 10.4018/978-1-7998-8561-0.ch001

tion 2 includes the definition of some algorithms. Section 3 highlights a listing of applications. Finally, the last section concludes the chapter.

2. PRINCIPLES OF SOME NATURE-INSPIRED ALGORITHMS

2.1. Genetic Algorithms

The genetic algorithms (GAs) are one of the most popular, oldest, and pioneer nature-inspired optimization algorithms. John Henry Holland developed the GAs in 1975 (Holland, 1975) based on genetic operators: chromosomes, parents, children, population, selection, crossover, and mutation. Many types of each operator can be used to solve a particular problem. The detailed principles can be found in Refs. (Holland, 1975; Sivanandam & Deepa, 2008; Sumathi & Paneerselvam, 2010).

2.2. Ant colony Optimization

The ant colony optimization (ACO) was developed by Marco Dorigo in 1992 (M. Dorigo, 1992). It is based on the ants' behavior for foraging. The ants left a pheromone on the ground and based on its intensity and evaporation process, the optimal path will be identified. It allows finding the shortest path. The detailed principles can be found in Refs. (M. Dorigo, 1992; Marco Dorigo, Maniezzo, & Colorni, 1996; M. A. Mellal & Williams, 2018).

2.3. Particle Swarm Optimization

The particle swarm optimization (PSO) is inspired by the principles of the moving style of some species, such as birds and fishes. It was developed by Kennedy and Eberhart in 1995 (Kennedy & Eberhart, 1995). Each element of the swarm, called a particle, has a position and a velocity. The particles move randomly and the best positions are computed and updated. Comprehensive details about the PSO can be found in Refs. (Clerc, 2006; Kennedy & Eberhart, 1995; M. A. Mellal & Williams, 2018).

2.4. Grey Wolf Optimizer

The grey wolf optimizer (GWO) was developed by Mirjalili et al. in 2014 (Mirjalili, Mirjalili, & Lewis, 2014) and is inspired by the grey wolves. This algorithm is reputed to be efficient and fast. In the GWO, the wolves are divided into four hierarchies, called alpha, beta, delta, and omega. These wolves follow some rules to hunt. Details can be found in Refs. (Mohamed Arezki Mellal, Frik, & Boutiche, 2021; Mirjalili et al., 2014; Panda & Das, 2019).

2.5. Plant Propagation Algorithm

The plant propagation algorithm (PPA), also called the strawberry algorithm, is inspired by the propagation mechanism of some plants, such as the strawberry plant. It was developed by Salhi and Fraga in 2011 (Salhi & Fraga, 2011), based on the runners of these plants. This algorithm is characterized by a few parameters to be tuned. A comprehensive presentation on the PPA can be found in Refs. (Salhi & Fraga, 2011; Sulaiman, Salhi, Selamoglu, & Kirikchi, 2014).

2.6. Cuckoo Optimization Algorithm

The cuckoo optimization algorithm (COA) was developed by Ramin Rajabioun in 2011 to deal with PID tuning in control systems (Rajabioun, 2011). The basic principles are based on the cuckoos' birds in terms of feeding, reproduction, and migration to the best habitat. The cuckoos are particular birds, as they don't build their proposer nests and lay their eggs in the nests of other bird species. The COA has been largely and successfully implemented to solve various engineering problems. More information about the COA can be found in Refs. (M. A. Mellal & Williams, 2017; Rajabioun, 2011).

Table 1. Some applications of the cited nature-inspired computing techniques.

Technique	Applications
GAs	- PV-Wind-Battery Storage (Adefarati et al., 2019). - Hybrid wind and solar renewable energy (Geleta & Manshahia, 2021). - Design of hybrid renewable energy systems (Ismail, Moghavvemi, & Mahlia, 2014). - Design and placement of wind turbines (Grady, Hussaini, & Abdullah, 2005; Mellal & Pecht, 2020). - Flow shop scheduling (Umam, Mustafid, & Suryono, 2021). - Image processing (Guo, Peng, & Tang, 2016). - Replacement of obsolete components (Mellal, Adjerid, Benazzouz, Berrazouane, & Williams, 2013).
ACO	- Energy storage (Deshun, Yumeng, Qiong, Jinhua, & Jelei, 2018). - Photovoltaic power forecasting (Pan et al., 2020). - Frequency control in hydropower plants (Singh & Verma, 2020). - Economic load dispatch (Srivastava & Singh, 2020). - Obstacle avoidance (Yang, Yang, Li, Zhang, & Kang, 2021). - Maintenance optimization (Zhou, Zhang, Lin, & Ma, 2013). - COVID-19 image segmentation (Liu et al., 2021).
PSO	- Sizing of renewable energy systems (Alshammari & Asumadu, 2020). - Scheduling in microgrid (Das, De, & Mandal, 2021). - Design of PV cells (Fan et al., 2022). - Economic load dispatch (Srivastava & Singh, 2020). - Machining processes (Quarto, D'Urso, & Giardini, 2022). - Camera calibration (Lü, Meng, Long, & Wang, 2021). - Reliability optimization (Mellal & Zio, 2019b).
GWO	- Hybrid wind and solar renewable energy (Geleta & Manshahia, 2021). - Biomass gasification for electricity generation (Musharavati, Khoshnevisan, Alirahmi, Ahmadi, & Khanmohammadi, 2022). - Parameters of PV (Xavier, Pradeep, Premkumar, & Kumar, 2021). - Reliability optimization (Mellal, Frik, et al., 2021). - Design of car side safety system (Hamadache & Mellal, 2021). - Budget allocation (Fu, Xiao, Lee, & Huang, 2021).
PPA	- Distribution network (Waqar et al., 2020). - Release management of water from reservoir (Asvini & Amudha, 2016). - Reliability and availability optimization (Mellal & Salhi, 2021a, 2021b; Mellal, Salhi, & Márquez, 2021).
COA	- Combined heat and power economic dispatch Mellal & Williams, 2015, 2020). - PID controller (Rajabioun, 2011). - Machining processes (Mellal & Williams, 2016). - Reliability and availability optimization (Mellal & Zio, 2019a; Mellal, Al-Dahidi, & Williams, 2020; Mellal & Zio, 2020). - Replacement of obsolete components (Mellal, Adjerid, Benazzouz, 2012; Mellal & Williams, 2019).
ANNs	- Forecasting in renewable energy systems (Azadeh, Babazadeh, & Asadzadeh, 2013; Purwanto, Hermawan, Suherman, Widodo, & Iksan, 2021; Wu et al., 2012; Wu, Park, Choi, Cha, & Lee, 2009). - Transportation system (Choi & Kim, 2021). - Virtual reality (Alkadri et al., 2021).

2.7. Artificial Neural Networks

The artificial neural networks (ANNs) are one of the oldest nature-inspired computing techniques, inspired by the functioning mechanism of the neural networks of the human brain. This technique was initialized in 1943 by McCulloch and Pitts (McCulloch & Pitts, 1943). It involves inputs and outputs through layers for learning from data and it is a very good alternative for problems without a specific mathematical model. Details and construction of the ANNs can be found in Refs. (Gurney, 1997; Park & Lek, 2016).

3. SOME APPLICATIONS

During the last decades, the development of computing machines has contributed to the improvement and development of nature-inspired computing techniques. These techniques have been applied to a wide range of engineering problems and showed their effectiveness. Table 1 illustrates some applications of the cited techniques.

4. CONCLUSION

Nature-inspired computing techniques are very diverse and have a great impact on solving various and complex engineering problems. Most of these techniques include stochastic models which allow to explore and exploit the searching space and to find adequate solutions with a reduced computing time and cost. The choice of the algorithm depends on the problem and the target of the used. In the literature, a daily evolution of these algorithms can be observed to tackle some special problems and to deal with their respective limitations.

REFERENCES

Adefarati, T., Potgieter, S., Bansal, R. C., Naidoo, R., Rizzo, R., & Sanjeevikumar, P. (2019). Optimization of PV-Wind-battery storage microgrid system utilizing a genetic algorithm. *ICCEP 2019 - 7th International Conference on Clean Electrical Power: Renewable Energy Resources Impact*, 633–638. 10.1109/ICCEP.2019.8890204

Alkadri, S., Ledwos, N., Mirchi, N., Reich, A., Yilmaz, R., Driscoll, M., & Del Maestro, R. F. (2021). Utilizing a multilayer perceptron artificial neural network to assess a virtual reality surgical procedure. *Computers in Biology and Medicine*, *136*, 104770. doi:10.1016/j.compbiomed.2021.104770 PMID:34426170

Alshammari, N., & Asumadu, J. (2020). Optimum unit sizing of hybrid renewable energy system utilizing harmony search, Jaya and particle swarm optimization algorithms. *Sustainable Cities and Society*, *60*, 102255. doi:10.1016/j.scs.2020.102255

Asvini, M. S., & Amudha, T. (2016). An efficient methodology for reservoir release optimization using plant propagation algorithm. *Procedia Computer Science*, *93*, 1061–1069. doi:10.1016/j.procs.2016.07.310

Azadeh, A., Babazadeh, R., & Asadzadeh, S. M. (2013). Optimum estimation and forecasting of renewable energy consumption by artificial neural networks. *Renewable & Sustainable Energy Reviews*, *27*, 605–612. doi:10.1016/j.rser.2013.07.007

Choi, S., & Kim, Y. J. (2021). Artificial neural network models for airport capacity prediction. *Journal of Air Transport Management*, *97*, 102146. doi:10.1016/j.jairtraman.2021.102146

Clerc, M. (2006). Particle swarm optimization. Particle Swarm Optimization. doi:10.1002/9780470612163

Das, G., De, M., & Mandal, K. K. (2021). Multi-objective optimization of hybrid renewable energy system by using novel autonomic soft computing techniques. *Computers & Electrical Engineering*, *94*, 107350. doi:10.1016/j.compeleceng.2021.107350

Deshun, W., Yumeng, Z., Qiong, T., Jinhua, X., & Jelei, Y. (2018). Research on planning and configuration of multi-objective energy storage system solved by improved ant colony algorithm. *China International Conference on Electricity Distribution*, 2279–2283. 10.1109/CICED.2018.8592157

Dorigo, M. (1992). *Optimization, learning and natural algorithms* (PhD Thesis). Politecnico di Milano, Italy.

Dorigo, M., Maniezzo, V., & Colorni, A. (1996). Ant system: Optimization by a colony of cooperating agents. *IEEE Transactions on Systems, Man, and Cybernetics. Part B, Cybernetics*, *26*(1), 29–41. doi:10.1109/3477.484436 PMID:18263004

Fan, Y., Wang, P., Heidari, A. A., Chen, H., HamzaTurabieh, & Mafarja, M. (2022). Random reselection particle swarm optimization for optimal design of solar photovoltaic modules. *Energy*, *239*, 121865. doi:10.1016/j.energy.2021.121865

Fu, Y., Xiao, H., Lee, L. H., & Huang, M. (2021). Stochastic optimization using grey wolf optimization with optimal computing budget allocation. *Applied Soft Computing*, *103*, 107154. doi:10.1016/j.asoc.2021.107154

Geleta, D. K., & Manshahia, M. S. (2021). A hybrid of grey wolf optimization and genetic algorithm for optimization of hybrid wind and solar renewable energy system. *Journal of the Operations Research Society of China*, 1–14. doi:10.1007/s40305-021-00341-0

Grady, S. A., Hussaini, M. Y., & Abdullah, M. M. (2005). Placement of wind turbines using genetic algorithms. *Renewable Energy*, *30*(2), 259–270. doi:10.1016/j.renene.2004.05.007

Guo, F., Peng, H., & Tang, J. (2016). Genetic algorithm-based parameter selection approach to single image defogging. *Information Processing Letters*, *116*(10), 595–602. doi:10.1016/j.ipl.2016.04.013

Gurney, K. (1997). *An introduction to neural networks*. CRC Press. doi:10.4324/9780203451519

Hamadache, I., & Mellal, M. A. (2021). Design optimization of car side safety system by particle swarm optimization and grey wolf optimizer. In M. A. Mellal & G. M. Pecht (Eds.), *Nature-Inspired Computing Paradigms in Systems: Reliability, Availability, Maintainability, Safety and Cost (RAMS+C) and Prognostics and Health Management (PHM)*. Elsevier. doi:10.1016/B978-0-12-823749-6.00006-4

Holland, J. H. (1975). Adaptation in Natural and Artificial Systems. University of Michigan Press. doi:10.1137/1018105

Ismail, M. S., Moghavvemi, M., & Mahlia, T. M. I. (2014). Genetic algorithm based optimization on modeling and design of hybrid renewable energy systems. *Energy Conversion and Management, 85*, 120–130. doi:10.1016/j.enconman.2014.05.064

Kennedy, J., & Eberhart, R. (1995). Particle swarm optimization. *Neural Networks, 1995. Proceedings, IEEE International Conference On, 4*, 1942–1948. 10.1109/ICNN.1995.488968

Liu, L., Zhao, D., Yu, F., Heidari, A. A., Li, C., Ouyang, J., Chen, H., Mafarja, M., Turabieh, H., & Pan, J. (2021). Ant colony optimization with Cauchy and greedy Levy mutations for multilevel COVID 19 X-ray image segmentation. *Computers in Biology and Medicine, 136*, 104609. doi:10.1016/j.compbiomed.2021.104609 PMID:34293587

Lü, X., Meng, L., Long, L., & Wang, P. (2021). Comprehensive improvement of camera calibration based on mutation particle swarm optimization. *Measurement, 110303*. Advance online publication. doi:10.1016/j.measurement.2021.110303

McCulloch, W. S., & Pitts, W. (1943). A logical calculus of the ideas immanent in nervous activity. *The Bulletin of Mathematical Biophysics, 5*(4), 115–133. doi:10.1007/BF02478259

Mellal, M. A., & Williams, E. J. (2019). Optimal replacement strategy of obsolete industrial components under fuzzy data. *Proceedings of the Institution of Mechanical Engineers. Part I: Journal of Systems and Control Engineering*. doi:10.1177/0959651819853483

Mellal, M. A., Frik, A., & Boutiche, R. (2021). Reliability optimization of power plant safety system using grey wolf optimizer and shuffled frog-leaping algorithm. In Nature-Inspired Computing Paradigms in Systems: Reliability, Availability, Maintainability, Safety and Cost (RAMS+C) and Prognostics and Health Management (PHM) (pp. 1–13). doi:10.1016/B978-0-12-823749-6.00008-8

Mellal, M. A., Salhi, A., & Márquez, F. P. G. (2021). System availability and cost optimization under failure dependencies by flower pollination and plant propagation algorithms. *Fifteenth International Conference on Management Science and Engineering, 78*, 469–476. 10.1007/978-3-030-79203-9_36

Mellal, M. A., Adjerid, S., & Benazzouz, D. W. E. J. (2012). Fault detection-isolation and tolerance control for maintenance operators: A bond graph methodology. CIMGLE'2012, Oran, Algeria.

Mellal, M. A., Adjerid, S., Benazzouz, D., Berrazouane, S., & Williams, E. J. (2013). Obsolescence optimization of electronic and mechatronic components by considering dependability and energy consumption. *Journal of Central South University, 20*(5), 1221–1225. doi:10.100711771-013-1605-9

Mellal, M. A., Al-Dahidi, S., & Williams, E. J. (2020). System reliability optimization with heterogeneous components using hosted cuckoo optimization algorithm. *Reliability Engineering & System Safety, 203*, 107110. doi:10.1016/j.ress.2020.107110

Mellal, M. A., & Pecht, M. (2020). A multi-objective design optimization framework for wind turbines under altitude consideration. *Energy Conversion and Management, 222*, 113212. doi:10.1016/j.enconman.2020.113212

Mellal, M. A., & Salhi, A. (2021a). Multi-objective system design optimization via PPA and a fuzzy method. *International Journal of Fuzzy Systems*, *23*(5), 1–9. doi:10.100740815-021-01068-z

Mellal, M. A., & Salhi, A. (2021b, April 30). System reliability-redundancy allocation by the multiobjective plant propagation algorithm. *International Journal of Quality & Reliability Management*. Advance online publication. doi:10.1108/IJQRM-10-2018-0285

Mellal, M. A., & Williams, E. J. (2015). Cuckoo optimization algorithm with penalty function for combined heat and power economic dispatch problem. *Energy*, *93*, 1711–1718. doi:10.1016/j.energy.2015.10.006

Mellal, M. A., & Williams, E. J. (2016). Total production time minimization of a multi-pass milling process via cuckoo optimization algorithm. *International Journal of Advanced Manufacturing Technology*, *87*(1-4), 747–754. Advance online publication. doi:10.100700170-016-8498-3

Mellal, M. A., & Williams, E. J. (2017). The cuckoo optimization algorithm and its applications. In *Handbook of Neural Computation* (pp. 269–277). Elsevier. doi:10.1016/B978-0-12-811318-9.00014-4

Mellal, M. A., & Williams, E. J. (2018). A survey on ant colony optimization, particle swarm optimization, and cuckoo algorithms. In *Handbook of Research on Emergent Applications of Optimization Algorithms*. IGI Global. doi:10.4018/978-1-5225-2990-3.ch002

Mellal, M. A., & Williams, E. J. (2020). Cuckoo optimization algorithm with penalty function and binary approach for combined heat and power economic dispatch problem. *Energy Reports*, *6*, 2720–2723. doi:10.1016/j.egyr.2020.10.004

Mellal, M. A., & Zio, E. (2019a). An adaptive cuckoo optimization algorithm for system design optimization under failure dependencies. *Journal of Risk and Reliability*, *233*(6), 1099–1105. doi:10.1177/1748006X19863639

Mellal, M. A., & Zio, E. (2019b). An adaptive particle swarm optimization method for multi-objective system reliability optimization. *Journal of Risk and Reliability*, *233*(6), 990–1001. doi:10.1177/1748006X19852814

Mellal, M. A., & Zio, E. (2020). System reliability-redundancy optimization with cold-standby strategy by an enhanced nest cuckoo optimization algorithm. *Reliability Engineering & System Safety*, *201*, 106973. doi:10.1016/j.ress.2020.106973

Mirjalili, S., Mirjalili, S. M., & Lewis, A. (2014). Grey Wolf Optimizer. *Advances in Engineering Software*, *69*, 46–61. doi:10.1016/j.advengsoft.2013.12.007

Musharavati, F., Khoshnevisan, A., Alirahmi, S. M., Ahmadi, P., & Khanmohammadi, S. (2022). Multi-objective optimization of a biomass gasification to generate electricity and desalinated water using Grey Wolf Optimizer and artificial neural network. *Chemosphere*, *287*, 131980. doi:10.1016/j.chemosphere.2021.131980 PMID:34509018

Pan, M., Li, C., Gao, R., Huang, Y., You, H., Gu, T., & Qin, F. (2020). Photovoltaic power forecasting based on a support vector machine with improved ant colony optimization. *Journal of Cleaner Production*, *277*, 123948. doi:10.1016/j.jclepro.2020.123948

Panda, M., & Das, B. (2019). Grey wolf optimizer and its applications: A survey. *Lecture Notes in Electrical Engineering*, *556*, 179–194. doi:10.1007/978-981-13-7091-5_17

Park, Y. S., & Lek, S. (2016). Artificial neural networks: Multilayer perceptron for ecological modeling. In Developments in Environmental Modelling (Vol. 28, pp. 123–140). doi:10.1016/B978-0-444-63623-2.00007-4

Purwanto, H., Suherman, W. D. A., & Iksan, N. (2021). Renewable energy generation forecasting on smart home micro grid using deep neural network. *International Conference on Artificial Intelligence and Mechatronics Systems*. 10.1109/AIMS52415.2021.9466089

Quarto, M., D'Urso, G., & Giardini, C. (2022). Micro-EDM optimization through particle swarm algorithm and artificial neural network. *Precision Engineering*, *73*, 63–70. doi:10.1016/j.precisioneng.2021.08.018

Rajabioun, R. (2011). Cuckoo optimization algorithm. *Applied Soft Computing*, *11*(8), 5508–5518. doi:10.1016/j.asoc.2011.05.008

Salhi, A., & Fraga, E. S. (2011). Nature-inspired optimisation approaches and the new plant propagation algorithm. *The International Conference on Numerical Analysis and Optimization.*

Singh, O., & Verma, A. (2020). Frequency control for stand-alone hydro power plants using ant colony optimization. *IEEE International Conference on Advent Trends in Multidisciplinary Research and Innovation*. 10.1109/ICATMRI51801.2020.9398417

Sivanandam, S. N., & Deepa, S. N. (2008). *Introduction to Genetic Algorithms*. Springer Berlin Heidelberg.

Srivastava, A., & Singh, S. (2020). Implementation of ant colony optimization and particle swarm optimization in economic load dispatch problem using renewable source. *2020 IEEE 17th India Council International Conference*. 10.1109/INDICON49873.2020.9342190

Sulaiman, M., Salhi, A., Selamoglu, B. I., & Kirikchi, O. B. (2014). A plant propagation algorithm for constrained engineering optimisation problems. *Mathematical Problems in Engineering*, *2014*, 1–10. Advance online publication. doi:10.1155/2014/627416

Sumathi, S., & Paneerselvam, S. (2010). *Computational intelligence paradigms : Theory & applications using MATLAB*. CRC Press. doi:10.1201/9781439809037

Umam, M. S., Mustafid, M., & Suryono, S. (2021). A hybrid genetic algorithm and tabu search for minimizing makespan in flow shop scheduling problem. *Journal of King Saud University - Computer and Information Sciences*. doi:10.1016/j.jksuci.2021.08.025

Waqar, A., Subramaniam, U., Farzana, K., Elavarasan, R. M., Habib, H. U. R., Zahid, M., & Hossain, E. (2020). Analysis of optimal deployment of several DGs in distribution networks using plant propagation algorithm. *IEEE Access: Practical Innovations, Open Solutions*, *8*, 175546–175562. doi:10.1109/ACCESS.2020.3025782

Wu, J., Zhao, L., Xie, N., Gao, L., Gao, W., Dai, X., & Zhang, J. (2012). Historical weather data supported hybrid renewable energy forecasting using artificial neural network (ANN). *Energy Procedia*, *14*, 1035–1040. doi:10.1016/j.egypro.2011.12.1051

Wu, L., Park, J., Choi, J., Cha, J., & Lee, K. Y. (2009). A study on wind speed prediction using artificial neural network at Jeju Island in Korea. *Transmission and Distribution Conference and Exposition: Asia and Pacific*. doi:10.1109/TD-ASIA.2009.5356873

Xavier, F. J., Pradeep, A., Premkumar, M., & Kumar, C. (2021). Orthogonal learning-based Gray Wolf Optimizer for identifying the uncertain parameters of various photovoltaic models. *Optik (Stuttgart)*, *247*, 167973. doi:10.1016/j.ijleo.2021.167973

Yang, X., Yang, X., Li, Y., Zhang, J., & Kang, Y. (2021). Obstacle avoidance in the improved social force model based on ant colony optimization during pedestrian evacuation. *Physica A*, *583*, 126256. doi:10.1016/j.physa.2021.126256

Zhou, Y., Zhang, Z., Lin, T. R., & Ma, L. (2013). Maintenance optimisation of a multi-state series-parallel system considering economic dependence and state-dependent inspection intervals. *Reliability Engineering & System Safety*, *111*, 248–259. doi:10.1016/j.ress.2012.10.006

Chapter 2
Nature-Inspired Approach Using Seasonal Comparison of Wind Speed With Spectral and Statistical Analysis

Tahir Cetin Akinci
https://orcid.org/0000-0002-4657-6617
Istanbul Technical University, Turkey & University of California, Riverside, USA

Ramazan Caglar
Istanbul Technical University, Turkey

Gokhan Erdemir
https://orcid.org/0000-0003-4095-6333
Istanbul Sabahhattin Zaim University, Turkey

Aydin Tarik Zengin
Istanbul Sabahattin Zaim University, Turkey

Serhat Seker
Istanbul Technical University, Turkey

ABSTRACT

Seasonal analysis of wind speed includes elements of its evaluation and analysis for wind energy production in complex geographical areas. These analyses require wind energy systems to be set up, integrated, operated, and designed according to seasonal differences. Istanbul wind speed data were collected hourly and analyzed seasonally. When the results of the analysis are examined, no significant increase in seasonal transitions was observed, while certain changes were observed between summer and winter. Here, statistical analysis, Weibull distribution function, and signal processing-based PSD analysis for wind speed is performed. In addition, correlation analysis was made between the seasons. Although significant results were obtained in signal-based analyses, results were obtained for seasonal transitions in correlation analyses. Seasonal spectral densities were calculated in the spectral analysis of wind speed data. This study has important implications in terms of extraction of seasonal characteristics of wind speed, resource assessment, operation, investment, and feasibility.

DOI: 10.4018/978-1-7998-8561-0.ch002

1. INTRODUCTION

The rise of global warming to high levels in the world, the increase in environmental pollution by industrialization and the inability to meet the energy demand cause fossil fuels to be replaced by renewable energy [Akanwa, and Joe-Ikechebelu, 2012). Among the renewable energy sources, there are many types of energy, especially wind, solar geothermal, hydrogen and biomass energy (Panwar at al., 2011; Gosunpro, 2021). Renewable energy sources and non-renewable energy sources are given in Figure 1 below.

Figure 1. Classification of Energy Resources

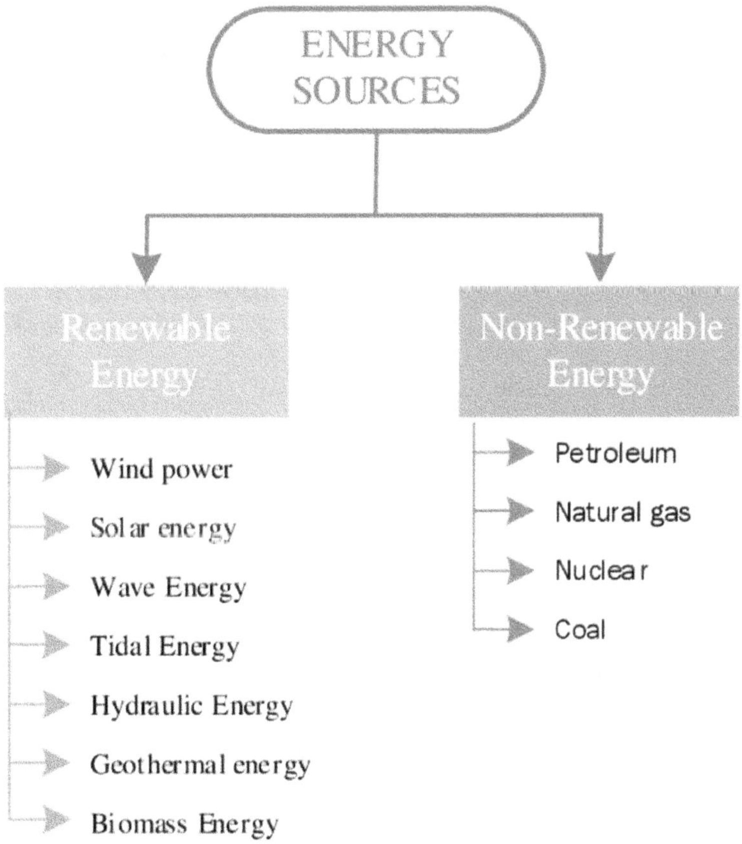

The use of renewable energy resources continues to increase in the world (Gielen at al, 2019; Wuebbles, and Sanyal, 2015; Kumar, 2020). The main reasons for preferring renewable energy production are; being friendly with nature and being inexhaustible (economical). Even initial investment costs in renewable energy sources can often be lower than non-renewable energy generation (Singh, Nyuur, and Richmond, 2018). Countries are increasing their investments in renewable energy resources. In Figure 2, investment statistics for renewable energy generation in the last ten years are given. In addition, governments aim to establish 488 gigawatts of hydroelectric capacity by 2030 in their energy investment programs (Fs-unep, 2020).

Figure 2. Investment in renewable energy sources (2010-2019) (Fs-unep, 2020).

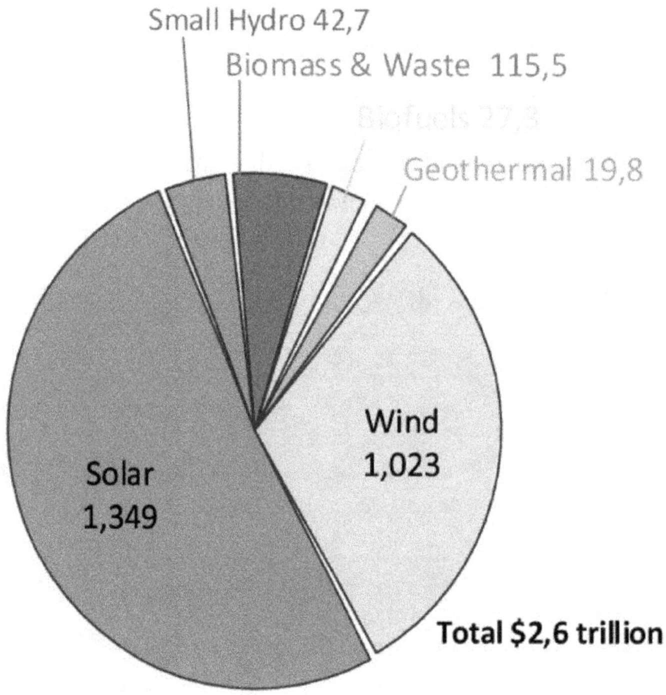

Wind energy is the most widely used type of energy among renewable energy sources (Onar and, Khaligh, 2015; Balat, 2005; Ilkilic, 2012). Advantages of wind energy: It reduces foreign dependency on energy, contributes to the national economy, is obtained naturally, and turbines cause little damage to the environment where they are located, they do not pollute the air since there is no fuel consumption. Disadvantages of wind energy: Since energy generation is directly connected to the wind, it changes seasonally, the noise generated by wind turbines causes noise pollution, disrupts the natural flight routes of birds, may cause electromagnetic pollution, the initial installation costs of the turbines are high, the turbines cover large areas in the area where they are installed (Ilkilic, 2011; Kuik, Branger, and Quirion, 2018; Energy, 2021; Advantages, 2021; Advantages, 2020).

Energy is the key to prosperity, economic development and technological developments in the world. In the energy generation process; One of the most important criteria is that it is clean, does not pollute the environment and does not harm nature. In addition, energy production centres should not be too far from consumption centres, which is important in terms of keeping energy transmission costs and maintenance costs at the lowest levels economically (Ghasemian, 2020).

Power Spectral Density (PSD)

Fourier transform is the main technique used in the frequency analysis. Power Spectral Density (PSD) is deterministic. Also, PSD is very practical, in the calculation of certain types of random signals and Fourier transform implementation. The power spectral density concept is formulated as below (Akinci at al, 2016; PSD, 2021; Akinci at al, 2014):

$$S_x(f) = \lim_{T\to\infty} E\left\{\frac{1}{2T}\left|\int_{-T}^{T} x(t)e^{-j2\pi ft}dt\right|^2\right\} \quad (1)$$

Where E is a mathematical operator of expected value, T is also period. In other writing styles, it is shown as:

$$S_x(f) = \int_{-T}^{T} R_x(\tau)e^{-j2\pi ft}dt \quad (2)$$

Here, $R_x(\tau)$ is also auto-correlation function which is given with the following relationship:

$$R_x(\tau) = E\{x(t)x*(t+\tau)\} \quad (3)$$

Applications

Figure 3 is a map of Turkey's Global given wind speed. In addition, the statistical and probabilistic distributions of Istanbul wind speed data were analyzed in detail in the applications. This application area is connected with the spectral analysis of seasonal wind speed data, as well as statistical analysis, by means of the annual measurement. For this purpose, data is divided into four seasons and power spectral density calculation is applied to extract the frequency properties depending on the seasonal behaviour. Here the frequency values are used as normalized values. As seen in Figure 4, the wind speed changes are presented in the form of the four seasons like spring, summer, autumn and winter times respectively.

Figure 3. Global Wind Atlas Mean Wind Speed Map Turkey (Global, 2021).

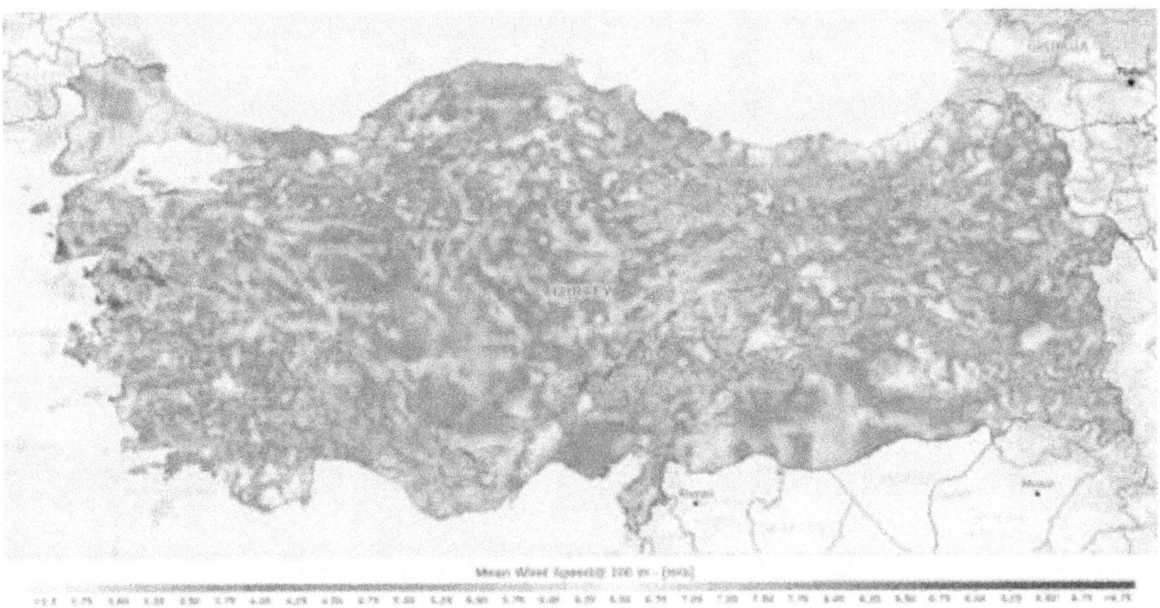

Figure 4. Seasonal Wind Speed Variations.

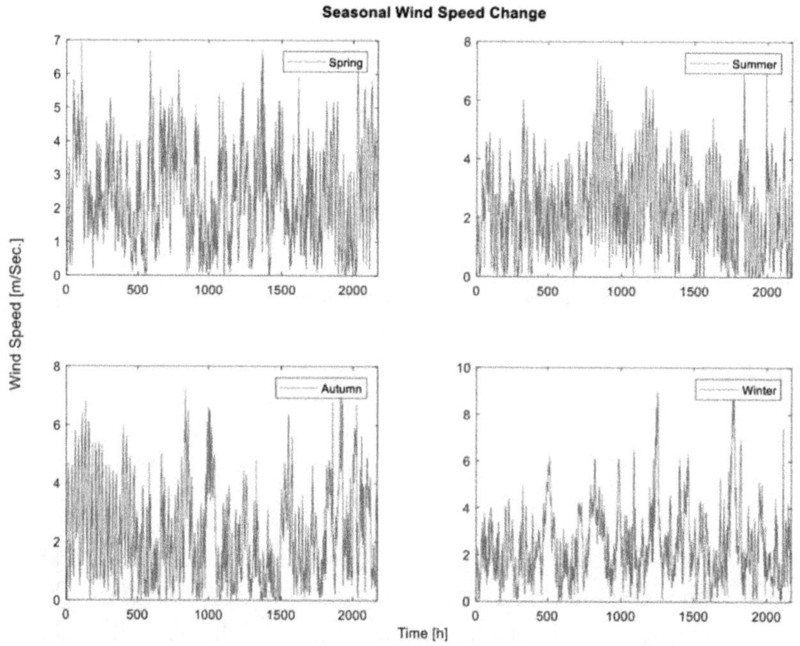

Figure 5. Power Spectrum Density for Wind Speed Data.

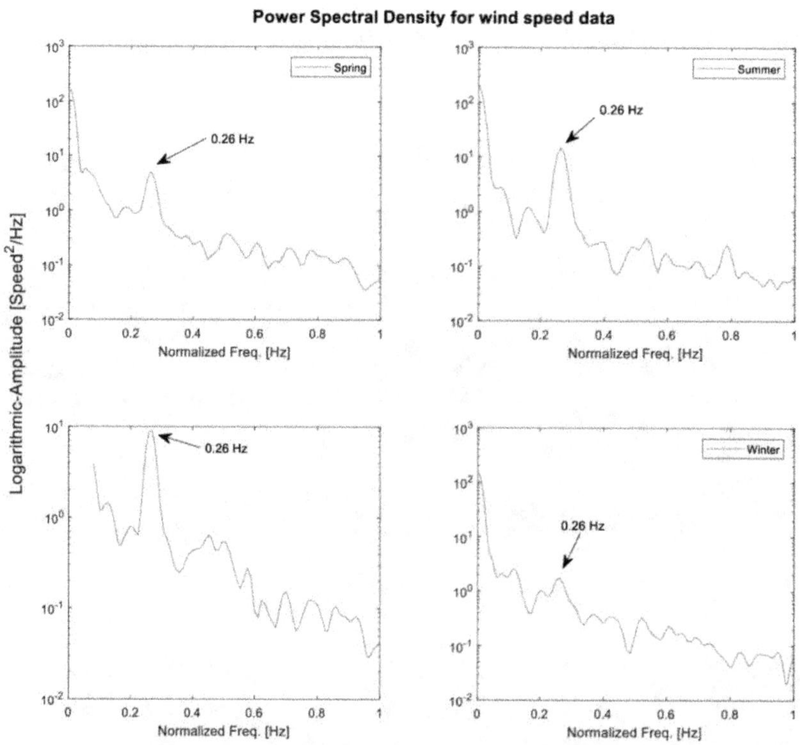

In terms of the power spectral density application, it is presented in Figure 4. Here the indicated frequency components are focused on the 0.26 Hz and 0.6 Hz. Especially, the frequency of 0.26 Hz is common property for four seasons but it is more dominant for spring and summer times. The second one, which is at 0.6 Hz, is highly related to autumn and winter seasons as seen in Figure 4.

In Figure 6, semi-logarithmic PSD is shown for all four seasons. From here it can be seen that they follow each other in all four seasons. This means that the wind frequency does not vary greatly seasonally.

Figure 7 shows the histograms for each season. As known, the statistical distribution of the wind speed data is compatible with the Weibull distribution. For this reason, the maximum points of the distributions are related to the mean wind speeds. Depending on the figure 5 results, these mean winds are 2.4131 m/s, 2.2882 m/s, 2.2429 m/s and 2.4094 m/s for spring, summer, autumn and winter times respectively. Details on statistical distributions are given in Table 1.

Table 1. Statistics on Wind Speed

	Spring	Summer	Autumn	Winter
Mean	2.4131	2.2882	2.2429	2.4094
Standard	1.4334	1.4874	1.5807	1.5505
Skewness	0.3458	0.5007	0.6196	1.1777
Kurtosis	2.3322	2.7228	2.5607	4.8667

Figure 6. Semi Logarithmic PSD for Wind Speed Data.

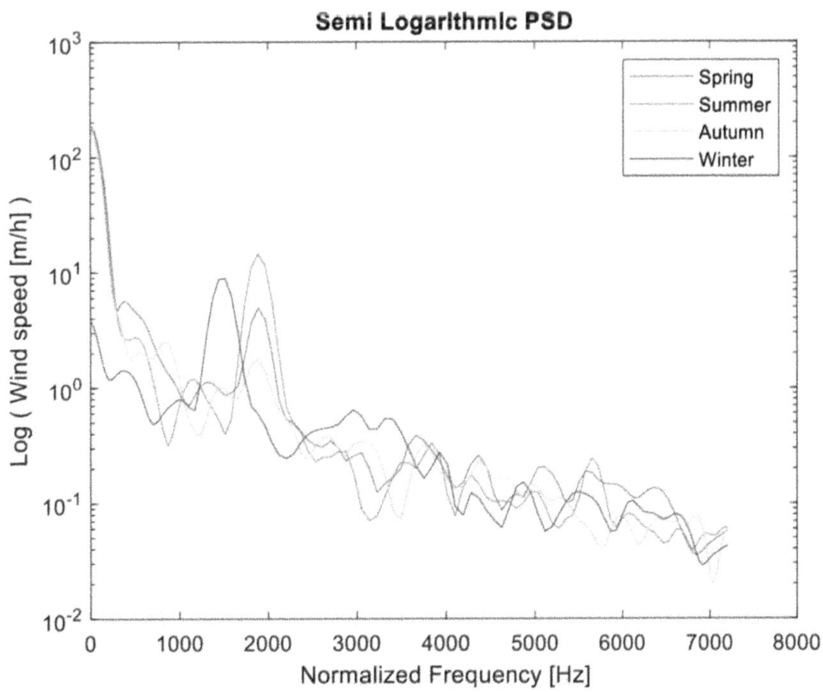

Figure 7. Comparison of the Wind Speed Histograms

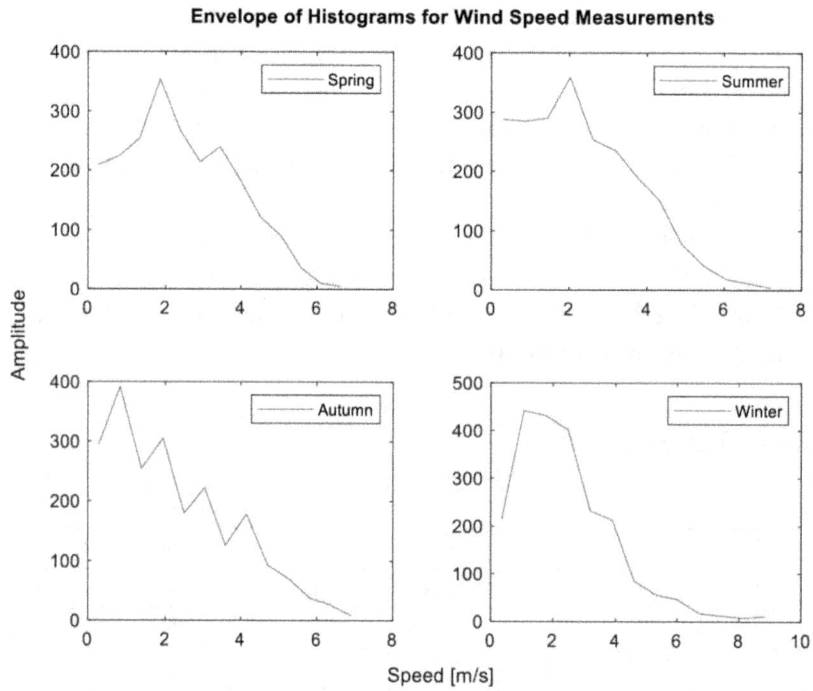

In addition to the statistical distributions, seasonal comparisons of wind speed are given in Figure 8. Here, it can be observed that the highest amplitude values occur in the winter season.

Figure 8. Seasonal comparison of histograms of Wind Speed

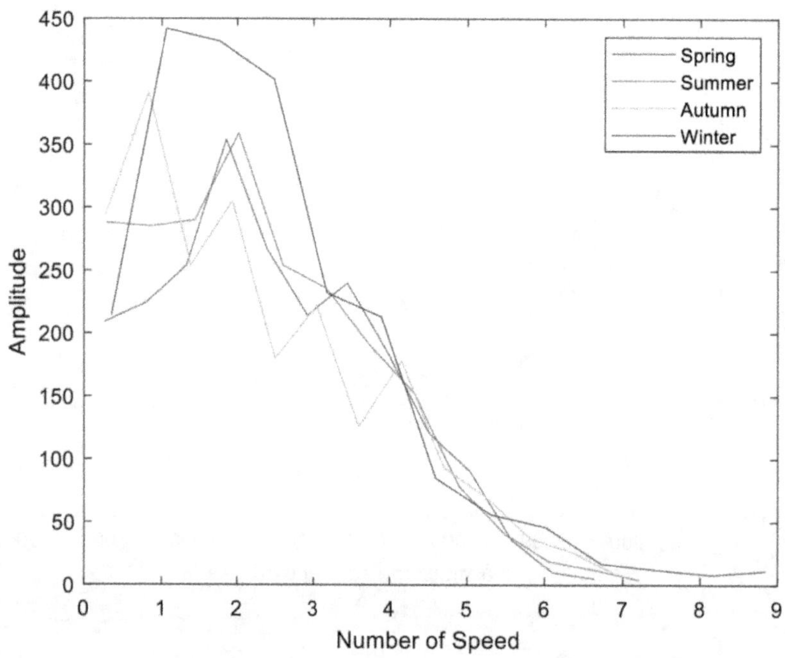

Figure 9. Seasonal change of rose charts of wind speeds

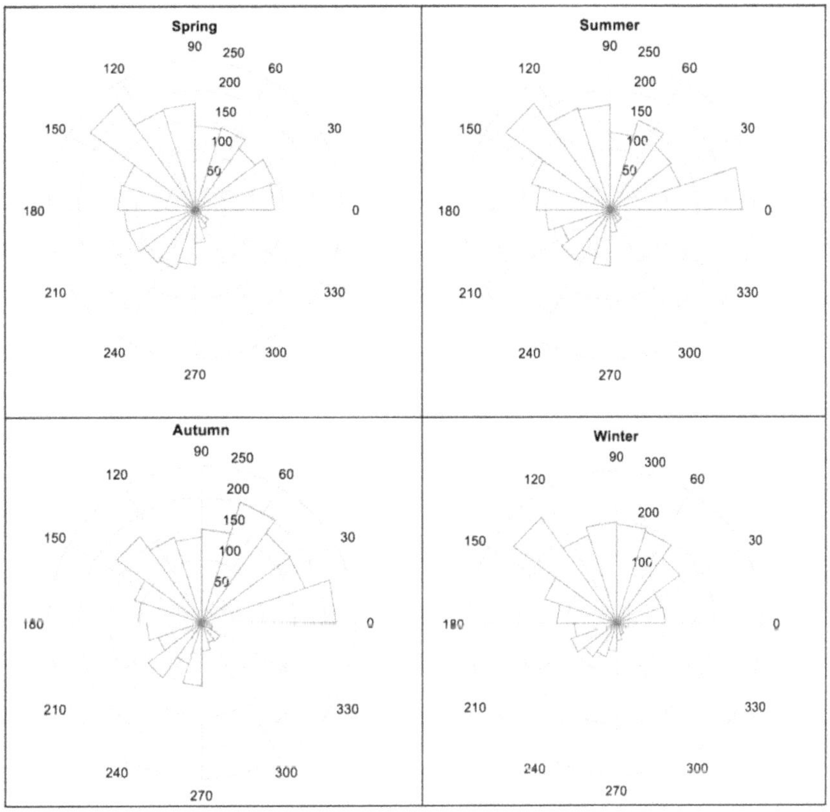

Figure 10. Seasonal Histogram Changes of Wind Speed

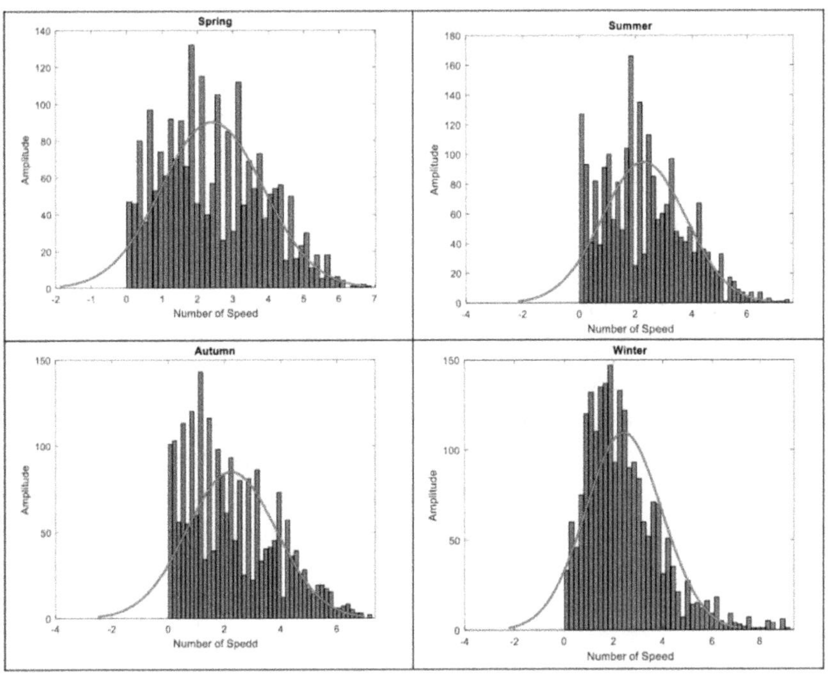

In Figure 9, the seasonal variation of the rose plot of wind speeds is given. This change is consistent with the histogram distributions given in Figure 10.

Figure 11. Probability plot of Normal Distribution

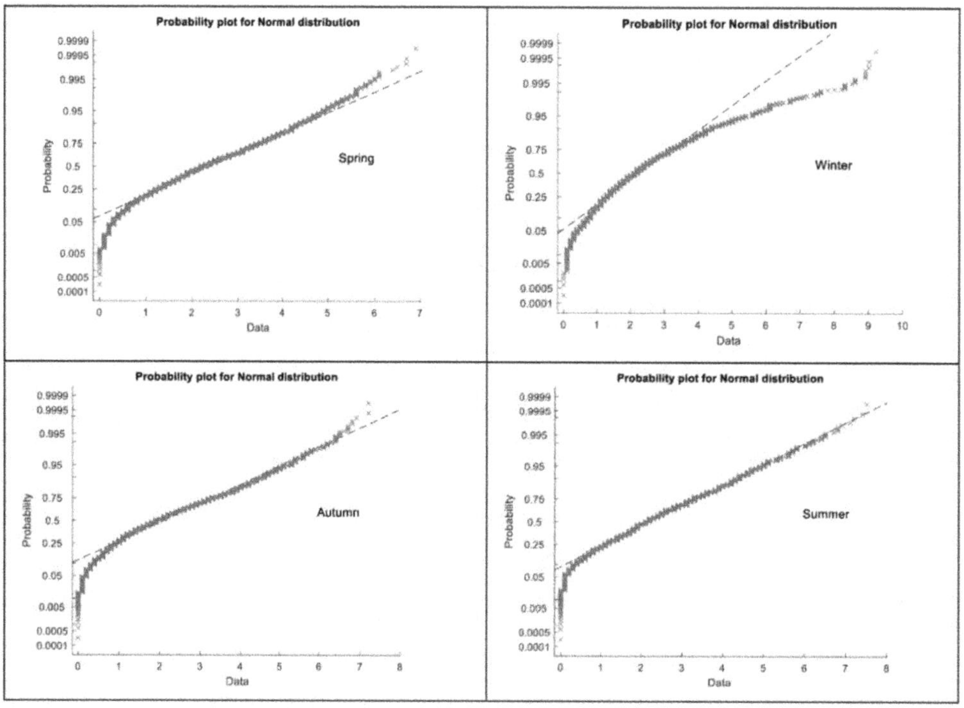

Figure 11 shows the Probability plot of Normal Distribution in the seasonal variation of Istanbul speed speeds. While there is a normal distribution in the summer season, it can be seen that this distribution has reached 0.99 values at 9. Very value in winter.

CONCLUSION

Wind speed data of the Istanbul region in Turkey have been analyzed to extract the seasonal wind characteristics for the year 2004. For these purposes Power Spectral Density calculation which is a Fourier transform approach was calculated to find the frequency characteristics. As a result of the frequency analysis, the joint frequency values for four seasons are at 0.26 Hz as a normalized value. But for winter and autumn times, an additional frequency and fluctuations appear. This is an expected value because these seasons are the windiest seasons than the summer and spring. Also, from the histograms with the Weibull distribution, the number of windy hours in winter times are around twice of the others. In this study, detailed statistical analyzes were given with Istanbul wind speed, and it was observed that rose charts and histogram graphics were compatible with the Probability plot of Normal Distribution.

REFERENCES

Advantages. (2020). https://www.energy.gov/eere/wind/advantages-and-challenges-wind-energy

Advantages. (2021). https://www.clean-energy-ideas.com/wind/wind-energy/advantages-and-disadvantages-of-wind-energy/

Akanwa, A. O., & Joe-Ikechebelu, N. (2019). The Developing World's Contribution to Global Warming and the Resulting Consequences of Climate Change in These Regions: A Nigerian Case Study. In Global Warming and Climate Change. IntechOpen.

Akinci, T. C., Nogay, H. S., Guseinoviene, E., Dikun, J., & Seker, S. (2016). Application of ANN for Short Term Forecasting of Wind Power Density. *15th International Scientific Conference Renewable Energy & Innovative Technologies*, 157-163.

Balat, H. (2005). Wind energy potential in Turkey. *Energy Exploration & Exploitation, 23*(1), 51-59.

Fs-unep. (2021). https://www.fs-unep-centre.org/wp-content/uploads/2020/06/GTR_2020.pdf

Ghasemian, S., Faridzad, A., Abbaszadeh, P., Taklif, A., Ghasemi, A., & Hafezi, R. (2020). An overview of global energy scenarios by 2040: Identifying the driving forces using cross-impact analysis method. *International Journal of Environmental Science and Technology*, 1–24.

Gielen, D., Boshell, F., Saygin, D., Bazilian, M. D., Wagner, N., & Gorini, R. (2019). The role of renewable energy in the global energy transformation. *Energy Strategy Reviews, 24*, 38–50.

Global. (2021). https://globalwindatlas.info/en/area/Turkey?print=true

Guseinoviene, E., Senulis, A., Seker, S., & Akinci, T. C. (2014, March). Statistical and continuous wavelet analysis of wind speed data in Mardin-Turkey. In *2014 Ninth International Conference on Ecological Vehicles and Renewable Energies (EVER)* (pp. 1-5). IEEE.

İlkiliç, C. (2012). Wind energy and assessment of wind energy potential in Turkey. *Renewable & Sustainable Energy Reviews, 16*(2), 1165–1173.

Kuik, O., Branger, F., & Quirion, P. (2019). Competitive advantage in the renewable energy industry: Evidence from a gravity model. *Renewable Energy, 131*, 472–481.

Kumar, M. (2020). *Social, economic, and environmental impacts of renewable energy resources*. Wind Solar Hybrid Renewable Energy System.

Lakatos, L., Hevessy, G., & Kovács, J. (2011). Advantages and disadvantages of solar energy and wind-power utilization. *World Futures, 67*(6), 395–408.

Onar, O. C., & Khaligh, A. (2015). Energy Sources. In Alternative Energy in Power Electronics (pp. 81-154). Butterworth-Heinemann.

Panwar, N. L., Kaushik, S. C., & Kothari, S. (2011). Role of renewable energy sources in environmental protection: A review. *Renewable & Sustainable Energy Reviews, 15*(3), 1513–1524.

PSD. (2021). http://faculty.etsu.edu/blanton/lab_3_psd.doc

Singh, N., Nyuur, R., & Richmond, B. (2019). Renewable energy development as a driver of economic growth: Evidence from multivariate panel data analysis. *Sustainability*, *11*(8), 2418.

The Guide to Renewable vs. Non-renewable Energy Sources. (n.d.). https://www.gosunpro.com/news/renewable-vs-non-renewable-energy-sources/

Wuebbles, D. J., & Sanyal, S. (2015). Air quality in a cleaner energy world. *Current Pollution Reports*, *1*(2), 117–129.

Chapter 3
Nature-Inspired Algorithm Applied to a Renewable Energy-Integrating Hydro-Thermal Power Plant

Sunanda Hazra
https://orcid.org/0000-0002-0612-9521
Central Institute of Petrochemicals Engineering and Technology, India

Provas Kumar Roy
https://orcid.org/0000-0002-3433-5808
Kalyani Government Engineering College, India

ABSTRACT

Due to the rising requirement on energy sources and the global doubts for using fossil fuel because of its consequences on the climate changes and the global warming caused by hazardous gases, the scientific research has shifted to the renewable energy. To minimize the usage of thermal power generation plants and to meet the rising load demand, a thermal-integrated wind-hydro-system is taking an important role in renewable power systems. A proficient nature-inspired optimization is proposed for solving economic and emission dispatch for the hydro-thermal-wind (HTW) scheduling problem. Further, the opposition-based learning have been incorporated with the chemical reaction optimization for improving the performance of the algorithm. To investigate the performance of oppositional chemical reaction optimization algorithm, the algorithm is tested on two different cases. Along with this, some statistical tests have also been performed. The results obtained by the OCRO algorithm are compared with other recently proposed methods to establish its robustness.

DOI: 10.4018/978-1-7998-8561-0.ch003

I INTRODUCTION

In latest trends, a few factors like rising in the globe residents have led to a spectacular hike in ordered of energy all across the every places. Certainly, this energy expenditure creates an extreme utilization of fossil fuels to adjust the energy claimed. Although, using fossil fuels such as gas, coal, petroleum, and few non-renewable energy sources have been previously used to generate electric energy and it results in extraordinary rising in ecological pollutants. In such situations, the energy resources with less emission generation has seem to be suitable alternatives to alleviate the environmental impacts, particularly in the electric power region. To reduce the habit of thermal power generation plants and to meet up the growing load demand, thermal integrated wind-hydro-system is delightful system and it plays an important task in renewable power system that have been represented by Hazra et al. (2019), and Li et al. (2014). In this manuscript, hydro thermal scheduling incorporating wind energy has been discussed and successfully been solved using three efficient meta-heuristics algorithms as well as power system operation and generation using conventional and non-conventional energy sources has been discussed. So, the proposed research work is very significant topic for the power system researchers. Conventional algorithms do not perform satisfactorily for non-linear optimization problem. Since, the proposed research work is non-linear in the presence of uncertain wind speed; the conventional algorithms will give local optimal solution instead of global optimal solution. The proposed research work is one of the promising topics for power system operation because by using the renewable energy sources the society can be protected from the effect of dangerous greenhouse gases as well as the power can be generated at cheap rate and it helps the consumer to get electricity at affordable price. Moreover, in this research work, few efficient meta-heuristics optimization algorithms are used to obtain optimal performance of renewable energy based power system.

II BACKGROUND

In a few decades ago, Newton's method (Lee et al. 1998) and dynamic programming (Farag, et al. 1995) etc. are used for the cost minimization. Traditional techniques have the complexities in non-linear constraints as well as considerable time-consuming effect. For that reason, previously developed techniques do not deal with sufficiently for solving economic optimization problems. The methods mentioned above suffer from poor local optimal optimization and slow convergence rate. To overcome this drawback, many populations based methods such as chemical reaction optimization (CRO) (Hazra & Roy, 2015), krill herd algorithm (Mandal et al. 2014), oppositional moth flame optimization (OMFO) (Hazra & Roy, 2019), quasi-oppositional chemical reaction optimization (QOCRO) (Hazra & Roy, 2019), grasshopper optimisation algorithm (GOA) (Hazra & Roy, 2020) etc. are represented. Chen et al. (1993) proposed distribution management-oriented renewable energy generation using novel interval. Aghaei et al. (2013) suggested programming framework over the 24hour time span based on wind power in a scenario-based stochastic dynamic economic and emission load transmit problem. Hetzer et al. (2008) briefly discussed the wind power underestimation cost and overestimation cost of available generation for renewable power. Bai et al. (2016) projected an artificial bee colony (ABC) to compact with the uncertainty of wind power for solving load dispatch problem. Panigrahi et al. (2010) discuss about wind resources as the stochastic nature type and for that reason wind generation output is difficult to predict. Earlier, the problem has been measured for several times as the progress of load dispatch, but now a day's research focuses on wind energy units together with exact cost functions. Most of these works are used as valid statistics distribution to characterize the inconsistency of wind and it is known as Weibull distribution (Shi et al. 2012). Moreover, Yu et al. (2007) explained different PSO techniques to state conditional hydrothermal scheduling (HTS) preparation. Hazra & Roy (2021) implemented MFO algorithm for solving the renewable energy integrating load transmit and HTS problem (Hazra et al. 2020).

III MAIN FOCUS OF THE ARTICLE

For optimal performance of the system, opposition-based learning (OBL) was planned by Tizhoosh (2005). Opposition based learning (OBL) is included with the fundamental CRO algorithm (Lam & Li, 2010). OCRO associated with the arrangement and breaking of chemical bonds in a chemical reaction. OCRO trim down the computational trouble and the opposite candidate solution has an advanced opportunity to be nearer to the global optimum solution than a random candidate solution. The aim of this investigation is to construct entire consumption of the unpredictable capability of wind power using hydropower. Optimal generation scheduling has been ensured skillful process of thermal generators and to build entire use of green energy by reducing consumption fossil fuel. In this manuscript, complete utilization of the changeable capability of hydropower using wind power fluctuations and fabricate optimal generation scheduling of thermal power considering single objective economic dispatch are solved using CRO and OCRO methods. Furthermore, to endorse the supremacy of the proposed technique, its simulation outcome is compared with the recently obtainable diverse optimization techniques and the corresponding results have been presented in the literature.

The remaining portions of this paper are prepared as follows. Section IV formulates mathematical problem formulation and the mathematical modelling of wind power. Section V presents system constraints of wind-based CEED problem. The CRO and QOBL techniques are briefly elaborated in Section VI. Section VII describes the algorithm phase of the QOCRO applied to wind oriented HTS problem. Wide numerical simulations of the few techniques for different cases are specified in Section VIII. Section IX describes the future research directions. The conclusion is summarized in section X.

IV PROBLEM FORMULATIONS

a. Objective Function

By utilizing the available renewable wind and hydro possessions the forecast horizon are considered. The novel item of this entire literature is to diminish the effective cost by appropriately substituting thermal and the wind power generations. The overall wind power (Younes et al. 2014). and thermal power (Hazra et al. 2015; Hazra & Roy 2021) cost can be written using the subsequent equations:

$$Minimize T_{\$} = \left[\sum_{i=1}^{N_T} \left(a_i P_{o,i}^{\ 2} + b_i P_{o,i} + c_i + \left| d_i \times \sin\left\{ e_i \times \left(P_{o,i}^{\min} - P_{o,i} \right) \right\} \right| \right) \right] + \\ + T_{wind(\$)} + \sum_{i=1}^{N_W} \left(f_i + g_i P_{o,i} + h_i P_{o,i}^{\ 2} \right) \kappa_{tx} \tag{1}$$

Where $T_\$$ is the whole generation cost. a_i b_i and c_i are fuel cost coefficient. d_i and e_i are the non-smooth thermal generating valve point co-efficient of fossil fuel unit. $P_{o,i}$ is the ith thermal generating unit. $P_{o,i}^{\min}$ is the least amount generation of thermal power. N_T is the thermal generating station number count. i is the thermal power generation index. f_i, g_i and h_i are the coefficients of emission. κ_{tx} is fuel pollutant factor of emission. $T_{wind(\$)}$ is the total working cost of renewable power. Underestimation and overestimation cost which are related to wind power might be present when the schedule wind power doesn't equivalent to the authentic value. If the genuine wind power is a lesser quantity than the scheduled value, then the wind power is termed as overestimated. In such condition, the employee needs to obtain a little energy from the separate source to adjust the load necessity. Cost of wind power (Hazra & Roy, 2021) is considered as follows:

$$T_{wind(\$)} = X^{ov,j} \int_0^{P_{n,j}} \left(P_{n,j} - W_u\right) f_w(w) dw + X^{un,j} \int_{P_{n,j}}^{P_{r,j}} \left(W_u - P_{o,j}\right) f_w(w) dw \quad (2)$$

Where, P_{nj} is output power from j^{th} wind Park. W_u is wind power output. P_{rj} is rated wind power output. $X^{un,j}$ is underestimation cost coefficient. $f_w(w)$ is probability density functions of output power. X^{ov} is overestimation cost coefficient.

b. Hourly Wind Speed and Probabilistic Analysis Output Modeling

The Weibull's pdf (Hetzer et al. 2008) is used for the wind speed estimation and described by the following equations:

$$f_V(V_W) = \frac{k}{c}\left[\left(\frac{v_W}{c}\right)^{k-1}\right]\exp\left[-\left(\frac{v_W}{c}\right)^k\right] \quad (3)$$

v_W is the wind speed which is a random variable. $c(c>0)$ is the scale factor and k $(k>0)$ is the shape factor. Weibull distribution is followed by the cumulative distribution Function (CDF):

$$F_V(V_W) = 1 - \exp\left[-\left(\frac{v_W}{c}\right)^k\right] \quad (4)$$

An easy model has been used to establish the relation between wind speed and wind power (WP). It is explained by the following equations:

$$W_P = \begin{cases} 0; \ \left(V_W < v_w^{in} \ or \ V_W > v_w^{out}\right) \\ 0.5\rho A_s v_W^3; \left(v_w^{in} \le V_W > v_w^r\right) \\ w_{rated}; \left(v_w^r \le V_W > v_w^{out}\right) \end{cases} \quad (5)$$

v_W is the wind speed and it is considered as a random variable. ρ is the air density (kg/m^3). As is the cross-sectional area. In the interval $v_w^{in} \leq V_W > v_w^{rated}$, the probability density function of wind power is described using the theory of random variables:

$$f_{pd}(P_r) = \frac{k}{3c^k}\left(\frac{2}{\rho A}\right)^{\frac{k}{3}} P_r^{\left(\frac{k}{3}-1\right)} \exp\left[-\frac{1}{c^k}\left(\frac{2P_r}{\rho A}\right)^{\frac{k}{3}}\right] \tag{6}$$

V SYSTEM CONSTRAINTS

a. Power Balance Constraints

Total load demand (Hetzar et al. 2008) is equal to the summation of entire power generation from thermal generating station, renewable wind power generation and hydro plants. It is represented as follows:

$$\sum_{j=1}^{N_W} P_{o,j}^{th} + \sum_{i=1}^{N_T} P_{o,i}^{wi} + \sum_{m=1}^{N_H} P_{hy,m} = P_d \tag{7}$$

$P_{o,j}^{th}$ is the output power from j^{th} thermal generating stations. $P_{o,i}^{wi}$ is the output power from i^{th}wind generating stations. $P_{o,m}^{hy}$ is the output power from m^{th} hydro generating stations. Hydropower output is denoted by $P_{hy,m}$. Hydropower output (P_{hy}) can be represented as:

$$P_{hy},m = \lambda_{1m}(V_m)^2 + \lambda_{2m}(Q_m)^2 + \lambda_{3m}V_m Q_m + \lambda_{4m}V_m + \lambda_{5m}Q_m + \lambda_{6m} \tag{8}$$

Where, $\lambda_{1m}, \lambda_{2m}, \lambda_{3m}, \lambda_{4m}, \lambda_{5m}$ *and* λ_{6m} are the power generation coefficients of mth hydro power plant. Q_m and V_m are the water discharge and volumes of reservoir storage of mth hydro power plant.

b. Limitation Constraints

Hydro power, renewable wind and fossil fuel generation output power have to uphold least amount and highest range (Yao et al. 2012). It is described by the following form:

$$P_{o,j}^{th,\min} \leq P_{o,j}^{th} \leq P_{o,j}^{th,\max} \quad j=1,2,\ldots\ldots,N_T \tag{9}$$

$$V_m^{\min} \leq V_m \leq V_m^{\max} \quad i=1,2,\ldots\ldots,N_W \tag{10}$$

$$P_{hy}^{\min} \leq P_{hy},m \leq P_{hy}^{\max} \quad m=1,2,\ldots\ldots,N_H \tag{11}$$

c. Reservoir Storage Volumes Limits

$$V_m^{\min} \le V_m \le V_m^{\max} \tag{12}$$

Where, $V_m^{\min}$ and $V_m^{\max}$ is the range boundary of storage volume of the mth reservoir.

d. Water Discharge Limits

$$Q_m^{\min} \le Q_m \le Q_m^{\max} \tag{13}$$

Where, $Q_m^{\min}$ and $Q_m^{\max}$ are the water least and highest discharge limit of the mth hydro power generation unit.

e. Water Dynamic Balance Constraints

Inflow, reservoir storage and spillage have been considered at the preceding incident for hydro plant. Hydraulic continuity equations at time interval t is represented as follows:

$$V_m^t = V_m^{t-1} + I_m^t - Q_m^t - S_m^t + \sum_{n=1}^{N_{sm}} \left(Q_{n,t-\tau_{nsm}} + S_{n,t-\tau_{nsm}} \right) \tag{14}$$

τ_{nsm} is the time delay between its upstream plant sm and hydro plant m. I_m^t and S_m^t are the inflow and spillage of the mth hydro unit at the mth interval. N_{sm} is the level of the upstream unit.

f. Final and Initial Reservoir Reserve Volumes Constraints

Initial and final reservoir storage of every unit should satisfy this constraint.

$$V_m^0 = V_m^{bn} \tag{15}$$

$$V_m^t = V_m^{end} \tag{16}$$

Where V_m^{bn} and V_m^{end} are the opening and ending tank storage restrictions of hydro unit m. V_m^0 and V_m^t are the reservoir storage of hydro unit m at interval 0 and t.

VI CHEMICAL REACTION OPTIMIZATION

Chemical reaction optimization (CRO) is a freshly introduced evolutionary calculation technique. It is initiated by Lam et al. (2010). Due to the configuration and breaking of chemical bonds, a chemical reaction has been occurred. So it is concluding that CRO is depended on a chemical change and it is occurs by the motion of electrons of the molecules.

a. Single Molecular Effect

In this on-wall ineffective collision, a molecule touches the trunk wall and then spring back. The change is allowed only if (Hazra & Roy, 2019):

$$KE_{(M_X)} + PE_{(M_X)} \geq PE_{(M_Y)} \tag{17}$$

Where, original molecule kinetic and potential energy is expressed by $KE_{(M_X)}$ and $Pe_{(M_X)}$. Recently produced molecule potential energy is represented by $PE_{(M_Y)}$. The molecular construction of the unique molecule is M_X and it produced a few resultant molecules M_{X1} and M_{X2}. A molecule hits the trunk wall and then decomposes at least two pieces. It is called chemical reaction decomposition. Decomposition occurs if the subsequent circumstance satisfy and it is expressed using the below mentioned equation (Hazra & Roy, 2019).

$$KE'_{(M_X)} + PE'_{(M_X)} \geq PE'_{(M_{X1})} + PE'_{(M_{X2})} \tag{18}$$

b. Multiple Molecular Effects

In Inter-molecular ineffective collision exists when two molecules collide with each other and then jump away. $M_{X1'}$ and $M_{X2'}$ and have been generated from the original molecules $M_{X1}$and M_{X2}.

The molecule changes occur if the following condition holds (Hazra & Roy, 2019):

$$\left[KE_{(M_{X1})} + PE_{(M_{X1})} + PE_{(M_{X2})} + KE_{(M_{X2})}\right] \geq \left[PE_{(M_{X1'})} + PE_{(M_{X2'})}\right] \tag{19}$$

For synthesis, when two or more reactants combine to enlarge a compound. Formation of a fresh molecule $M_{n'}$ from two lively molecules M_{X1} and M_{X2} is permitted when the following situation holds (Hazra & Roy, 2019):

$$\left[KE_{(M_{X1})} + PE_{(M_{X1})} + PE_{(M_{X2})} + KE_{(M_{X2})}\right] \geq PE_{(M_{n'})} \tag{20}$$

c. Opposition-Based Learning

Opposition-based learning (OBL) is a new idea in optimization aptitude, which considers existing approximation and its opposite estimate. At the similar cases, to get an enhanced estimate for an existing candidate resolution, an opposite number is considered by explaining the mirror point of the solution from the search space interior. It has been validated that an opposition candidate solution has an improved opportunity to be closer to the worldwide optimum result than a random candidate result.

d. Opposition Number

In an one-dimensional investigation region, if X_r be any actual number between $[q_om,q_on]$, its opposite number X_o is represented as below (Hazra & Roy 2019):

$$X_o = [q_om + q_on - X_r] \tag{21}$$

e. Opposition Point

Similarly, for n-dimensional search space, the opposite point may be scientifically presented as follows (Hazra & Roy, 2019):

$$D_{o,j}= [(q_om)_j + (q_on)_j - D_j];\ D_j\in[q_om,\ q_on];\ j=1,2,\ldots\ldots,d \tag{22}$$

VII OCRO APPLIED TO WIND ENERGY INCORPORATING HTS

The course of action for validating the OCRO method in solving wind energy incorporating HTS problem can be processed by the subsequent steps:

Step1: Set the system data (hydro, thermal and wind related individual parameter). Population size is described as the entire count of molecular organization. Working limits and the active power of the orientation unit is scientifically calculated by the power balance using equation (9)- equation (11). Re computation has been occurred if any of the generators violates the operating boundary ranges.

Step2: Opposite population $O^{\#}_{i,j}$ is shaped using equation (23) and initialize every molecule of a given molecule set (Hazra & Roy, 2019).

$$O^{\#}_{i,j} = F_j + E_j - P_{i,j} \tag{23}$$

Where, P_i,j is the jth independent variables of the ith vector of the population. $i=1,2,\ldots,P_S$ and $j=1,2,\ldots,P_C$.

Step3: Access the fitness value (Total generation cost) of the current and opposite populations. The fitness value denotes the potential energy PE assessment of every molecule.

Step4: Applying single molecular reaction and multiple molecular reaction and check whether it satisfies the equations, described in equation (17) and equation (20), respectively. If satisfied, the independent variables of every non-elite solution are customized.

Step5: The infeasible outcomes are replaced by arbitrarily produced innovative solutions set by checking the practicability of recently generated solutions.

Step6: The active power generation of the generating units is restructured using jumping rate. Populations of CRO algorithm is applied in the single molecular and multiple molecular phases for calculating the fitness values of opposite population.

Step7: Stop the search procedure if the executing criteria are maintained, and present the optimum result, else continue to the next iteration.

VIII SIMULATION RESULTS AND DISCUSSION

The projected OCRO method has been introduced to the single and multi-objective transmit for Hydro–Thermal-Wind (HTW) scheduling problem. The proposed algorithm has been tested on four hydro plants, three thermal and two wind plants with the economic problem for solving daily hydrothermal scheduling. In this study, four hydro and three thermal plants cost co-efficient are taken from (Liao et al.2013) and two wind farms parameters are taken from (Yao et al.2012). The agenda is urbanized with 4 GB RAM personal computer with core i^3 processor in MATLAB 12.0.

a. Case A: Fuel Cost Minimization Without Renewable Energy

To minimize the fuel cost, OCRO is implemented. The outcome results for single objective cost minimization particularly the optimal hydropower generation of each unit, and total thermal power generation using OCRO for each hour for cost minimization without renewable energy are presented in Table 1. The obtained results in terms of statistical data using CRO and OCRO for fuel cost minimization without renewable energy is compared with the former active methods and exposed in Table 2 and it is also revealed that the mathematical outcome using OCRO is the best. From the simulation outcome it is clearly visible that costs gained using OCRO and CRO is 41517.55 $/day and 41528.90. The cost are most outstanding and smallest amount than fuel cost of 41751.15$, 41856.50$, 42032.35$ obtained by DGSA (Yu & Li 2015), MHDE (Lakshminarasimman & Subramanian 2008), GSA (Yu & Li 2015), and 42187.49$, 42385.88$, 42440.57$, obtained by QTLBO (Roy et al. 2014), TLBO (Roy 2013), CSA (Swain et al.2011). Figure 1 displays the cost convergence outline for cost minimization by the proposed OCRO and CRO method.

Table 1. Simulation results of without renewable energy systems using OCRO

Hour	*Hydro Power generation (MW)*				*Thermal Power generation(MW)*	Hour	*Hydro Power generation (MW)*				*Thermal Power generation (MW)*
	Unit1	**Unit2**	**Unit3**	**Unit4**	**All Unit**		**Unit1**	**Unit2**	**Unit3**	**Unit4**	**All Unit**
1	78.18	62.03	57.19	183.87	368.71	13	70.95	47.11	43.99	248.14	699.79
2	96.15	71.21	0	160.16	452.47	14	77.49	48.12	33.19	246.63	624.54
3	80.57	69.22	27.71	148.22	374.26	15	78.08	76.25	52.34	261.98	541.33
4	69.82	65.59	0	146.86	367.71	16	58.21	50.55	47.71	277.88	625.62
5	80.95	68.20	0	149.75	371.08	17	65.88	47.29	52.63	252.33	631.84
6	78.29	60.47	42.53	161.00	457.69	18	80.81	59.24	52.28	300.80	626.85
7	59.05	63.61	36.54	238.41	552.37	19	69.60	47.59	54.81	272.32	625.66
8	52.38	46.98	37.41	250.87	622.33	20	96.79	57.26	55.78	297.95	542.19
9	80.17	48.19	19.95	225.43	716.23	21	67.83	52.37	57.15	283.12	449.51
10	81.32	73.51	48.59	249.68	626.88	22	76.02	57.72	58.03	292.34	375.87
11	91.75	61.01	48.96	259.97	638.29	23	56.03	52.65	59.38	283.31	398.60
12	76.80	56.99	46.80	252.72	716.68	24	57.15	53.40	51.32	271.23	366.88

Table 2. Statistical comparison of different algorithms for without renewable energy systems

Methods	Best cost ($/day)	Mean cost ($/day)	Worst cost($/day)
OCRO	**41517.55**	**41531.24**	**41540.55**
CRO	**41528.90**	**41553.95**	**41577.02**
DGSA (Yu & Li 2015)	41751.15	41989.02	41821.49
MHDE (Lakshminarasimman & Subramanian 2008)	41856.50	NA	NA
GSA (Yu & Li 2015)	42032.35	42561.53	42292.12
QTLBO (Roy et al. 2014)	42187.49	42202.75	42193.46
TLBO (Roy 2013)	42385.88	42441.36	42407.23
CSA (Swain et al.2011)	42440.57	NA	NA

Figure 1.

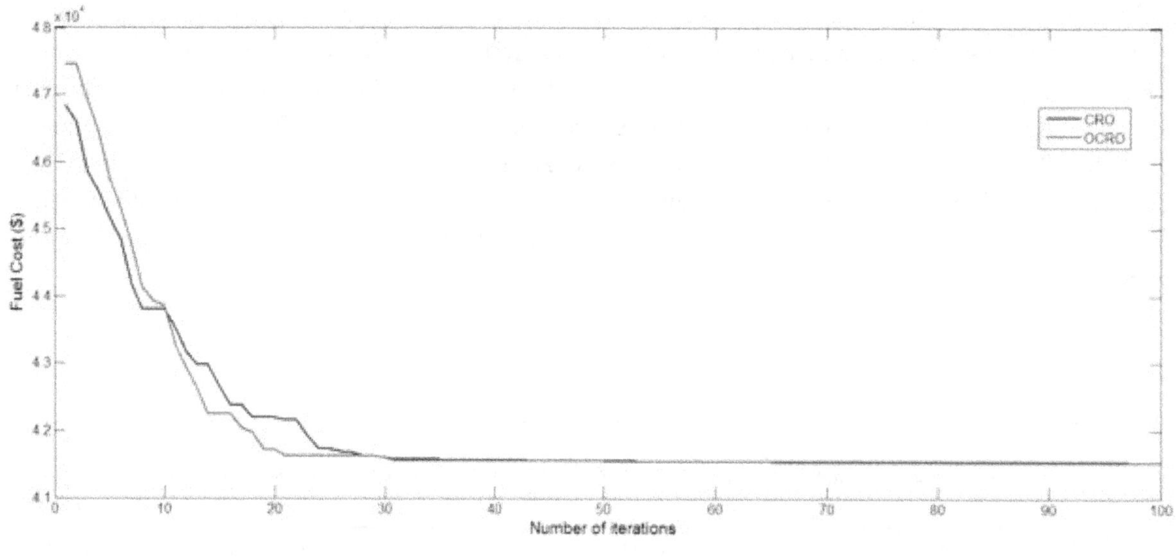

b. Case B: Fuel Cost Minimization Incorporating Wind Energy

Responding to environmental problems, two renewable wind plants have been incorporated and new system is become more complex. An optimal result obtained by OCRO i.e. total thermal power generations of three thermal plant, total hydropower generation of four hydropower plant for 24 hours, wind power generation of two wind park are listed in Table 3. Statistical results are listed in Table 4 and the achieved result is also compared with previous techniques. It is also confirmed that to achieving the minimum cost, proposed OCRO method is the best. It is also shown that after considering renewable energy, the fuel cost by OCRO are diminished from 41517.55 $/day to 36809.52 $/day respectively. A comparison of generation cost without and with using wind energy is shown in Figure 2. Figure 3 and Figure 4 indicates the hourly water storage volumes and wind power generation of each unit using OCRO. Figure 5 describes the comparison of total thermal, hydro, and wind power generations.

Table 3. Simulation results of renewable energy added systems using OCRO

Hour	Wind Power Generations (MW)		*Hydro Power generation (MW)*	*Thermal Power generation (MW)*	Hour	Wind Power Generations (MW)		*Hydro Power generation (MW)*	*Thermal Power generation(MW)*
	Unit1	Unit2	All Unit	All Unit		Unit1	Unit2	All Unit	All Unit
1	87.46	41.33	319.41	196.52	13	80.90	44.69	431.24	628.46
2	68.47	22.38	307.66	367.54	14	75.50	59.85	414.54	453.98
3	42.48	54.46	346.34	294.02	15	89.27	59.51	457.28	454.56
4	68.15	29.43	408.17	196.32	16	70.61	44.08	359.17	540.30
5	74.98	37.28	355.01	279.41	17	64.21	42.84	437.17	537.26
6	89.93	46.81	388.16	293.38	18	76.20	37.26	437.39	543.28
7	81.20	33.42	355.78	451.22	19	77.96	59.76	450.64	541.43
8	73.01	38.43	367.79	545.85	20	60.75	57.95	424.47	545.57
9	88.38	60.00	433.88	539.74	21	88.14	13.12	408.99	362.66
10	87.52	32.58	355.99	625.91	22	85.36	51.96	190.12	277.56
11	83.96	56.42	455.36	537.84	23	85.68	26.98	152.33	286.47
12	45.07	45.74	392.28	631.16	24	1.30	2.12	78.44	366.84

Figure 2.

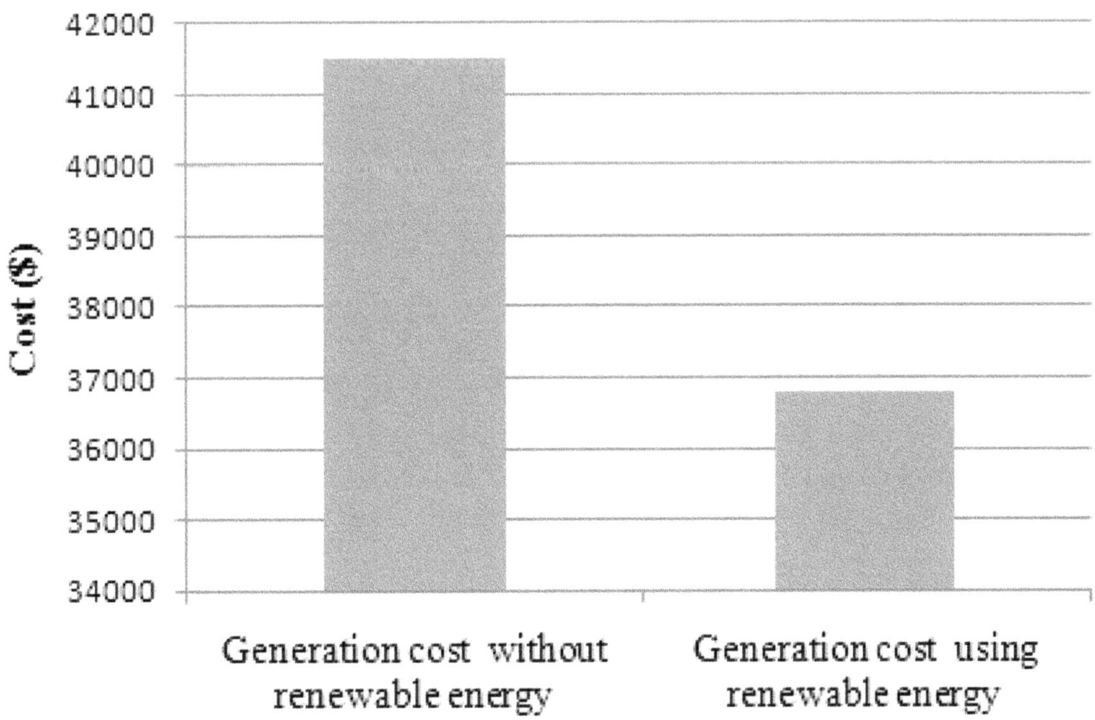

Table 4. Statistical comparison of algorithms for renewable energy added systems

Methods	Best cost ($/day)	Mean cost ($/day)	Worst cost($/day)
OCRO	36809.5282	36819.6505	36827.5282
CRO	36832.04	36848.0994	36860.0040

Figure 3.

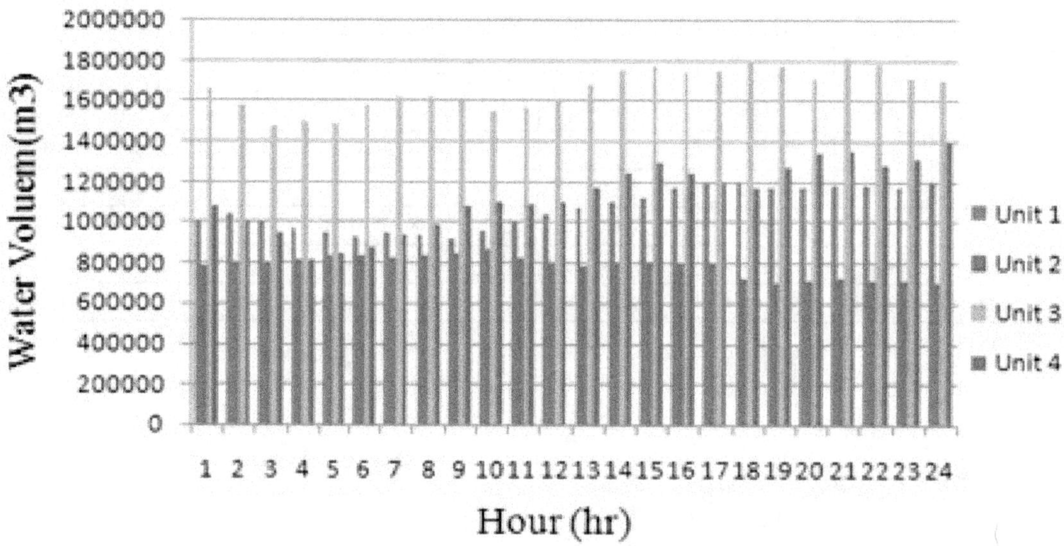

Figure 4.

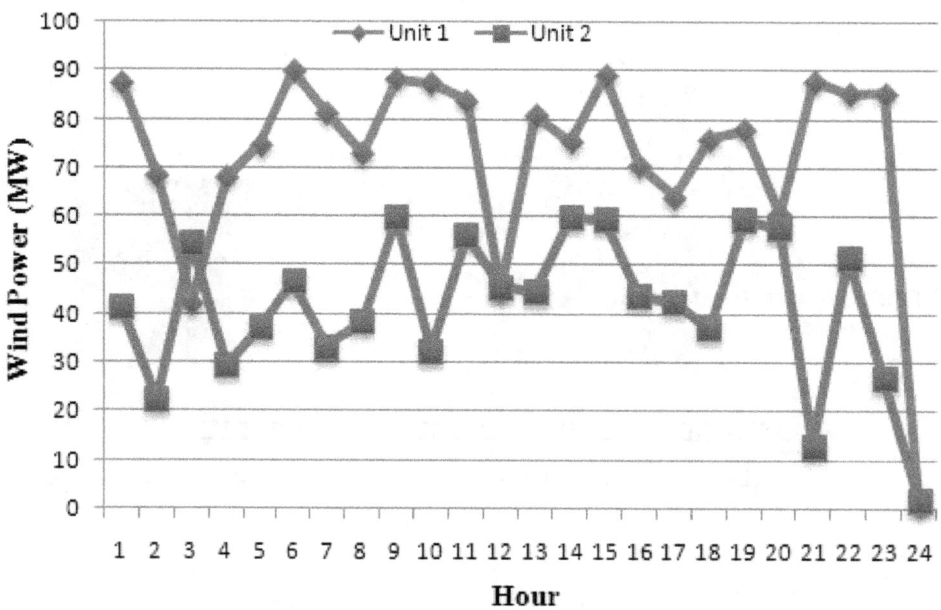

Figure 5.

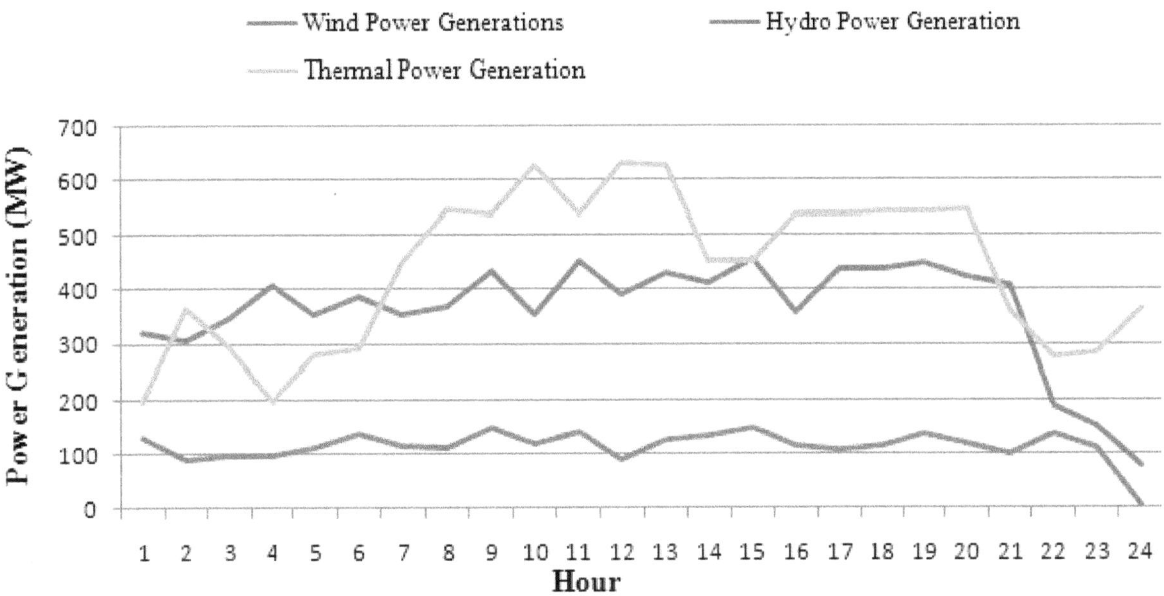

IX FUTURE RESEARCH DIRECTIONS

In future, Geothermal, Ocean, Hydrogen, Biomass will be included in the research. The effectiveness of the proposed control algorithms will be examined in large scale system.

X CONCLUSION

Using an influential novel search technique, the hydro-thermal-wind scheduling problem is solved for the cost and emission minimization objectives. It is modelled using the chemical reaction of molecule. The projected technique depending on opposition criteria handles multifaceted practical constraints very effectively. Moreover, a number of statistical analysis have been processed to discover the usefulness of OCRO. The proposed OCRO have been tested on three test systems and it has been confirmed that OCRO provide better results than DGSA, MHDE, GSA, QTLBO, TLBO, and CSA. It is proved that OCRO is more robust than others and it is also best technique in terms of producing quality solution. The exploitation of energy due to rising inhabitants, financial development and living standard are increasing reasonably at an elevated rate. In the context of current trends in environmental directive and energy segment, such revision becomes more relevant to the clean energy sources and green atmosphere. A related enhancement is also accepted in the future, for hybrid scheduling with a multiplicity of renewable power, unit commitment, e-vehicle charging with the mixture of renewable energy generation. In this manuscript, hydro thermal scheduling incorporating wind energy has been discussed and successfully been solved using efficient meta-heuristics algorithms as well as power system operation and generation using conventional and non-conventional energy sources has been discussed. The proposed research work is one of the promising topics for power system operation because by using the renewable energy sources the society can be protected from the effect of dangerous greenhouse gases as well as the power can be generated at cheap rate and it helps the consumer to get electricity at affordable price.

REFERENCES

Aghaei, J., Niknam, T., Azizipanah-Abarghooee, R., & Arroyo, J. M. (2013). Scenario-based dynamic economic emission dispatch considering load and wind power uncertainties. *International Journal of Electrical Power & Energy Systems*, *47*, 351–367. doi:10.1016/j.ijepes.2012.10.069

Bai, W., & Lee, Y. (2016). Modified Optimal Power Flow on Storage Devices and Wind Power Integrated System. *IEEE International Conference on Power and Energy Society General Meeting (PESGM)*. doi: 10.1109/PESGM.2016.7742033

Chen, C. L., & Wang, S. C. (1993). Branch-and bound scheduling for thermal generating units. *IEEE Transactions on Energy Conversion*, *8*(2), 184–189. doi:10.1109/60.222703

Farag, A., Al-Baiyat, S., & Cheng, T. C. (1995). Economic load dispatch multi-objective optimization procedures using linear programming techniques. *IEEE Transactions on Power Systems*, *10*(2), 731–738. doi:10.1109/59.387910

Hazra, S., Pal, T., & Roy, P. K. (2019). Renewable Energy Based Economic Emission Load Dispatch Using Grasshopper Optimization Algorithm. *International Journal of Swarm Intelligence Research*, *10*(1), 38–57. doi:10.4018/IJSIR.2019010103

Hazra, S., & Roy, P. K. (2015). Economic Load Dispatch Considering Non-smooth Cost Functions Using Predator–Prey Optimization. *Springer Conference on Advances in Intelligent Systems and Computing*, 67–78. doi: 10.1007/978-81-322-2268-2_8

Hazra, S., & Roy, P. K. (2019). Evolutionary Oppositional Moth Flame Optimization for Renewable and Sustainable Wind Energy Based Economic Dispatch. *International Journal of Applied Evolutionary Computation*, *10*(4), 65–84. doi:10.4018/IJAEC.2019100104

Hazra, S., & Roy, P. K. (2019). Quasi-oppositional chemical reaction optimization for combined economic emission dispatch in power system considering wind power uncertainties. *Renewable Energy Focus*, *31*, 45–62. doi:10.1016/j.ref.2019.10.005

Hazra, S., & Roy, P. K. (2020). Newly developed Swarm Intelligence Algorithms (GOA) applied on real-world non conventional energy based load dispatch Problems. *Advancements of Swarm Intelligence Algorithms for Solving Real-World Problems*, 1-26. doi:10.4018/978-1-7998-9152-9.ch035

Hazra, S., & Roy, P. K. (2021). Metaheuristic Moth-Flame Optimization Applied on Renewable Wind Energy Incorporating Load Transmit Penetration. *International Journal of Applied Metaheuristic Computing*, *12*(1), 185–210. doi:10.4018/IJAMC.2021010110

Hazra, S., Roy, P. K., Hazra, S., & Roy, P. K. (2020). Optimal dispatch using moth-flame optimization for hydro-thermal-wind scheduling problem. *International Transactions on Electrical Energy Systems*, *30*(8), e12460. doi:10.1002/2050-7038.12460

Hazra, S., Roy, P. K., & Sinha, A. (2015). An efficient evolutionary algorithm applied to economic load dispatch problem. *International Conference on Computer, Communication, Control and Information Technology*, 1–6. 10.1109/C3IT.2015.7060129

Hetzer, J., Yu, D. C., & Bhattrarai, K. (2008). An economic dispatch model incorporating wind power. *IEEE Transactions on Energy Conversion*, *23*(2), 603–611. doi:10.1109/TEC.2007.914171

Lakshminarasimman, L., & Subramanian, S. (2008). A modified hybrid differential evolution for short-term scheduling of hydrothermal power systems with cascaded reservoirs. *Energy Conversion and Management*, *49*(10), 2513–2521. doi:10.1016/j.enconman.2008.05.021

Lam, A. Y. S., & Li, V. O. K. (2010). Chemical-reaction-inspired metaheuristic for optimization. *IEEE Transactions on Evolutionary Computation*, *14*(3), 381–399. doi:10.1109/TEVC.2009.2033580

Lee, K. Y., Yome, A. S., & Park, J. H. (1998). Adaptive Hopðeld neural networks for economic load dispatch. *IEEE Transactions on Power Systems*, *13*(2), 519–526. doi:10.1109/59.667377

Li, Y. Z., Wu, Q. H., Li, M. S., & Zhan, J. P. (2014). Mean-variance model for power system economic dispatch with wind power integrated. *Energy*, *72*, 510–520. doi:10.1016/j.energy.2014.05.073

Liao, X., Zhou, J., Ouyang, S., Zhang, R., & Zhang, Y. (2013). An adaptive chaotic artificial bee colony algorithm for short-term hydrothermal generation scheduling. *International Journal of Electrical Power & Energy Systems*, *53*, 34–42. doi:10.1016/j.ijepes.2013.04.004

Mandal, B., Roy, P. K., & Mandal, S. (2014). Economic load dispatch using krill herd algorithm. *International Journal of Electrical Power & Energy Systems*, *57*, 1–10. doi:10.1016/j.ijepes.2013.11.016

Panigrahi, B. K., Pandi, V. R., & Das, S. (2010). Multi-objective fuzzy dominance based bacterial foraging algorithm to solve economic emission dispatch problem. *Energy*, *35*(12), 4761–4770. doi:10.1016/j.energy.2010.09.014

Roy, P. K. (2013). Teaching learning based optimization for short-term hydrothermal scheduling problem considering valve point effect and prohibited discharge constraint. *International Journal of Electrical Power & Energy Systems*, *53*, 10–19. doi:10.1016/j.ijepes.2013.03.024

Roy, P. K., & Hazra, S. (2015). Economic emission dispatch for wind- fossil fuel based power system using chemical reaction optimization. *International Transaction on Electrical Energy Systems*, *25*(12), 3248–3274. doi:10.1002/etep.2033

Roy, P. K., Paul, C., & Sultana, S. (2014). Oppositional teaching learning based optimization approach for combined heat and power dispatch. *International Journal of Electrical Power & Energy Systems*, *57*, 392–403. doi:10.1016/j.ijepes.2013.12.006

Shi, L., Wang, C., Yao, L., Ni, Y., & Bazargan, M. (2012). Optimal power flow solution incorporating wind power. *IEEE Systems Journal*, *6*(2), 233–241. doi:10.1109/JSYST.2011.2162896

Swain, R. K., Barisal, A. K., Hota, P. K., & Chakrabarti, R. (2011). Short-term hydrothermal scheduling using clonal selection algorithm. *International Journal of Electrical Power & Energy Systems*, *33*(3), 647–656. doi:10.1016/j.ijepes.2010.11.016

Tizhoosh, H. (2005). Opposition-based learning: A new scheme for machine intelligence. *Proceedings of the International Conference on Computational Intelligence for Modelling Control and Automation*, 695–701. 10.1109/CIMCA.2005.1631345

Yao, F., Dong, Z. Y., Meng, K., Xu, Z., Iu, H. H., & Wong, K. P. (2012). Quantum-inspired particle swarm optimization for power system operations considering wind power uncertainty and carbon tax in Australia 2012. *IEEE Transactions on Industrial Informatics*, *8*(4), 880–888. doi:10.1109/TII.2012.2210431

Younes, M., Khodja, F., & Kherfane, R. L. (2014). Multi-objective economic emission dispatch solution using hybrid FFA (firefly algorithm) and considering wind power penetration. *Energy*, *67*, 1595–11606. doi:10.1016/j.energy.2013.12.043

Yu, B., Yuan, X., & Wang, J. (2007). Short-term hydro–thermal scheduling using particle swarm optimization method. *Energy Conversion and Management*, *48*(7), 1902–1908. doi:10.1016/j.enconman.2007.01.034

Yu, J. J. Q., & Li, V. O. K. (2015). A social spider algorithm for global optimization. *Applied Soft Computing*, *30*, 614–627. doi:10.1016/j.asoc.2015.02.014

Chapter 4
Ant Colony Optimization Algorithm for Electrical Power Systems Applications:
A Literature Review

Ragab A. El-Sehiemy
https://orcid.org/0000-0002-3340-4031
Kafrelsheikh University, Egypt

Almoataz Y. Abdelaziz
https://orcid.org/0000-0001-5903-5257
Ain Shams University, Egypt

ABSTRACT

Optimization has been an active area of research for several decades. As many real-world optimization problems become increasingly complex, better optimization algorithms are always needed. Recently, meta-heuristic global optimization algorithms have become a popular choice for solving complex and intricate problems, which are otherwise difficult to solve by traditional methods. This chapter reviews the recent applications of ant colony optimization (ACO) algorithm in the field of electrical power systems. Also, the progress of the ACO algorithm and its recent developments are discussed. This chapter covers the aspects like (1) basics of ACO algorithm, (2) progress of ACO algorithm, (3) classification of electrical power system applications, and (4) future of ACO for modern power systems application.

DOI: 10.4018/978-1-7998-8561-0.ch004

Box. List of symbols

Symbol	Definition
τ_{ij}	pheromone trail deposit ed between city i and j by ant k,
α and β	two parameters which influence the relative weight of pheromone trail and heuristic guide function,
η_{ij}	visibility or sight and equal to the inverse of the distance or (= $1/d_{ij}$).
η_{ij}	transition cost between city i and j
q	cities that will be visited after city i,
N_i^k	a tabu list in the memory of ant that recodes the cities visited to avoid stagnations
$\tau_{ij}(t+1)$	pheromone after one tour or iteration
ρ	pheromone evaporation
ε	elite path weighting constant
$\tau_o = 1/d_{ij}$	incremental value of pheromone of each ant
λ	large positive constant
d_{best}	shortest tour distance.
P_{gj} and Q_{gj}	active/reactive power outputs from the generator bus j
P_{dj} and Q_{dj}	active and reactive power demand at bus j,
V_i and V_j	voltages at sending end i and receiving end j,
Y_{ij} and θ_{ij}	admittance magnitude and angle between buses i and j
δ_i and δ_j	phase angles of voltages at buses i and j,
P_{DGj} and Q_{DGj}	active and reactive power injections at location j,
QCj	the reactive power injection at location j.
VSI_j	the voltage stability index of bus j,
Vi,	the voltage magnitude of sending end bus i
P_j and Q_j	the total active and reactive power load fed through bus j,
R_{ij} and X_{ij}	the resistance and reactance of the line connected buses i and j, respectively.
w_1, w_2 and w_3	weighting factors.
P_{Loss}	the total real power loss
P_i and Q_i	the net active and reactive power at bus i,
N_b	the system buses number.
R_{ij}	line resistance between buses i and j,
VD	total voltage deviation
VSI	voltage stability index
G_{ij} and B_{ij}	mutual conductance and susceptance between bus i and j,
N_{PQ}	load buses number
P_{L_i} and Q_{L_i}	active and reactive power demand at bus i
Q_{C_i}	capacitive or inductive power of existing VAR source installed at bus i.
N_{pv}	total number of voltage-controlled buses;
T_k	tapping change of a transformer
N_t	total number of on-load tap changing transformers.

continued on following page

Box. Continued

Symbol	**Definition**
S^{flow}	apparent power flow; N_L is all transmission lines in the system;
F_t	non-linear objective function of power generation cost
a_i, b_i and c_i	coefficients of power generation cost function;
e_i and f_i and is	fuel cost coefficients of the *i*th unit with valve-points effects
NG	number of generation buses.
$F_i(x)$, $F_k(x)$, $F_j(x)$	minimum number of PMU channels, costs and the total fitness function
N_b,	number of system buses,
w_k	the cost of PMUs weighting factor based on the cost of PMU channel,
x_i, x_j and x_k	the control variables vector in a binary or logic form [0 or 1],
N_C	the total number of existing VAR sources;
N_g	refers to the total number of generators.
Y_{LL} *and* Y_{LG}	*are sub-matrices of Y-Bus matrix.*
PF_k *and* PF_k^{max}	power flow in line *k and its* maximum value in line *k*
$D_{k,i}$	sensitivity parameters of the power flows related to the power generations.
F_1	refers to the total fuel costs of generators in $/h;
PG_i	is the power generation at bus *i*,
PLj	the load demand at load bus *j*,
NL	the number of load buses,
P_{losses}	total power losses in the system
min and max operator	the maximum and minimum limits of power generation
ρ	is the evaporation rate,
m	is the number of ants
$\Delta\tau i_j$	is the quantity of pheromone laid on edge e_{ij} by ant *k*
Co	is a constant
L_k	is the length of the tour constructed by ant *k*.
$\eta_{ij} = \frac{1}{d_{ij}}$	is the heuristic information
d_{ij}	is the distance between cities *i* and *j*.
L_{best}	is the length of the tour of the best ant *k*.
φÎ(0,1)	is the pheromone decay coefficient
$\tau 0$	is the initial value of the pheromone.

1. BASICS OF ANT COLONY OPTIMIZER

At first, the ACO algorithms were proposed in (Dorigo & Gambardella, 1997a; Dorigo & Gambardella, 1997b). Applications of the ACO to combinatorial optimization problems such as traveling salesman problem (TSP) (Jun-man & Yi, 2012) and quadratic assignment problem (QAP). The ACO algorithms emulate the behavior of real ants that are members of a family of social insects (Dorigo, 2008; Dorigo & Blum, 2005). A literature survey on the concept and the formulations of ACO was presented in (Dorigo et al., 1999; Dorigo & Stützle, 2009). Ant algorithms for discrete optimization problem were presented (Dorigo et al., 1999; Dréo & Siarry, 2002; Mathur et al., 2000) and for continuous function optimization were developed in (Dréo & Siarry, 2002; Gomez et al., 2004; Mathur et al., 2000). The mathematical model of ACO algorithm involve random number of pheromones that are is placed in each pass after each ant finalizes its tour. Other ants attract to the shortest route according to the probabilistic transition rule that depends on the amount of pheromone deposited and a heuristic guide function. Therefore, the probabilistic transition rule of ant k to go from city i to city j can be expressed as in TSP (Dorigo & Gambardella, 1997a; Dorigo & Gambardella, 1997b; Jun-man & Yi, 2012) as:

$$P_{ij}^{k}(t) = \frac{\left[\tau_{ij}(t)\right]^{\alpha}\left[\eta_{ij}(t)\right]^{\beta}}{\sum_{q}\left[\tau_{iq}(t)\right]^{\alpha}\left[\eta_{iq}(t)\right]^{\beta}}; j,q \in N_{i}^{k} \tag{1}$$

After completing each tour, the local pheromone update is resolute for each ant depending on the route of each ant as in equation (2).

$$\tau_{ij}(t+1) = (1-\rho)\tau_{ij}(t) + \rho\tau_{o} \tag{2}$$

After all ants attracted to the shortest route, a global pheromone update is considered to show the influence of the new addition deposits by the other ants that attractive to the best tour as:

$$\tau_{ij}(t+1) = (1-\rho)\tau_{ij}(t) + \varepsilon\Delta\tau\mathrm{ij}(\mathrm{t}) \tag{3}$$

Then, $\Delta\tau_{ij}$ is the amount of pheromone for elite path as:

$$\Delta\tau_{ij}(\mathrm{t}) = \lambda/d_{best} \tag{4}$$

2. FAMILY OF ACO ALGORITHMS

In the literature, there are many variants of the ACO algorithm. Samples of these variants which aim at improving the performance of the basic version. Hybridizations are the combination of the basic algorithm with a local search. The original ant system and the most successful variants are presented. Description of these variants is presented in the following subsections.

2.1 Ant System (AS)

The main merits of AS is that the pheromone values are updated at each iteration by all the m ants in the iteration. The associated pheromone τ_{ij} with the edge joining cities i and j is computed from Equation (5) as:

$$\tau_{ij} = (1-\rho).\tau_{ij} + \sum_{k=1}^{m} \Delta\tau_{ij}^{k} \tag{5}$$

$$\Delta\tau_{ij}^{k} = \begin{cases} Co/L_k & \text{if the ant } k \text{ used } e_{ij} \text{ in its tour,} \\ 0 & \text{otherwise} \end{cases} \tag{6}$$

for the solution construction, ants select the following city to be visited through a stochastic mechanism. When ant k is in city i and has so far constructed the partial solution x the probability of going to city j is defined by:

$$p_{ij} = \begin{cases} \dfrac{\tau_{ij}^{\alpha}.\eta_{ij}^{\beta}}{\sum\{k \in \{1,..,V\} \mid v_k \notin Ta\}\tau_{ik}^{\alpha}.\eta_{ik}^{\beta}}, & \forall j \in \{1,...,V\}, v_j \notin Ta \\ 0 & \text{otherwise} \end{cases} \tag{7}$$

2.2 MAX –MIN Ant System (MMAS)

MAX–MIN AS is an improvement over the original AS. Its characterizing elements are that only the best ant updates the pheromone trails and that the value of the pheromone is bound. The pheromone update is implemented as follows:

$$\tau_{ij} = \left[(1-\rho).\tau_{ij} + \Delta\tau_{ij}^{best}\right]_{\tau_{min}}^{\tau_{max}}, \tag{8}$$

The operator $[x]_b^a$ is defined as:

$$[x]_b^a = \begin{cases} a & \text{if } x > a \\ b & \text{if } x < b \\ x & \text{otherwise} \end{cases} \tag{9}$$

and $\Delta\tau_{ij}^{best}$ is:

$$\Delta\tau_{ij}^{best} = \begin{cases} 1/L_{best} & \text{if } e_{ij} \text{ belongs to the best tour,} \\ 0 & \text{otherwise} \end{cases} \tag{10}$$

Concerning the lower and upper bounds on the pheromone values that are typically tuned for the considered problem. Nonetheless, some guidelines have been provided for defining the upper and lower bounds based on analytical considerations.

2.3 Ant Colony System (ACS)

Each ant applies the pheromone only to the last edge traversed:

$$\tau_{ij} = (1 - \varphi).\tau_{ij} + \varphi.\tau_0 \tag{11}$$

The update is slightly different:

$$\tau_{ij} = \begin{cases} (1-\rho).\tau_{ij} + \rho.\Delta\tau_{ij} & \text{if } e_{ij} \text{ belongs to best tour} \\ \tau_{ij} & \text{otherwise} \end{cases} \tag{12}$$

As in MMAS, $\Delta\tau_{ij} = 1/L_{best}$, where L_{best} can be either L_{ib} or L_{bs}.

The probability for an ant to move from city i to city j depends on a random variable q uniformly distributed over [0, 1] and a parameter q_0; if $q \leq q_0$, then

$$j = \begin{cases} \arg\max \forall j \in \{1,...,V\},\ v_j \notin Ta,\ \left\{\tau_{ik}^{\alpha}.\eta_{ik}^{\beta}\right\},\ \{k \in \{1,..,V\} \mid v_k \notin Ta\} \\ \text{Equation (12) is used} \qquad\qquad \text{otherwise} \end{cases} \tag{13}$$

2.4 Model Induced Max-Min Ant System

In this model, the induced Max-Min AS (MIMMAS) algorithm is applied for asymmetric TSP (ATSP). The MIMM-ACO has two aspects.

1. Adjusted transition probabilities are developed by replacing static biased weighting factors with dynamic ones. The dynamic weighting factor is closely dependent on partial solution that ant has constructed. Idea behind it is that it favors the choice of edges with small residual cost instead of small actual cost. As byproduct, non-optimal arcs will be identified at each step of tour construction using dual information derived from solving associated assignment problem (AP) and these arcs will be discarded from future consideration.
2. A terminal condition is analytically determined on the basis of the considered pheromone state. The result comes with a necessary condition for obtaining one optimal solution.

Algorithm of MIMM-ACO is based on the computation of AP for residual cost and the PATCH algorithm repairs the AP solutions below the lower bound. This PATCH algorithm will return first candidate solution for MIMM-ACO (s_1). Then, the termination criteria minimum pheromone value are computed from Equation (14) as:

$$\tau_{\min} = \frac{1}{\left(\frac{1}{2}\right) f(s_1) + \left(\frac{1}{2}\right) Z_{AP}^{*}} \tag{14}$$

Then, the initial minimum pheromone is computed dynamically and chosen. Then, the tour will be constructed and then the local search will get used using 2-OPT heuristics.

3. CLASSIFICATION OF ELECTRICAL POWER SYSTEM APPLICATIONS

In power system engineering, researchers found more interests in application of ACO for various power system applications. Table 1 summarizes the developments of ACO for transmission and distribution grids. Most of the referred problems are characterized by the binary decision variables as distribution systems configuration, switching of protective devices and PMU placement for full observability problem. A number of problems that were solved by ACO algorithm that have the binary and continuous variables such as unit commitment, preventive maintenance, Fact's devices, design problem, generation and transmission expansion problem. The main merit of the ACO algorithm is the capability of solving discrete and continuous problem. General description of various optimization problems in power system is expressed in Table 1 for different non-linear complex power system problems.

Table 1. Applications of ACO variants to power systems problems

Reference	Topic	Main features
(Annaluru, Das, & Pahwa, 2004)	Reconfiguration and capacitor placement in distribution systems	The considered formulations deal with the binary design variable for the location and consider the real representation for sizing of allocated capacitors.
(Song et al., 1999)	Combined heat and power economic dispatch	In this work, the improved ant colony search algorithm was developed to solve the combined heat and power scheduling in power systems.
(Hou, 2003; Hou, 2002; Thanathip, 2004)	Economic load dispatch	The economic load dispatch problem deals with continuous variables that represent the scheduling of committed generators to achieve minimum production costs.
(Huang, 2001; Pothiya et al., 2010; Yu & Song, 2001)	Generation scheduling	The generation scheduling aims to find the optimal settings of generation units that minimize the total operational costs.
(Vlachogiannis et al., 2005)	Constrained load flow problem	This problem deals with solving equality constraints at certain loading condition.
(Falaghi et al., 2009; Tippachon & Rerkpreedapong, 2009; Wang & Singh, 2008)	Switches and protective devices	The allocation of switchable and protective devices deal with the binary design variable
(Sum-Im, & Ongsakul, 2003; Simon, Padhy, & Anand, 2006; Chen, 2008; Chandrasekaran, & Simon, 2012; Columbus, Chandrasekaran, & Simon, 2012)	Unit commitment	The unit commitment problems deal with the binary design variable for the location and consider the real representation for sizing of committed generation scheduling.
(Samrout et al., 2005)	Preventive maintenance	The problem identifies the maintenance scheduling of generating units and combines binary and real variables.
(Abdelaziz & Reham, 2012; Falaghi & Haghifam, 2007; Georgilakis & Hatziargyriou, 2013; Niknam, 2005; Niknam et al., 2005; Sheidaei et al., 2008)	Allocation of distributed generation	The allocation of distributed generation has binary design variable for the location and real design variables for sizing of the distributed generation.

continued on following page

Table 1. Continued

Reference	Topic	Main features
(Abbasy & Hosseini, 2007; Abou El-Ela et al., 2011; El-Sehiemy et al., 2012; Gardel, 2006; Ketabi et al., 2010; Lin, 2003; Liu et al., 2010; Mouwafi et al., 2014; Ren et al., 2003)	Power market and pricing of reactive power systems Voltage collapse problem	These references deal with the binary design variable for the location and consider the real representation for sizing of allocated reactive power resources that prevent voltage collapse problem. The pricing of reactive power is also considered an urgent topic in power markets.
(Chen & Tang, 2005; Dong, 2007; Hu et al., 2005; Sun et al., 2006)	Distribution network planning	The planning of distribution network combines both binary and continuous design variables.
(Eroğlu & Seçkiner, 2012; Karaboga, 2009)	Design problems	The design problem combines between binary and continuous design variable to obtain the best control variables that minimize the considered technical and economical objective functions such as reduction of total harmonic distortion level for the digital filter.
(Gasbaoui & Allaoua, 2009)	Optimal power flow	The problem aims at finding the optimal settings of active and reactive control variables that minimizing the generation costs as an economical aspect and transmission power losses minimization, the voltage profile improvement at generation buses and enhancing voltage stability index at load buses.
(Chang et al., 1999; Chen et al., 2006; Teng & Liu, 2003)	Fault estimation and location	In this type of problems, the fault in distribution section is determined.
(Foong et al., 2008)	Power plant maintenance scheduling	The power plant maintenance problem finds the optimal scheduling for generation/ transmission lines maintenance periods considering the minimum nutriment loads.
(Abdelsalam et al., 2014; Abou El-Ela, Kinawy, El-Sehiemy, & Mouwafi, 2014; Abou El-Ela, Kinawy, El-Sehiemy, & Mouwafi, 2014; Abou El-Ela, Kinawy, Mouwafi et al, 2014; Mouwafi & Ragab, 2016)	PMU placement	The PMU placement problem has binary design variables.
(da Silva, 2010; Kannan & Mary Raja Slochanal, 2005; Rahmani et al., 2013)	Power system planning involving generation and Transmission expansion planning	The expansion problem identifies additional generation units and transmission lines that are needed to meet the expected load over short/long planning time horizon.
(Wang, Chiou, & Liu, 2009)	Power system stabilizer	This problem is concerned with the optimal design of power system stabilizer to preserve the power system transient stability.
(Kefayat et al., 2015)	Optimal placement and sizing of distributed energy resources	The optimal placement and sizing of distributed energy resources problem constitutes both binary and real design variables.
(Lu et al., 2013)	FACTS allocation	Various FACTS allocation problems have both discrete and real continuous variables.
(Berbaoui, 2010)	Shunt active power filter system	The ACO algorithm is combined with fuzzy logic to find the optimal configuration of shunt active power filter system.
(Nakawiro & Erlich, 2009)	Load shedding	This problem aims at determining the optimal load to be shed with preserving the voltage stability

4. MATHEMATICAL FORMULATION OF DIFFERENT POWER SYSTEMS PROBLEMS

In this section, the main aim is to present the mathematical formulations of number of power system problems. At first, the optimization problem is expressed as a constrained optimization problem as:

Minimize $f(x)$ (15)

subject to: $g(x)=0$ (16)

$$h(x) \leq 0 \tag{17}$$

where, $f(x)$ is the objective function such as generators fuel costs, transmission line losses etc, $g(x)$ represents the equality constraints, $h(x)$ represents the inequality constraints, and x is the vector of the control variables that may be generator real power outputs, generator voltages, switchable reactive power and transformer tap setting.

4.1 Constrained Economic Load Dispatch Problem

The constrained economic load dispatch (CELD) problem is a non-linear problem. It aims at finding the optimal power generation outputs. The basic objective function in CEED problem minimizes the total fuel costs of the committed generators and preserving all operating requirements Generally, the fuel costs are represented by quadratic functions with superimposed sine components that represent the rippling of steam valve opening effects produced that can be expressed in (18):

$$MinF_t = \sum_{i=1}^{NG} f_i\left(PG_i\right) = \sum_{i=1}^{NG} a_i + b_i PG_i + c_i PG_i^2 + \left|e_i \times \sin\left(f_i \times \left(PG_i^{\min} - PG_i\right)\right)\right| \tag{18}$$

The objective function (18) is subjected to the following constraints:

4.1.1 Equality Constraints

(a) Power Balance Constraint

The generators real power output should be cover the total load demand plus transmission line losses as:

$$\sum_{i=1}^{NG} PG_i = \sum_{j=1}^{NL} PD_j + P_{losses} \tag{19}$$

4.1.2 Inequality Constraints

(a) Generator Real Power Output Limits

The generator real power output is kept within the feasible limits as:

$$PG_i^{\min} \leq PG_i \leq PG_i^{\max} \tag{20}$$

(b) Power Flow Constraints

The power flow in each transmission is bounded by the maximum power flow limits in the line as:

$$\left|PF_k\right| = \left|D_{k,i} PG_i\right| \prec PF_k^{\max} \tag{21}$$

4.2 Optimal Power Flow (OPF) Problem

The objective of the OPF is to optimally identify the power system control variables, while satisfying various equality and inequality constraints. This is mathematically stated as follows:

4.2.1 Problem Objectives

Minimization of fuel cost: Fuel cost for any generator is traditionally modeled as polynomial quadratic function as:

$$F_1 = \sum_{i=1}^{N_g} a_i Pg_i^{\,2} + b_i Pg_i + c_i \ \$/\text{h} \tag{22}$$

Minimization of system power losses: The minimization of system real power losses F_2 (MW) can be calculated as follows:

$$F_2 = \sum_{i,j \in N_b} g_{ij} \left(V_i^2 + V_j^2 - 2V_i V_j \cos\theta_{ij} \right) \tag{23}$$

Voltage profile improvement: Load voltage represents important indicators of system security and service quality by minimizing the voltage deviation (VD) of the load buses than the flat voltage. The VD objective is expressed as:

$$F_3 = VD = \sum_{i=1}^{N_{Load}} \left| V_i - 1 \right| \tag{24}$$

Voltage stability enhancement: For heavy stressed power systems, voltage stability becomes a significant issue. A specific objective function is considered in this case that aims to minimizing the maximum stability index (L-index). The L-index is employed on the basis of static power flow computations every load bus. Its range is from Zero to 1 to reflect no load and voltage collapse cases. Computation of L-index is carried at each load bus as:

$$L_j = \left| 1 - \sum_{i=1}^{N_g} F_{ji} \frac{V_i}{V_j} \angle(\theta_{ij} + \delta_i - \delta_j) \right| \tag{25}$$

$$F_{ji} = -[Y_{LL}]^{-1}[Y_{LG}] \tag{26}$$

The maximum L-index should be minimized to improve the voltage stability (F_4) as:

$$F_4 = \text{Max}(L_j)\ j=1,2,\ldots\ldots,N_b \tag{27}$$

4.2.2 Constraints

4.2.2.1 Equality Constraints

Two load flow equations are used to represent the equality constraints as:

$$Q_{gi} - Q_{Li} + Q_{Ci} - V_i \sum_{j=1}^{N_b} V_j (G_{ij} \sin\theta_{ij} - B_{ij} \cos\theta_{ij}) = 0, \quad i = 1,2,...N_{PQ} \tag{28}$$

$$P_{gi} - P_{Li} - V_i \sum_{j=1}^{N_b} V_j (G_{ij} \cos\ \theta_{ij} + B_{ij} \sin\ \theta_{ij}) = 0, \quad i = 1,2,......N_b - slack \tag{29}$$

4.2.2.2 Inequality Constraints

Furthermore, the power system has to satisfy inequality constraints corresponding to the operational variables as:

$$P_{g\,i}^{\min} \le P_{g\,i} \le P_{g\,i}^{\max}, \quad i = 1,2,......N_g \tag{30}$$

$$V_i^{\min} \le V_i \le V_i^{\max}, \quad i = 1,2,......N_b \tag{31}$$

$$T_k^{\min} \le T_k \le T_k^{\max}, \quad k = 1,2,......N_t \tag{32}$$

$$0 \le Q_{C\,e} \le Q_{C\,e}^{\max}, \quad e = 1,2,......N_C \tag{33}$$

$$Q_{g\,i}^{\min} \le Q_{g\,i} \le Q_{g\,i}^{\max}, \quad i = 1,2,......N_{pv} \tag{34}$$

$$\left|S_L^{flow}\right| \le S_L^{\max}, \ L = 1,2,......N_L \tag{35}$$

4.3 Optimal Placement Problem (OPP) Phasor Measurement Units

The solution of OPP with minimum PMU channels' number and the cost of PMUs to make the power system complete observability can be expressed as constrained multi-objective optimization problem as:

$$\text{Min } F(x) = \text{Min}(F_i(x) + F_j(x) + F_k(x)) \tag{36a}$$

$$= \sum_{i=1}^{N_b} w_i x_i + \sum_{j=1}^{N_b} w_j x_j + \sum_{k=1}^{N_b} w_k x_k \tag{36b}$$

Subject to:

$$g(x) = \sum_{i=1, j=i}^{N_b} A_{ij} x_i \geq b \tag{37}$$

$$N_{PMU} < N_{PMU\max} \tag{38}$$

$$N_{Ch_i} \leq N_{Ch_{i\max}} \tag{39}$$

If x_i equals 0 refers to no existing of PMU at bus i while 1 refers to installing of PMU at that bus, which are defined in (40).

$$x_i = \begin{cases} 1 & \text{if a } PMU \text{ installed at bus } i \\ 0 & \text{otherwise} \end{cases} \tag{40}$$

In (37), $g(x)$ is the observability constraint which must be verified at each bus in the system. On the other hand, if a PMU is installed at bus i, so all the buses connected to it don't need installing PMU at them, because phasors at these buses can be determined using Ohm's law. A_{ij} is the connectivity matrix which can be constructed based on the line data of the system by replacing the values of the line data with binary logic, so it can be expressed as:

$$A_{ij} = \begin{cases} 1 & z_{ij} \geq 0 \\ 0 & z_{ij} = \infty \end{cases} \tag{41}$$

The bound b in (37) refers to the minimum limit of measurement redundancy matrix of length N_b and may be taken as:

$$b = \begin{cases} 1 & \text{for normal operating condition} \\ 2 & \text{for emergency operating condition} \end{cases} \tag{42}$$

Form (42), if b is 1, which means each bus is observed one time at least and this condition is suitable at normal condition. While, if the value of b is 2 that means each bus is observed twice at least. Therefore, this condition is suitable at emergency condition such as any single line outage or any single PMU loss.

In (38), N_{PMU} is the number of required PMUs to complete system observability, while $N_{PMU\max}$ is the maximum number of PMUs equals to the number of total buses. In (39), N_{Chi} is the number of PMU channels at bus i, while $N_{Chi\max}$ is the maximum number of PMU channels at that bus equals to the number of lines connected to bus i plus one channel refers to the voltage measurement at that bus.

4.4 Distributed generation (DG)/Capacitor Placement Problem (CPP) in Distribution Systems

The optimal DGs and capacitors placement problem with single and multi-objective functions are based on various scenarios of DGs and capacitors placements subject to equality and inequality constraints. Mathematical formulation for different objective functions is presented as follows:

4.4.1 Individual Objective Functions

The purpose of optimal DGs and capacitors placement in radial distribution systems is to reduce the total power loss, VD and the inverse of total VSI.

The main merit of optimal DGs and capacitors placement in distribution systems is to minimize the real power loss. Mathematically, the real power loss can be expressed as:

$$f_1 = Min\, P_{Loss} = \sum_{i=1}^{N_b}\sum_{j=1}^{N_b}\left[\alpha_{ij}\left(P_iP_j + Q_iQ_j\right) + \beta_{ij}\left(Q_iP_j - P_iQ_j\right)\right] \tag{43}$$

$$\alpha_{ij} = \frac{R_{ij}}{V_iV_j}\cos\left(\delta_i - \delta_j\right),\ \ \beta_{ij} = \frac{R_{ij}}{V_iV_j}\sin\left(\delta_i - \delta_j\right)$$

The distribution system suffers from the higher decline in the voltage levels with move away from the substation especially for radial distribution system. The objective function for improving the voltage profile by minimization of the voltage deviation can be formulated as:

$$f_2 = Min\, VD = \sum_{i=1}^{N_b}\left(V_i - V_{rated}\right)^2 \tag{44}$$

The flat voltage at each bus that equals 1.0 p.u.

Another important aspect is the maximization the VSI. The weakest voltage bus is identified by VSI that can lead to voltage instability and indicate the sensitive buses to voltage collapse under certain condition such as load increasing in the future. Therefore, it is necessary to maintain VSI within a specific limit. When DGs and/or capacitors are connected to distribution system, the VSI will be changed, where this index can be calculated at all buses in the distribution system. Consider a two-node radial distribution system. Buses i and j are considered as the sending and receiving end buses. After the load flow calculation, the VSI for all the receiving end buses in the distribution system can be formulated as:

$$VSI_j = \left|V_i\right|^4 - 4\left|V_i\right|^2\left[P_jR_{ij} + Q_jX_{ij}\right] - 4\left[P_jX_{ij} - Q_jR_{ij}\right]^2 \tag{45}$$

The objective function for improving the VSI can be expressed as:

$$f_3 = Min\frac{1}{VSI} = Min\left(\frac{1}{\sum_{j=1}^{N_b} VSI_j}\right) \tag{46}$$

For achieving the voltage stability criteria, the objective function f_3 should be minimized.

4.4.2. Multi-Objective Function Formulation

The multi-objective function (F) minimizes the total power loss, the total VD and the inverse of total VSI at the same time. Therefore, it can be formulated as:

$$\begin{aligned} \min F &= \left(w_1 f_1 + w_2 f_2 + w_3 f_3\right) \\ \min F &= \left(w_1 * P_{Loss} + w_2 * VD + w_3 * \frac{1}{VSI}\right) \end{aligned} \tag{47}$$

The weighting factors are selected according to the priority of the objective functions.

4.4.3 System Constraints

The above mentioned single and multi-objective functions in equations (48-57). Equations (48) and (49) give the power balance constraints. Equation (50) gives the constraint of voltage constraint with the aim of preserving the voltage at each bus within their minimum and maximum permissible limits. The capability of transmission power flow constraint in each line (PF_k) is presented in Equation (51). The transmission system must be operated below the maximum power flow limit (PF_k^{max}). The overall system power factor ($pf_{overall}$) must be greater than or equal to the minimum limit of overall power factor (pf_{min}) as presented in (52). The constraints that are presented in (53) and (54) aim to reduce respectively the number of DGs and capacitors, Therefore, the optimal number of DGs (N_{DG}) is limited by the maximum number of possible locations ($N_{DG}^{\max}$). Also, the optimal number of capacitors (N_C) must be is limited by their maximum number of possible locations ($N_C^{\max}$). The active and reactive power injections by DGs and the reactive power injected by capacitors are kept with their boundries as presented in Equations (55)-(57).

$$P_{gj} - P_{dj} - V_j \sum_{j=1}^{N_b} V_i Y_{ij} \cos(\delta_j - \delta_i - \theta_{ij}) = 0 \tag{48}$$

$$Q_{gj} - Q_{dj} - V_j \sum_{j=1}^{N_b} V_i Y_{ij} \sin(\delta_j - \delta_i - \theta_{ij}) = 0 \tag{49}$$

$$V_i^{\min} \leq V_i \leq V_i^{\max} \tag{50}$$

$$\left|PF_k\right| < PF_k^{\max} \tag{51}$$

$$\left|pf_{overall}\right| \geq pf_{overall}^{\min} \tag{52}$$

$$N_{DG} \leq N_{DG}^{\max} \tag{53}$$

$$N_C \leq N_C^{\max} \tag{54}$$

$$P_{DGj}^{\min} \leq P_{DGj} \leq P_{DGj}^{\max} \tag{55}$$

$$Q_{DGj}^{\min} \leq Q_{DGj} \leq Q_{DGj}^{\max} \tag{56}$$

$$Q_{Cj}^{\min} \leq Q_{Cj} \leq Q_{Cj}^{\max} \tag{57}$$

5. FUTURE OF ACO FOR POWER SYSTEM APPLICATIONS

The future applications of ACO variants to power systems application is rich field especially in the following modern topics:

1. Integration of distribution generation into distribution systems.
2. Economic studies in Microgrid.
3. Design of electrical apparatus.
4. Transmission switching problem.
5. Unit commitment problems considering transmission commitment.
6. Capacitor placement for enhancing the voltage profile with different types of energy renewable sources.
7. Fault location in optimal framework.
8. Advanced restoration mechanisms.
9. Power quality problems involving allocation of harmonic meters and allocation of active filters into distribution systems.
10. Congestion management in the open access electricity markets.
11. Energy hubs
12. Combined heat and power dispatch problem with renewable energy resources.

In the view point of algorithm improvements, the future is open for hybridization between ACO algorithm with a number of information-based algorithm such as:

1. Rough theory.
2. Graph theory.
3. Neuterosofic.

6. CONCLUSION

This chapter reviews the recent progress of one recent optimization algorithm, namely ant colony optimization algorithm, to power system applications. It is found that the ACO variants are efficiently applied for binary and real design variables' problems. The future of developing of ACO variants is open to add the merits of fuzzy logic, rough theory and Neuterosofic to enhance the routine of ACO algorithm.

REFERENCES

Abbasy, A., & Hosseini, S. H. (2007). *Ant colony optimization-based approach to optimal reactive power dispatch: a comparison of various ant systems. In Power Engineering Society Conference and Exposition in Africa, 2007. PowerAfrica'07*. IEEE.

Abdelaziz, A. Y., Osama, R. A., & El-Khodary, S. M. (2012). Reconfiguration of distribution systems for loss reduction using the hyper-cube ant colony optimisation algorithm. *IET Generation, Transmission & Distribution*, *6*(2), 176–187. doi:10.1049/iet-gtd.2011.0281

Abdelaziz, A. Y., & Reham, A. (2012). Application of ant colony optimization and harmony search algorithms to reconfiguration of radial distribution networks with distributed generations. *Journal of Bioinformatics and Intelligent Control*, *1*(1), 86–94. doi:10.1166/jbic.2012.1007

Abdelsalam, H. A., Abdelaziz, A. Y., & Mukherjee, V. (2014). Optimal PMU placement in a distribution network considering network reconfiguration. *2014 International Conference on Circuits, Power and Computing Technologies, ICCPCT 2014*, 191–196. 10.1109/ICCPCT.2014.7054997

Abou El-Ela, A., Kinawy, A., El-Sehiemy, R., & Mouwafi, M. (2014). Ant colony optimizer for phasor measurement units placement. *International Review of Applied Sciences and Engineering*, *5*(2), 127–134. doi:10.1556/irase.5.2014.2.4

Abou El-Ela, A. A., Kinawy, A. M., El-Sehiemy, R. A., & Mouwafi, M. T. (2011). Optimal reactive power dispatch using ant colony optimization algorithm. *Electrical Engineering*, *93*(2), 103–116. doi:10.100700202-011-0196-4

Abou El-Ela, A. A., Kinawy, A. M., El-Sehiemy, R. A., & Mouwafi, M. T. (2014). Optimal Placement of Phasor Measurement Units with Limited Channels Using Ant Colony Optimization Algorithm. Engineering Research Journal, 37(2), 191-197. doi:10.21608/erjm.2014.66918

Abou El-Ela, A. A., Kinawy, A. M., Mouwafi, M. T., & El-Sehiemy, R. A. (2014). Optimal Placement of Phasor Measurement Units for Power System Observability Using Ant Colony Optimization Algorithm. *16th International Middle- East Power Systems Conference (MEPCON'2014)*, 1-6.

Ahuja, A., & Pahwa, A. (2005). Using ant colony optimization for loss minimization in distribution networks. In *Proceedings of the 37th Annual North American Power Symposium*. IEEE. 10.1109/NAPS.2005.1560562

Annaluru, R., Das, S., & Pahwa, A. (2004). *Multi-level ant colony algorithm for optimal placement of capacitors in distribution systems. In Evolutionary Computation, 2004. CEC2004. Congress on* (Vol. 2). IEEE.

Berbaoui, B. (2010). Optimization of shunt active power filter system fuzzy logic controller based on ant colony algorithm. *Journal of Theoretical and Applied Information Technology*, *14*(1), 117–125.

Bouri, Zeblah, Ghoraf, Hadjeri, & Hamdaoui. (2005). Ant colony optimization to shunt capacitor allocation in radial distribution systems. *Acta Electrotechnica et Informatica*, *5*(1), 4.

Carpaneto, E., & Chicco, G. (2004). Ant-colony search-based minimum losses reconfiguration of distribution systems. In *Electrotechnical Conference, 2004. MELECON 2004. Proceedings of the 12th IEEE Mediterranean (Vol. 3)*. IEEE. 10.1109/MELCON.2004.1348215

Carpaneto, E., & Chicco, G. (2008). Distribution system minimum loss reconfiguration in the hyper-cube ant colony optimization framework. *Electric Power Systems Research*, *78*(12), 2037–2045. doi:10.1016/j.epsr.2008.06.009

Chandrasekaran, K., & Simon, P. (2012). Network and reliability constrained unit commitment problem using binary real coded firefly algorithm. *International Journal of Electrical Power & Energy Systems*, *43*(1), 921–832. doi:10.1016/j.ijepes.2012.06.004

Chang, C.-F. (2008). Reconfiguration and capacitor placement for loss reduction of distribution systems by ant colony search algorithm. *IEEE Transactions on Power Systems*, *23*(4), 1747–1755. doi:10.1109/TPWRS.2008.2002169

Chang, C. S., Tian, L., & Wen, F. S. (1999). A new approach to fault section estimation in power systems using Ant system. *Electric Power Systems Research*, *49*(1), 63–70. doi:10.1016/S0378-7796(98)00127-8

Chen, Ding, & Zhao. (2006). Ant colony algorithm for solving fault location in distribution networks. *Automation of Electric Power Systems*, *5*, 17.

Chen & Tang. (2005). Tabu search-ant colony optimization hybrid algorithm based distribution network planning. *Power System Technology*, *2*, 5.

Chen, Y. (2008). An Ant Colony Optimization and Particle Swarm Optimization Hybrid Algorithm for Unit Commitment Based on Operate Coding. *Power System Technology*, *6*, 14.

Columbus, C., Chandrasekaran, K., & Simon, S. P. (2012). Nodal ant colony optimization for solving profit based unit commitment problem for GENCOs. *Applied Soft Computing*, *12*(1), 145–160. doi:10.1016/j.asoc.2011.08.057

da Silva, A. M. (2010). Reliability worth applied to transmission expansion planning based on ant colony system. *International Journal of Electrical Power & Energy Systems*, *32*(10), 1077–1084. doi:10.1016/j.ijepes.2010.06.003

Dong, Y.-F. (2007). Combination of genetic algorithm and ant colony algorithm for distribution network planning. In *2007 International Conference on Machine Learning and Cybernetics* (Vol. 2). IEEE. 10.1109/ICMLC.2007.4370288

Dorigo, M. (2008). Ant Colony Optimization and Swarm Intelligence. In *6th International Conference, ANTS 2008, Brussels, Belgium, September 22-24, 2008, Proceedings (Vol. 5217).* Springer.

Dorigo, M., & Blum, C. (2005). Ant colony optimization theory: A survey. *Theoretical Computer Science, 344*(2), 243–278. doi:10.1016/j.tcs.2005.05.020

Dorigo, M., Di Caro, G., & Gambardella, L. M. (1999). Ant algorithms for discrete optimization. *Artificial Life, 5*(2), 137–172. doi:10.1162/106454699568728 PMID:10633574

Dorigo, M., & Gambardella, L. M. (1997a). Ant Colonies for the Traveling Salesman Problem. *Bio Systems, 43*(2), 73–81. doi:10.1016/S0303-2647(97)01708-5 PMID:9231906

Dorigo, M., & Gambardella, L. M. (1997b). Ant colony system: A cooperative learning approach to the traveling salesman problem. *IEEE Transactions on Evolutionary Computation, 1*(1), 53–66. doi:10.1109/4235.585892

Dorigo, M., & Stützle, T. (2009). *Ant colony optimization: overview and recent advances. Techreport.* IRIDIA, Universite Libre de Bruxelles.

Dréo, J., & Siarry, P. (2002). A new ant colony algorithm using the heterarchical concept aimed at optimization of multiminima continuous functions. In *International Workshop on Ant Algorithms*. Springer Berlin Heidelberg.

Ebrahimi, J., Hosseinian, S. H., & Gharehpetian, G. B. (2011). Unit Commitment Problem Solution Using Shuffled Frog Leaping Algorithm. *IEEE Transactions on Power Systems, 26*(2), 573–581. doi:10.1109/TPWRS.2010.2052639

El-Ela, A. A. A., El-Sehiemy, R. A., Kinawy, A.-M., & Mouwafi, M. T. (2016). Optimal capacitor placement in distribution systems for power loss reduction and voltage profile improvement. *IET Generation, Transmission & Distribution, 10*(5), 1209–1221. doi:10.1049/iet-gtd.2015.0799

El-Sehiemy, R. A., El Ela, A. A. A., Kinawy, A. M. M., & Mouwafia, M. T. (2012). An ant colony optimization algorithm-based emergency control strategy for voltage collapse mitigation. *International Review of Applied Sciences and Engineering, 3*(2), 147–156. doi:10.1556/irase.3.2012.2.8

Eroğlu, Y., & Seçkiner, S. U. (2012). Design of wind farm layout using ant colony algorithm. *Renewable Energy, 44*, 53–62. doi:10.1016/j.renene.2011.12.013

Falaghi, H., & Haghifam, M.-R. (2007). ACO based algorithm for [72 sources allocation and sizing in distribution systems. In Power Tech, 2007 IEEE. IEEE.

Falaghi, H., Haghifam, M.-R., & Singh, C. (2009). Ant colony optimization-based method for placement of sectionalizing switches in distribution networks using a fuzzy multiobjective approach. *IEEE Transactions on Power Delivery, 24*(1), 268–276. doi:10.1109/TPWRD.2008.2005656

Foong, W. K., Simpson, A. R., Maier, H. R., & Stolp, S. (2008). Ant colony optimization for power plant maintenance scheduling optimization—A five-station hydropower system. *Annals of Operations Research, 159*(1), 433–450. doi:10.100710479-007-0277-y

Gardel, P. (2006). Multiobjective reactive power compensation with an ant colony optimization algorithm. In *AC and DC Power Transmission, 2006. ACDC 2006. The 8thIEE International Conference on.* IET. 10.1049/cp:20060056

Gasbaoui, B., & Allaoua, B. (2009). Ant colony optimization applied on combinatorial problem for optimal power flow solution. *Leonardo Journal of Sciences, 14*, 1–17.

Georgilakis, P. S., & Hatziargyriou, N. D. (2013). Optimal distributed generation placement in power distribution networks: Models, methods, and future research. *IEEE Transactions on Power Systems, 28*(3), 3420–3428. doi:10.1109/TPWRS.2012.2237043

Gomez, J. F., Khodr, H. M., DeOliveira, P. M., Ocque, L., Yusta, J. M., Villasana, R., & Urdaneta, A. J. (2004). Ant colony system algorithm for the planning of primary distribution circuits. *IEEE Transactions on Power Systems, 19*(2), 996–1004. doi:10.1109/TPWRS.2004.825867

Guoqing, C. G. W. L. T. (2001). Distribution network reconfiguration for loss reduction using an ant colony optimization method. *Proccedings of thc Csu-cpsa, 2*.

Hao, J. (2002). Optimal Unit Commitment Based On Ant Colony Optimization Algorithm. *Power System Technology, 11*, 6.

Hou, Y-h. (2003). Economic dispatch of power systems based on generalized ant colony optimization method. *Proceedings of the CSEE, 3*, 13.

Hou, Y.-H. (2002). Generalized ant colony optimization for economic dispatch of power systems. In *Power System Technology, 2002. Proceedings. PowerCon 2002. International Conference on* (Vol. 1). IEEE.

Hu, B., Gu, J., & Wang, Y.-D. (2005). An ant colony optimization based method for power distribution network planning. *Relay, 33*(21), 54–57.

Huang, S.-J. (2001). Enhancement of hydroelectric generation scheduling using ant colony system based optimization approaches. *IEEE Transactions on Energy Conversion, 16*(3), 296–301. doi:10.1109/60.937211

Jiang, L. L., Maskell, D. L., & Patra, J. C. (2013). A novel ant colony optimization-based maximum power point tracking for photovoltaic systems under partially shaded conditions. *Energy and Building, 58*, 227–236. doi:10.1016/j.enbuild.2012.12.001

Jun-man, K., & Yi, Z. (2012). Application of an improved ant colony optimization on generalized traveling salesman problem. *Energy Procedia, 17*, 319–325. doi:10.1016/j.egypro.2012.02.101

Kannan, S., & Mary Raja Slochanal, S. (2005). Application and Comparison of Metaheuristic Techniques to Generation Expansion Planning Problem. *IEEE Transactions on Power Systems, 20*(1), 466–475. doi:10.1109/TPWRS.2004.840451

Karaboga, N. (2009). A new design method based on artificial bee colony algorithm for digital IIR filters. *Journal of the Franklin Institute, 346*(4), 328–348. doi:10.1016/j.jfranklin.2008.11.003

Kefayat, M., Lashkar Ara, A., & Nabavi Niaki, S. A. (2015). A hybrid of ant colony optimization and artificial bee colony algorithm for probabilistic optimal placement and sizing of distributed energy resources. *Energy Conversion and Management*, *92*, 149–161. doi:10.1016/j.enconman.2014.12.037

Ketabi, A., Alibabaee, A., & Feuillet, R. (2010). Application of the ant colony search algorithm to reactive power pricing in an open electricity market. *International Journal of Electrical Power & Energy Systems*, *32*(6), 622–628. doi:10.1016/j.ijepes.2009.11.019

Kumar, K. S., & Jayabarathi, T. (2012). Power system reconfiguration and loss minimization for distribution systems using bacterial foraging optimization algorithm. *International Journal of Electrical Power & Energy Systems*, *36*(1), 13–17. doi:10.1016/j.ijepes.2011.10.016

Lin, Z-h. (2003). Generalized ant colony optimization algorithm for reactive power optimization in power systems. *Journal of North China Electric Power University, 2*, 1.

Liu, Z., Wang, X., & Li, C. (2010). *Reactive power optimization in power system based on chaos ant colony algorithm. In CICED 2010 Proceedings*. IEEE.

Lu, C.-F., Hsu, C.-H., & Juang, C.-F. (2013). Coordinated control of flexible AC transmission system devices using an evolutionary fuzzy lead-lag controller with advanced continuous ant colony optimization. *IEEE Transactions on Power Systems*, *28*(1), 385–392. doi:10.1109/TPWRS.2012.2206410

Mathur, M., Karale, S. B., Priye, S., Jayaraman, V. K., & Kulkarni, B. D. (2000). Ant colony approach to continuous function optimization. *Industrial & Engineering Chemistry Research*, *39*(10), 3814–3822. doi:10.1021/ie990700g

Meziane, R., Massim, Y., Zeblah, A., Ghoraf, A., & Rahli, R. (2005). Reliability optimization using ant colony algorithm under performance and cost constraints. *Electric Power Systems Research*, *76*(1), 1–8. doi:10.1016/j.epsr.2005.02.008

Mouwafi, M. T., El-Sehiemy, R. A., Abou El-Ela, A. A., & Kinawy, A. M. (2016). Optimal placement of phasor measurement units with minimum availability of measuring channels in smart power systems. *Electric Power Systems Research*, *141*, 421–431. doi:10.1016/j.epsr.2016.07.029

Mouwafi, M. T., El-Sehiemy, R. A., Abou El-Ela, A. A. & Kinawy, A. M. (2014). A Complete Reactive Power Management Strategy Using Ant Colony Optimization Algorithm. *Emirates Journal for Engineering Research*, *19*(3), 19–31.

Nakawiro, W., & Erlich, I. (2009). Optimal load shedding for voltage stability enhancement by ant colony optimization. In *Intelligent System Applications to Power Systems, 2009. ISAP'09. 15th International Conference on* (pp. 1-6). IEEE. 10.1109/ISAP.2009.5352886

Niknam, T. (2005). A new approach based on ant algorithm for Volt/Var control in distribution network considering distributed generation. *Iranian Journal of Science & Technology, Transaction B*, *29*(B4), 1–15.

Niknam, T. (2008). A new approach based on ant colony optimization for daily Volt/Var control in distribution networks considering distributed generators. *Energy Conversion and Management*, *49*(12), 3417–3424. doi:10.1016/j.enconman.2008.08.015

Niknam, T., Ranjbar, A. M., & Shirani, A. R. (2005). A new approach for distribution state estimation based on ant colony algorithm with regard to distributed generation. *Journal of Intelligent & Fuzzy Systems*, *16*(2), 119–131.

Pothiya, S., Ngamroo, I., & Kongprawechnon, W. (2010). Ant colony optimization for economic dispatch problem with non-smooth cost functions. *International Journal of Electrical Power & Energy Systems*, *32*(5), 478–487. doi:10.1016/j.ijepes.2009.09.016

Rahmani, R., Yusof, R., Seyedmahmoudian, M., & Mekhilef, S. (2013). Hybrid technique of ant colony and particle swarm optimization for short term wind energy forecasting. *Journal of Wind Engineering and Industrial Aerodynamics*, *123*, 163–170. doi:10.1016/j.jweia.2013.10.004

Ren, X., Deng, Y., Zhao, C., & Zhao, D. P. (2003). Study on the algorithm for dynamic reactive power optimization of distribution systems. *Zhongguo Dianji Gongcheng Xuebao*, *23*(1), 31–36.

Saffar, A., Hooshmand, R., & Khodabakhshian, A. (2011). A new fuzzy optimal reconfiguration of distribution systems for loss reduction and load balancing using ant colony search-based algorithm. *Applied Soft Computing*, *11*(5), 4021–4028. doi:10.1016/j.asoc.2011.03.003

Samrout, M., Yalaoui, F., Châtelet, E., & Chebbo, N. (2005). New methods to minimize the preventive maintenance cost of series–parallel systems using ant colony optimization. *Reliability Engineering & System Safety*, *89*(3), 346–354. doi:10.1016/j.ress.2004.09.005

Sheidaei, F., Shadkam, M., & Zarei, M. (2008). Optimal Distributed Generation allocation in distribution systems employing ant colony to reduce losses. In *Universities Power Engineering Conference, 2008. UPEC 2008. 43rd International*. IEEE.

Shi, L., Hao, J., Zhou, J., & Xu, G. (2004). Ant colony optimization algorithm with random perturbation behavior to the problem of optimal unit commitment with probabilistic spinning reserve determination. *Electric Power Systems Research*, *69*(2), 295–303. doi:10.1016/j.epsr.2003.10.008

Simon, S. P., Padhy, N. P., & Anand, R. S. (2006). An ant colony system approach for unit commitment problem. *International Journal of Electrical Power & Energy Systems*, *28*(5), 315–323. doi:10.1016/j.ijepes.2005.12.004

Sisworahardjo, N. S., & El-Keib, A. A. (2002). Unit commitment using the ant colony search algorithm. In *Power Engineering 2002 Large Engineering Systems Conference on, LESCOPE 02*. IEEE. 10.1109/LESCPE.2002.1020658

Song, Y. H., Chou, C. S., & Stonham, T. J. (1999). Combined heat and power economic dispatch by improved ant colony search algorithm. *Electric Power Systems Research*, *52*(2), 115–121. doi:10.1016/S0378-7796(99)00011-5

Su, C.-T., Chang, C.-F., & Chiou, J.-P. (2005a). Optimal capacitor placement in distribution systems employing ant colony search algorithm. *Electric Power Components and Systems*, *33*(8), 931–946. doi:10.1080/15325000590909912

Su, C.-T., Chang, C.-F., & Chiou, J.-P. (2005b). Distribution network reconfiguration for loss reduction by ant colony search algorithm. *Electric Power Systems Research*, *75*(2), 190–199. doi:10.1016/j.epsr.2005.03.002

Sum-Im, T., & Ongsakul, W. (2003). Ant colony search algorithm for unit commitment. In *Industrial Technology, 2003 IEEE International Conference on* (Vol. 1). IEEE. 10.1109/ICIT.2003.1290244

Sun, Shang, & Niu. (2006). Application of Improved Ant Colony Optimization Algorithm in Distribution Network Planning. *Power System Technology, 15*, 21.

Swarnkar, A., Gupta, N., & Niazi, K. R. (2011). Adapted ant colony optimization for efficient reconfiguration of balanced and unbalanced distribution systems for loss minimization. *Swarm and Evolutionary Computation*, *1*(3), 129–137. doi:10.1016/j.swevo.2011.05.004

Teng, J.-H., & Liu, Y.-H. (2003). A novel ACS-based optimum switch relocation method. *IEEE Transactions on Power Systems*, *18*(1), 113–120. doi:10.1109/TPWRS.2002.807038

Thanathip. (2004). Economic dispatch by ant colony search algorithm. In *Cybernetics and Intelligent Systems, 2004 IEEE Conference on* (Vol. 1). IEEE.

Tippachon, W., & Rerkpreedapong, D. (2009). Multiobjective optimal placement of switches and protective devices in electric power distribution systems using ant colony optimization. *Electric Power Systems Research*, *79*(7), 1171–1178. doi:10.1016/j.epsr.2009.02.006

Vaisakh, K., & Srinivas, L. R. (2011). Evolving ant colony optimization based unit commitment. *Applied Soft Computing*, *11*(2), 2863–2870. doi:10.1016/j.asoc.2010.11.019

Vlachogiannis, J. G., Hatziargyriou, N. D., & Lee, K. Y. (2005). Ant colony system-based algorithm for constrained load flow problem. *IEEE Transactions on Power Systems*, *20*(3), 1241–1249. doi:10.1109/TPWRS.2005.851969

Vuppalapati, S. H. K., & Srivastava, A. (2010). Application of ant colony optimization for reconfiguration of shipboard power system. *International Journal of Engineering Science and Technology*, *2*(3), 119–131. doi:10.4314/ijest.v2i3.59181

Wang, B., Liu, D., & Li, X. (2009). An improved ant colony system in optimizing power system PMU placement problem. In *2009 Asia-Pacific Power and Energy Engineering Conference*. IEEE. 10.1109/APPEEC.2009.4918138

Wang, L., & Singh, C. (2008). Reliability-constrained optimum placement of reclosers and distributed generators in distribution networks using an ant colony system algorithm. *IEEE Transactions on Systems, Man and Cybernetics. Part C, Applications and Reviews*, *38*(6), 757–764. doi:10.1109/TSMCC.2008.2001573

Wang, S.-K., Chiou, J.-P., & Liu, C.-W. (2009). Parameters tuning of power system stabilizers using improved ant direction hybrid differential evolution. *International Journal of Electrical Power & Energy Systems*, *31*(1), 34–42. doi:10.1016/j.ijepes.2008.10.003

Wu, Y.-K., Lee, C.-Y., Liu, L.-C., & Tsai, S.-H. (2010). Study of reconfiguration for the distribution system with distributed generators. *IEEE Transactions on Power Delivery*, *25*(3), 1678–1685. doi:10.1109/TPWRD.2010.2046339

Yazdani, D., & Totonchi, A. (2012). Power system reconfiguration and loss minimization for an distribution systems using bacterial foraging optimization algorithm. *International Journal of Electrical Power & Energy Systems*, *36*(1), 13–17. doi:10.1016/j.ijepes.2011.10.016

Yu, I. K., & Song, Y. H. (2001). A novel short-term generation scheduling technique of thermal units using ant colony search algorithms. *International Journal of Electrical Power & Energy Systems*, *23*(6), 471–479. doi:10.1016/S0142-0615(00)00065-X

Zhang, Zhao, & Feng. (2008). A Novel Optimization Reconfiguration Algorithm for Power Supply Restoration of Distribution Network. *Power System Technology, 7*, 12.

Chapter 5
Application of Optimization to Sizing Renewable Energy Systems and Energy Management in Microgrids:
State of the Art and Trends

Khaled Dassa
University of Boumerdes, Algeria

Abdelmadjid Recioui
https://orcid.org/0000-0001-9028-3910
University of Boumerdes, Algeria

ABSTRACT

The smart grid is the aggregation of emerging technologies in both hardware and software along with practices to make the existing power grid more reliable and ultimately more beneficial to consumers. The smart grid concept is associated with the production of electricity from renewable energy sources (RES). For the distant isolated regions, microgrids (MG) with RES are offering a suitable solution for remote and isolated region electrification. The improper sizing would lead to huge investment cost which could have been avoided. The objective of this chapter is to review the state-of-the-art studies on the use of optimization techniques to renewable energy design and sizing. The chapter reviews the optimization techniques employed at different components of the microgrid including the energy sources, storage elements, and converters/inverters with their control systems.

INTRODUCTION

The traditional power grid is made up of synchronous machines, power transformers, transmission lines, transmission substations, distribution lines, distribution substations, and various types of loads that are

DOI: 10.4018/978-1-7998-8561-0.ch005

interconnected altogether. They are located far from the power consumption area and electric power is transmitted through long transmission lines. Traditional power grids have served us for decades because they are predictable and reliable however, Due to the growing concern about the climate change caused by greenhouse gas emissions and the escalating increase in electricity demand (2% per year until 2040) which exceeds the demand for any other form of final energy globally, there is a tendency to replace fossil fuels by renewable sources such as wind and solar (Baimel et al., 2016). Figure 1 shows the growth of renewable energy penetration into total energy generation over next three decades across the world. The Europe will take the maximum share of 44% of renewable generation in their total energy production (Ramesh Babu, 2017).

Figure 1. Percentage growth forecast in renewable energy generation

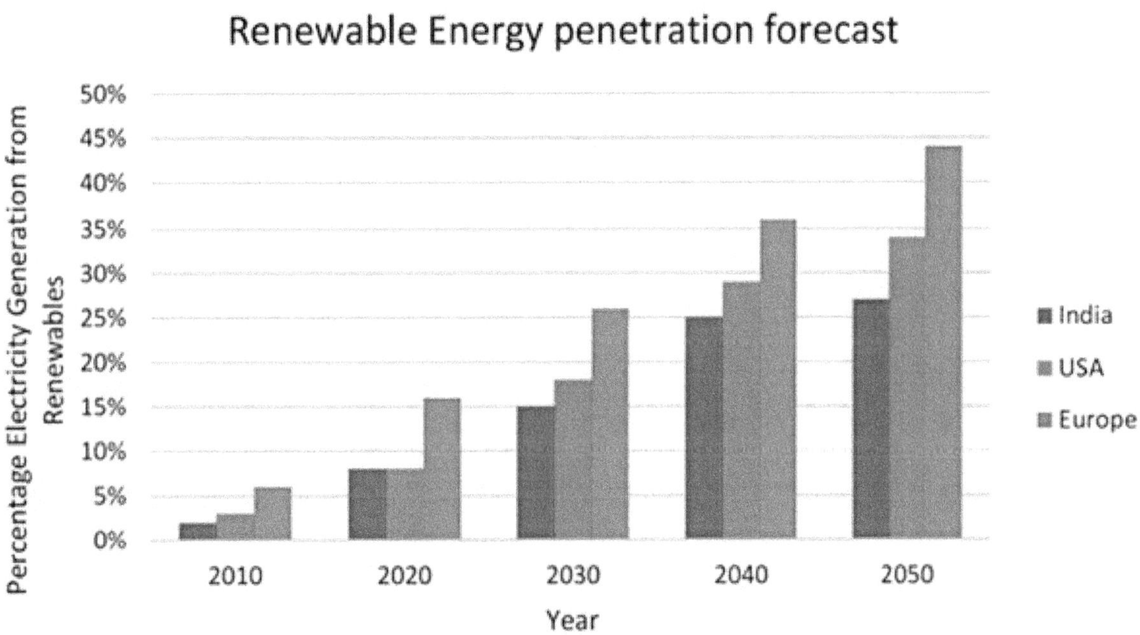

Built on top of an intelligent communications infrastructure, the smart grid is a combination of hardware, management, and reporting software. Consumers and utility corporations alike have capabilities to manage, monitor, and respond to energy challenges in the smart grid era. The transfer of electricity from the utility to the consumer becomes a two-way conversation, saving consumers money and energy while also providing greater transparency in terms of end-user usage and lowering carbon emissions. (Ramesh Babu, 2017).

A smart grid is an electricity network monitors and manages the delivery of power from all generation sources to satisfy the variable electricity demands of end-users using digital and other modern technology. Smart grids coordinate the demands and capacities of all generators, grid operators, end-users, and electrical market stakeholders to run the system as efficiently as possible, reducing costs and environmental consequences while maximizing system reliability, resilience, and stability. (Kingsley, Shongwe, & Joseph, 2018).

Figure 2. Smart grid illustration diagram

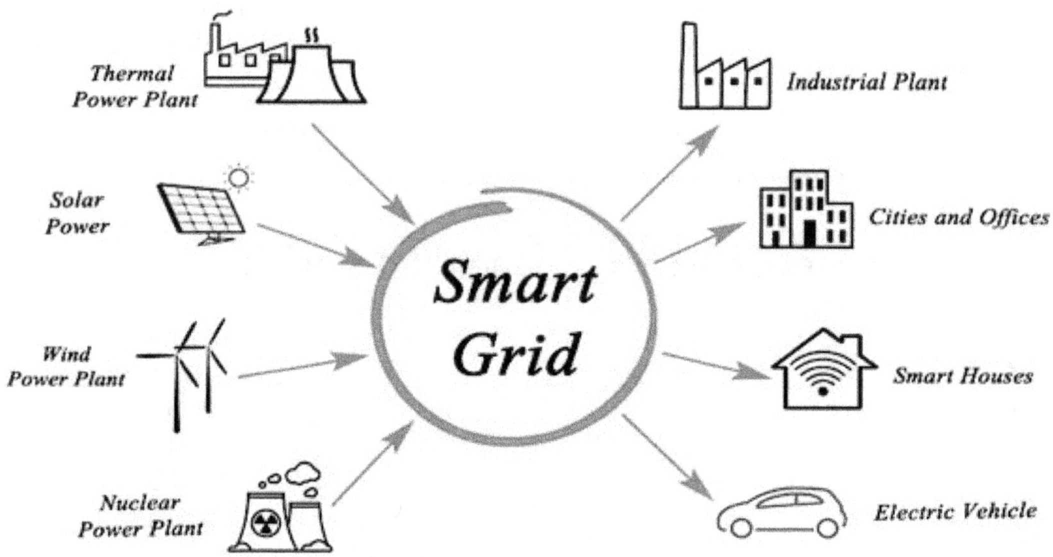

There are a lot of advantages from setting up Smart Grids (Ekanayaka, J. et al., 2012; Momoh J., 2012; Bayindir, R. et al., 2016). These include the following:

Technical Advantages

The complete implementation of Smart Grid would result in a number of technical advantages, including:

- Increased energy efficiency: This is accomplished through loss reduction, peak shaving (demand control), AMI adoption, and automated energy system operation.
- Increased grid reliability: This is accomplished by minimizing the frequency and duration of power outages.
- Increased operational efficiency: Achieved through active control, automation, and management services in distribution grids, as well as customer empowerment via home automation and smart appliance use.
- Enhanced security and safety: Enhanced security can be achieved through the use of sensors and automated operations.

Environmental Advantages

The following are some of the environmental advantages of Smart Grid deployment:

- Carbon emissions are reduced as a result of reduced grid losses, the integration of renewable and distributed generation, and the encouragement of efficient end-use by plug-in electric vehicles.
- Climate change benefits: As previously stated, the reduction in grid losses as a result of Smart Grid deployment, combined with the facilitation of electricity generation from renewable energy sources such as wind, solar, and hydro, has significant implications for CO2 emission reduction, which, in turn, improves climate change prospects.

Advantages to the Electricity Markets

Due to the dynamic interaction of the demand side of the market (consumers) with the power supply side (suppliers/providers) in a Smart Grid setting, the electricity price can be reduced compared to a conventional grid. Consumers would naturally choose the least expensive electricity source if information about electricity prices from various suppliers was made available in such an environment. As a result, there is strong competition in the electricity market, which benefits customers and helps to optimize the operation of the power system network.

MAIN FOCUS OF THE CHAPTER

Being part of the smart grid, Hybrid Renewable Energy Systems and microgrids constitute a market share that is worth investment due to the increasingly lower costs of PV and wind generators. In this chapter, it is attempted to present a comprehensive review on optimization of power systems built around hybrid renewable multi-source systems and microgrids. For a renewable energy deployment project to be economically feasible, correct and accurate sizing and optimization of the hybrid system components is needed. The review presented in this chapter would give a reasonably good idea about the tools that researchers have been using so far to optimize standalone or grid connected Hybrid Renewable Energy Systems. The chapter also stresses microgrid energy management with renewable energies. Two approaches exist. In its name, the centralized approach uses a computerized control center optimize the settings to each microgrid component. The decentralized management system optimizes the microgrid operation using partial information with each component evaluating its own optimal settings. Centralized management is often deployed via metaheuristics and decentralized management is commonly implemented in methods based on multi-agents.

Background

Microgrid is an important part of the smart grid concept. It's part of a larger grid that includes nearly all of the utility grid's components, but these ones are smaller. Microgrids are smaller size and can operate independently from the larger utility grid, whereas smart grids take place at a bigger utility level, such as massive transmission and distribution lines. (Han, Y., 2014).

Microgrids are small, local distribution systems that include a set of micro sources including micro-turbines, fuel cells, photovoltaic (PV) arrays, and wind turbines, as well as storage devices like flywheels, energy capacitors, and batteries, as well as controllable and uncontrollable loads. During faults or other external disruptions, it can be connected to the utility grid (grid mode) or operated independently and separated from the utility grid (island mode). As a result, they improve supply quality, allowing customers to acquire more efficient, less expensive, and cleaner energy. One of the important advantages of microgrids is their ability to store energy. Microgrids may also increase local reliability, minimize investment costs, reduce emissions, improve power quality, and reduce distribution network power losses. (Banerji, A.et al., 2013).

Various organizations have comparable definitions for microgrids, which include the concept of a system with numerous loads and generation, as well as the concept of islanding off the grid. Microgrids provide a number of advantages, including the ability to modernize the grid and integrate numerous Smart Grid technologies.

- Improving the integration of distributed and renewable energy sources, which can assist minimize peak load and losses by putting generation closer to demand.
- Meeting end-user needs by assuring energy supply for key loads, controlling power quality and reliability at the local level, and encouraging customer participation in the electricity supply through demand side management and community involvement.
- Providing auxiliary services to the bulk power system and supporting the microgrid by handling sensitive loads and the fluctuation of renewable sources locally. (Ton, Wang, W. M., & Wang, W. T. P., 2011).

Figure 3. Basic Microgrid architecture

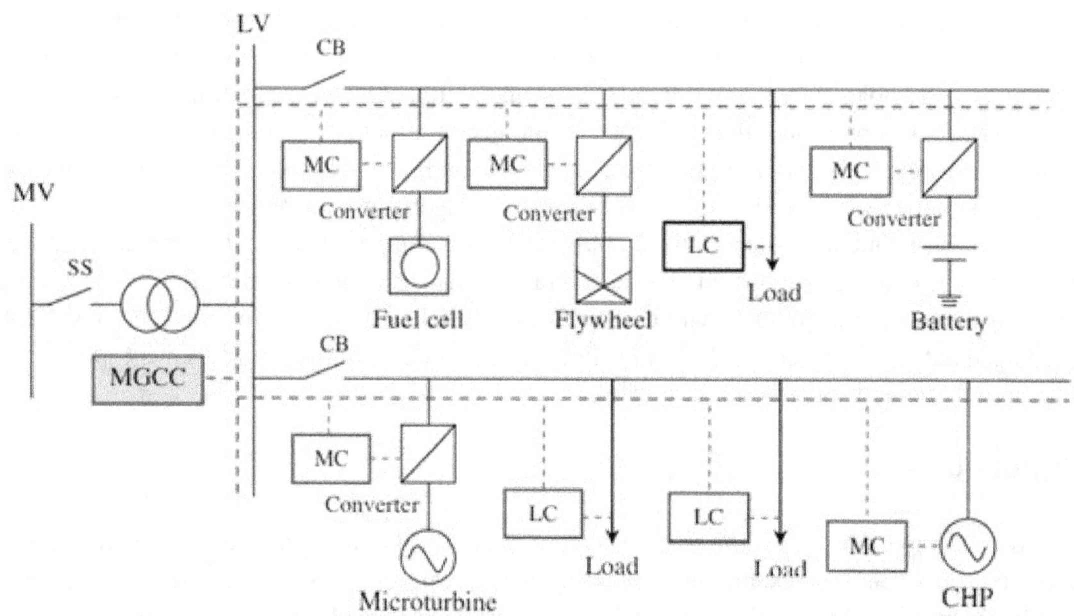

Components of Microgrid

A microgrid is made up of a central grid source, distributed generators, energy storage devices, power electronics, and a control system for managing the generator power supply.

- In a microgrid, distributed generators (DG) are the primary source of electricity generation. Renewable energy DGs and non-renewable energy DGs are two types of DGs that can be classified based on their technologies (Nojavan & Jermsittiparsert, 2020).
- Microgrids rely heavily on renewable energy sources like wind and solar energy. Due to the rising implementation of renewable energy sources and their intermittent nature, energy storage devices have become an unavoidable aspect of a microgrid. Batteries, flywheels, and other storage devices are examples.
- In microgrid systems, the electrical load is key to their operation and stability because in some applications, prioritizing the supply to critical loads is required. Microgrids can be used to power both residential and industrial loads, which are divided into sensitive and non-sensitive categories. (Sriyakul & Jermsittiparsert, 2020).
- In most DER microgrids, power converters are necessary to convert generated electricity to AC power suitable with appliances. (Bevrani, François & Ise, 2017; Jiang, He & Jermsittiparsert, 2020).

The global community has started to adopt the renewable energy as the primary source of power in order to address the issue of global warming and reduce the production of greenhouse gases due to burning of fossil fuel. Utilization of electrical energy is the key growth of any country economically. Huge investments are made by various countries in developing and implementing the renewable energy-based power production. Renewable energy also plays a vital role is electrifying the rural and remote areas where the transmission of grid power is impossible. The major renewable sources which produce the electricity are solar, wind, hydro, tides, geothermal, and biomass. Figure 4 shows the general production of electricity at present, worldwide (Ramesh Babu, 2017).

Figure 4. Global production of electricity in 2020 by share (World Energy Data, 2021)
Data: BP Statistical Review of World Energy 2021

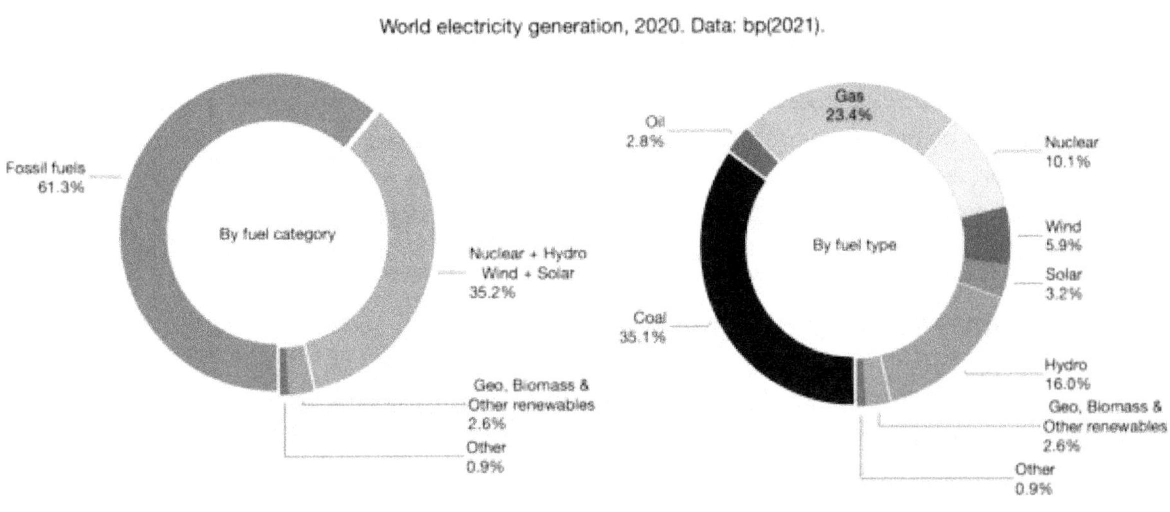

Although the use of renewable energy is increasing, we can distinguish many kinds of challenges. These challenges include:

Widespread Unpredictability

The inability to forecast whether the wind and sun will be available for energy generation an hour or a day later is known as unpredictability. Because grid operators employ unit commitment to manage the majority of electricity on the system, the hour-to-day variability is reduced. Unit commitment is the process of planning generation ahead of time, usually a day ahead of time, in order to fulfill the predicted load. As a result, when supply falls short of demand, the grid operator turns to ancillary services to make up the shortfall. (Kingsley, Shongwe & Joseph, 2018). As a result, when the process of unit commitment and reserve computation is performed based on hypothetical or random data in order to ensure reliability, a difficult challenge occurs.

Non-Controllable Variability

Variability refers to a non-consistent production in the context of renewable energy resources. It differs from unpredictability in that even if operators are able to accurately estimate wind and solar output, the output will still be varied, posing problems to the operator. Grid operators must deal with voltage and frequency fluctuations on a second to minute scale that, if left uncontrolled, can cause major damage to the system and its associated equipment. (Kingsley, Shongwe & Joseph, 2018).

Dependence on Location

Wind and solar energy resources are frequently found in remote locations far from where they are used. As a result, the building of adequate transmission infrastructure is critical for the grid integration of renewable energy. Transmission planning regulations vary greatly and are frequently influenced by regional politics. Energy production capacity can be found in one condition, then passed via another before being used in another. Because of the disparities in generation capacity, transmission capacity location, and load size changes between different sites, the development of renewable energy transmission is complicated, especially in terms of cost allocation.

Capital-Intensive Grid Upgrading

Upgrades to the grid may be necessary to accommodate wind and solar electricity. For example, new transmission lines or improvements to existing lines may be required if high-quality wind and solar resources are located far from demand centers. At the distribution level, rooftop PV may hasten the deterioration of distribution components such as low-voltage transformers, resulting in a higher return on RE expenditures. (Ramesh Babu, N., 2017).

Uncertain Renewable Energy Cost and Cash Flow

Smart grid solutions are emerging to address two specific challenges that have harmed RE project economics in the past: grid upgrading expenses awarded to renewable energy project developers and energy curtailment when full renewable energy production cannot be easily integrated into the power system. Both of these factors could cause the project's cash flows to deviate even more from expectations. The investment landscape for variable renewable energy becomes more uncertain as improvements grow more expensive or curtailments become more common, which can hinder total adoption. (Ramesh Babu, N., 2017).

Fortunately, smart solutions exist to mitigate these challenges

- **Enhanced Forecasting:** System operators can better predict and manage renewable energy variability and uncertainty thanks to widespread instrumentation and advanced computer models.
- **Smart Inverters:** Inverters and other power electronics can provide control to system operators as well as provide some grid support automatically.
- **Incorporated storage:** Storage can assist smooth out short-term changes in RE output and manage supply and demand mismatches.
- **Demand Response:** Demand-side contributions to balancing can be made using smart meters, intelligent appliances, and even industrial-scale loads.
- **Advanced energy management systems:** Advanced energy management systems, which give real-time, high-resolution visibility and control of power systems, can let grid operators postpone more expensive capital expenditures.
- **Grid-level storage:** Various types of large-scale storage can assist in reducing the demand for extra transmission capacity. (Speer, B. *et al.,* 2015).

The AMI and Renewable Energy Grid Integration

AMI stands for advanced metering infrastructure, which is an integrated system of communications networks, data management systems, and smart meters that allows utilities and customers to communicate in real time. Unlike traditional electricity meters, AMI works in three different ways. They can measure energy consumption with improved time resolution and send usage data to the utility on a regular basis. AMI enables a variety of smart grid technologies that support renewable energy (Recioui & Bentarzi, 2021). AMI is distinguished in terms of renewable energy and grid integration by:

- For reparation, regulation, and development, AMI can size renewable resource production, including auxiliary infrastructure.
- AMI can be used to incorporate distributed resources into DA systems as a communications setup.
- AMI enables improved electricity rating systems, which can boost distributed PV's financial viability.
- AMI can serve as a communication channel for DR, which works in tandem with PV. (Kingsley, Shongwe & Joseph, 2018).

Components of a Typical Hybrid Renewable Energy System

As a subordinate power source, a hybrid setup combining the electricity gathered from both the wind and the sun and stored in a battery can be far more reliable and realistic. Even if there is no sun or wind due to climate change, the system will be powered by the stored energy in the batteries. Hybrid systems are typically designed to provide the lowest feasible cost while simultaneously providing the highest level of reliability by taking into account all of the constraints that may affect the amount of generated electricity. Solar PV cells are becoming more expensive, making them unsuitable for larger-capacity systems. This is where the wind turbine comes into play, with its key advantage being its lower cost as compared to PV cells. The system requires a battery to store solar and wind energy generated during the day. The presence of wind during the night is an added benefit that boosts the system's reliability. A wind turbine is a more dynamic source of electricity. A wind turbine added to a system would protect batteries from deep discharges, extending their life. (Rajendra & Natarajan, 2006; Shaahid & Elhadidy, 2008; Khatib, Azah, & Kamaruzzaman, 2012).

Figure 5. System components of hybrid wind and PV systems

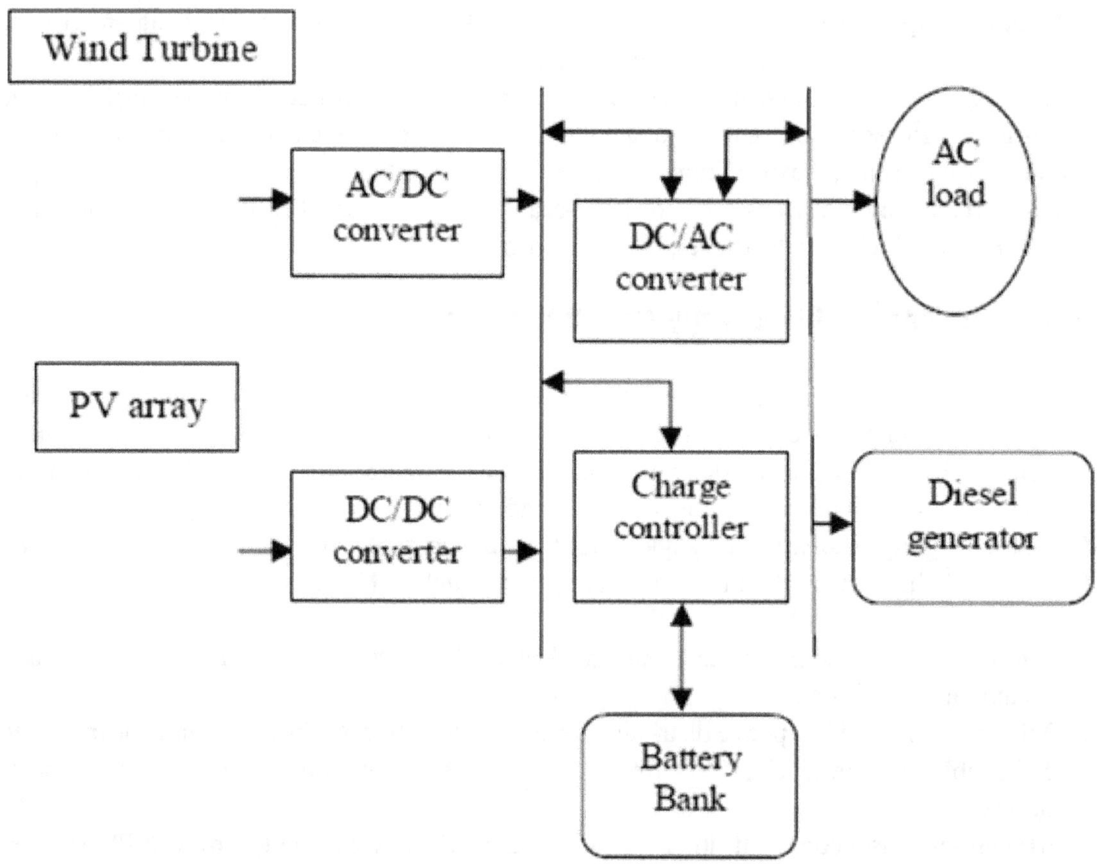

Looking at the components of a hybrid renewable energy system shown in Figure 5, a lot of works have dealt with the optimization of such a system as a whole or in part depending on the intended investment and the renewable energy resources in the area of interest. Also, the system optimization may be categorized into standalone systems for remote areas or grid connected for smart cities applications (Ding *et al.*, 2021). It should be noted that the optimization task is successful provided the renewable resources data are available and accurate. This issue has to be considered whenever a renewable energy system is to be deployed and designed for the investment to be worth it.

COMMON PERFORMANCE INDICES IN RENEWABLES SIZING PROBLEMS

In the literature, a number of key performance metrics have been used to analyze the efficiency and practicability of hybrid renewable energy systems. The following are the most commonly used metrics:

Loss of Power Supply Probability

The loss of power supply probability (LPSP) is the probability that insufficiency in energy supply occurs when the hybrid system is unable to satisfy the load demand (Mahesh & Sandhu, 2015; Yang, Lu, & Burnett, 2003). It indicates the portion of the operating time T in which the energy supplied by the energy system $E_{supplied}$ is insufficient to meet the load demand E_{load} and may be written as (Mahesh & Sandhu, 2015):

$$LPSP = \frac{\sum_{t=1}^{T} Powerfailuretime(E_{supplied}(t) < E_{load}(t))}{Totaltimeofoperation, T} \tag{1}$$

Expected Energy Not Supplied

The expected energy not supplied, EENS, is an indicator to measure the amount of energy not supplied by the power system when the load exceeds available generation (Mahesh & Sandhu, 2015, Chauhan & Saini, 2014). The EENS at any instant in time is given by the difference between the load level and total generation corresponding to that time instant (Khatod, Pant & Sharma, 2010)

$$EENS(t) = \begin{cases} E_{load}(t) - E_{supplied}(t) When E_{load}(t) > E_{supplied}(t) \\ 0 Otherwise \end{cases} \tag{2}$$

The EENS over the whole operation period is then:

$$EENS = \sum_{t=1}^{T} EENS(t) \tag{3}$$

Energy Index of Reliability

The energy index of reliability (EIR) is the fraction of the demand satisfied by an energy system and is directly related to EENS (Tina, Gagliano & Raiti, 2006) as:

$$EIR = 1 - \frac{EENS}{\sum_{t=1}^{T} E_{load}(t)} \tag{4}$$

Renewable Energy Fraction

The renewable energy fraction, REF, represents the fraction of the total energy delivered to the load that was generated from a renewable resource. It is typically used for renewable-based systems with diesel generators as backup to prevent power failure and is given as (Al-Shamma'a & Addoweesh, 2014):

$$REF = 1 - \frac{\sum_{t=1}^{T} E_{DG}(t)}{\sum_{t=1}^{T} E_{load}(t)} \tag{5}$$

Where E_{DG}(t) represents the energy supplied from diesel generators and $E_{load}(t)$ the load demand, respectively.

Loss of Load Probability (LOLP)

Loss of load (LOL) is defined as the HRES inability to meet the daily peak load. A LOL occurs whenever the system load exceeds the available generating capacity (Pillai, 2008). The overall probability that there will be a shortage of power (loss of power) is called loss of load probability (LOLP) which is expressed in terms of days per year, hours per day or percentage of time. The LOLP measure was first introduced by Calabrese (Calabrese, 1947). The LOLP is given by the following equations (Yang, Lu & Burnett, 2003):

$$LOLP = \frac{\sum_{t=1}^{T} Time(I_{supply} < I_{needed}(t))}{T} \tag{6}$$

Where

$$I_{needed}(t) = \frac{L(t) - P_W(t) - P_{PV}(t)}{V_L} \eta\left(I_{battery}(t)\right) \tag{7}$$

$$I_{needed}(t) = \min\left(I_{max} = \frac{0.2SOC}{"t}, \frac{SOC(t)\sigma - SOC_{min}}{"t}\right) \tag{8}$$

And I_{needed}(t) is the current required for the load at time t, I_{supply}(t) is the current supplied by HRES at time t, T is number of samples. V_L is the nominal voltage needed by the system, L(t) is the electrical load requirements at time t, $P_W(t)$ is the power generated by the wind turbine at time t, and $P_{PV}(t)$ is the power generated by PV modules at time t. If LOLP is low, it results in high cost of the system and vice versa.

Demand Satisfaction Criteria

Generally, the objective of the sizing problem is to determine the minimum cost design which is guarantees to meet the load demands throughout the year. A simple energy balance constraint over the entire time period of operation (also termed the demand satisfaction criteria (Maleki. & Askarzadeh, 2014; Maleki & Pourfayaz, 2015) or demand-supply criteria (Markvart, 1996) can be used to assess the performance. For hybrid energy systems, this is given by:

$$E_{supplied}(t) + E_{DF}(t) \geq E_{load}(t)\ \forall t \tag{9}$$

REVIEW ON HYBRID RENEWABLE ENERGY SYSTEM SIZING OPTIMIZATION

The main goals of optimizing a hybrid renewable energy system using any suitable technique are to identify system component values, set up an objective function containing variables and components, as well as realistic constraints that can affect the function, and satisfy the load demand economically and effectively. As a result, the optimization problem's objective function and its overall system components are discovered, subject to the following two points:

1. Keeping the overall net current cost of the system as low as possible;
2. Ensuring that the load is served in accordance with specific reliability requirements.

As indicated in Fig. 6, the hybrid renewable energy sizing approaches can be categorized into two main classes: optimization based on dedicated software and optimization using computational methods (Ammari et al, 2021). Different categorization approaches are suggested by a thorough literature analysis on the application of optimization techniques to solve the challenge of scaling renewable energy installations.

Figure 6. Classification of Hybrid Renewable Energy system sizing

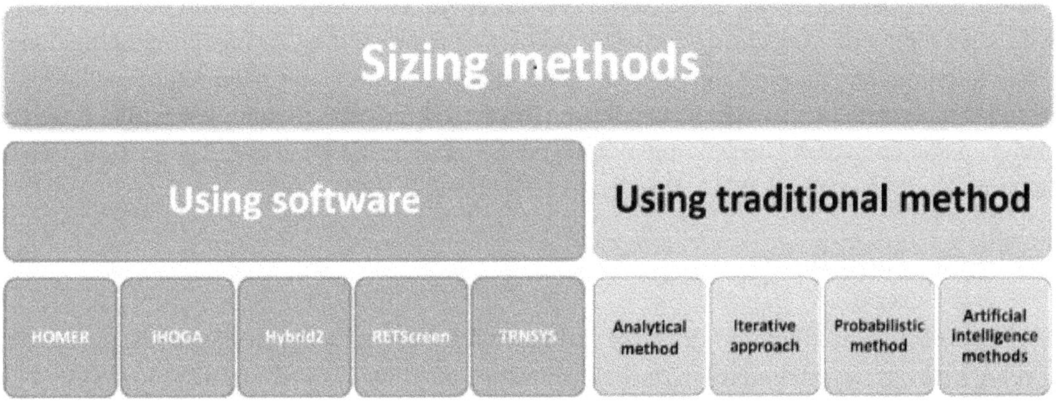

From the philosophical point of view of optimization, the renewable energy sizing works can be categorized as:

Graphical Solution

This method works better for situations involving two design variables because it allows you to see how one changes in relation to the other graphically. The limitations are shown graphically in the same graph. Once the objective function contours are drawn, the optimized point on the graph can be identified by inspecting the feasible region (Borowy & Salameh, 1996; Markvart, 1996). For a standalone hybrid Wind/PV system, Borowy and Salameh (Borowy & Salameh, 1996) devised a methodology for calculating the optimum size of a battery bank and PV array. In a hybrid solar/wind system, a graphical construction technique was utilized to discover the best battery and PV array combination. The loss of power supply probability (LPSP) was calculated using various combinations of PV array and battery sizes. The PV array against battery size is plotted using this LPSP, and the optimum option that minimizes the total system cost is obtained. Markvart et al. (Markvart, Fragaki, & Ross, 2006) used a long time series of solar radiation to establish the best sizing by superimposing contributions from climatic cycles with low daily solar radiation. A method for determining the optimal size of a hybrid PV/wind energy system was presented by Ai et al. (Ai et al., 2003). Annual Loss of Load Probability (LOLP) was estimated for various PV array and battery bank capacity values, and the best configuration (in terms of cost and LPSP) was discovered by drawing a tangent to the trade-off curve.

Probabilistic Approach

On an hourly basis or daily average power per month, the day of minimum PV power per month, and the day of minimum wind power per month, the optimal size of a hybrid PV/wind energy system can be estimated. The cost and time required to obtain environmental and load data are both minimal with this strategy (Tina, Gagliano & Raiti, 2006; Bagul, Salameh & Borowy,1996). Tina et al. (Tina, Gagliano, & Raiti, 2006) proposed a probabilistic approach based on the convolution technique to account for resource and load fluctuation. This eliminates the requirement for time-series data to evaluate a hybrid solar–wind system's long-term performance for both standalone and grid-connected applications. The energy index of reliability (EIR) directly related to energy expected not supplied (EENS) was used in the reliability study, with the time-value of energy being included as applicable in economic assessments of the system. In another case, the hourly average generation capacity approach was applied (Karaki, Chedid & Ramadan, 1999; Kellogg et al., 1996). The hourly average wind, insolation, and power consumption are used in this method to optimize system sizing. This computation is based on average annual monthly solar and wind statistics. The photovoltaic and wind components' sizes have been determined. The size of the PV and Wind generators was determined using the most unfavorable month technique. Based on the provided data, the unfavorable irradiation month and the unfavorable wind speed month are determined (El-Khadimi, Bchir & Zeroual, 2004; Kusakana & Vermaak, 2013).

Deterministic Algorithms

Unlike the probabilistic technique, every set of variable states is uniquely defined by model parameters and sets of prior states of these variables in the deterministic approach. As a result, there is always a unique solution for given parameters (Bhuiyan, & Ali, 2003). For a PV system installed in Nepal, Bhandari and Stadler calculated the system size and cost (Bhandari & Stadler, 2011). Rachid B. et al. (Belfkira, R. et al., 2008) developed a new sizing optimization methodology for a stand-alone hybrid renewable energy system. They used a deterministic method to reduce the system's life cycle cost while still ensuring energy availability. The ideal number of wind turbines, solar panels, and batteries has been found. (Lotfi, Farid, and Ghiamy, 2013) employed a deterministic approach technique to determine the best unit size for hybrid power generating systems that combine photovoltaic and wind energy. The weighting approach is used to create a discrete collection of Pareto optimum solutions numerically.

Classical Optimization Techniques

Iterative technique is a mathematical procedure that generates a series of improving approximation solutions for the optimization problem until a termination requirement is met, usually using a computer. When employing this method, the computation time grows exponentially as the number of optimization variables grows (Yang, Lu & Burnett, 2003; Yang, Lu & Zhou, 2007; Chedid & Rahman, 1997; Chedid, Karaki & Rifai, 2005). This approach was applied by Li et al. (Li, Wei, & Xiang, 2012) to improve PV-wind-battery HRES based on life cycle cost minimization. Kellogg et al. (Kellogg et al., 1998) employed an iterative optimization method to figure out how much generation capacity and storage a stand-alone, wind, PV, and hybrid wind/PV system would need. The utilization of renewable energy is compared to the construction of a power line extension from the nearest power line. (Diaf et al., 2008) presented an iterative optimization technique for the techno-economic optimization of a hybrid PV/wind energy system with/without an uninterruptible power supply to provide a certain load. The main optimization goals were to find the best HRES component size and the lowest LEC. The authors investigated the performance of HRES with and without an uninterruptible power supply, finding that system configuration influences LEC and battery state of charge (SOC), particularly at low windy sites. Furthermore, the authors confirmed that for the systems under investigation, the hybrid system is the optimum alternative. However, when determining the optimal system size, the authors did not use smart grid applications. Using the iterative technique, Kaabeche et al. (Kaabeche, Belhamel & Ibtiouen, 2011) and Yang et al. (Yang, Lu & Zhou, 2007) provided a sizing optimization model for a hybrid PV/wind/battery energy system. The LOLP and the LEC are used as optimization objectives in this model to determine the best system configuration. In (Hocaolu, Gerek, & Kurban, 2009), an iterative optimization methodology was presented to maximize the size of a hybrid PV/wind/battery energy system while keeping the system components' investment costs to a minimum. To find the maximum battery and minimum supply size, this methodology first assumes that the battery capacity is infinite, and then determines the optimal number of WT and PV arrays to supply the load demand with a specified LOLP. The authors looked at the impact of using real meteorological data vs generic analytical meteorological models for estimating system components. In (Sukumar et al., 2017), a linear optimization method for determining the appropriate size of an energy storage system was given. Many critical limits on the LPSP and REF indices were overlooked by the suggested technique. The authors of (Helal et al., 2017) used mixed integer nonlinear programming to examine an energy management system for a microgrid in an isolated community. Correa et al. (Correa, Marulanda, & Garces, 2016) suggested a virtual power plant-based energy management system. To reduce operating costs, they used linear programming to model the parts. The battery size has a significant impact on the microgrid's operational costs.

Artificial Intelligence Techniques

Artificial intelligence (AI) is a discipline of computer science concerned with the development of intelligent computers and software. Artificial neural networks (ANN), genetic algorithms (GA), fuzzy logic (FL), and hybrid systems are examples of AI techniques. Intelligent technologies enable us to create systems that are more practical and perform better than traditional approaches (Russell & Norvig, 2003).

Several research papers that use nature-inspired Computational Intelligence Techniques as a solution to HRES optimization have been published in the literature. The following sub-sections summarize these works.

Genetic Algorithms

Many studies have looked into the best design for a hybrid PV/Wind energy system in order to reduce the system's total net present cost and ensure that the load is reliably provided. Satish Kumar R. et al. (Ramoji, Bibhuti, & Kumar, 2014) used a genetic algorithm to build a system that took into account both economic and environmental factors. improved photovoltaic (PV)-wind hybrid energy house system with a storage battery (Nafeh, 2011). The design ensures that the load is properly fed to meet specified reliability criteria while keeping the loss of power supply probability (LPSP) below a predetermined threshold. (Ko et al., 2015) proposed a hybrid energy system (HES) that included both fossil fuel and renewable energy systems in order to bring the cost of renewable energy down to a tolerable level for customers. For the simultaneous multi-objective design of three objectives: life cycle cost, renewable energy integration, and greenhouse gas emissions reduction, an elitist non-dominated sorting genetic algorithm was applied. Koutroulis et al. (Koutroulis, Dionissia, & Kostas, 2006) optimized the kind and number of components in order to achieve the lowest cost while meeting the load requirement with zero load rejection. They employed a genetic algorithm to optimize the component sizes of a freestanding system made up of solar panels, wind turbines, and a battery bank. The authors of (Benatiallah, Kadia, & Dakyob, 2010) used the genetic algorithm to size freestanding wind power plants (GA). In the Algerian town of Bechar, the developed approach was used to size a wind system that is expected to supply power to a household load. The system is designed for applications that require fluctuating or constant power loads. The authors of (Koutroulis et al., 2006) provided an optimal sizing of stand-alone PV/Wind systems by selecting the appropriate number and kind of units from a list of commercially accessible system devices, guaranteeing that the total system cost is reduced. In compared to traditional methods such as dynamic programming and gradient techniques, they used Genetic Algorithm to reduce the cost function. (Al-Shamma'a & Addoweesh, 2012) used a Genetic Algorithm to optimize the configuration of a hybrid energy system (HES) (GA). The suggested method was used to analyze the HES that supplies energy to a distant town in Saudi Arabia's northern region. The best size is determined by the renewable energy fraction (REF) and the system's reliability (LPSP). The hybrid PV/WT/Bat/DG, hybrid WT/Bat/DG, and hybrid PV/WT/DG systems are the most cost-effective, as they have a lower COE than the other systems. Genetic algorithms were employed by Kamaruzzaman et al. (Sopian et al., 2008) to optimize the system components of a hybrid energy system that included a Pico hydro system, solar photovoltaic modules, a diesel generator, and battery sets. They want to employ renewable energy as much as possible while limiting the use of diesel generators. They used genetic algorithms programming to analyze both circumstances in order to reduce the overall net present cost for the best configuration. They took into account a variety of operating tactics when creating the vectors for the best plan. Satish Kumar

Ramoji et al. (Ramoji, 2014) provided a new methodology for sizing a hybrid PV/Wind energy system for best performance. The hybrid PV-wind system was designed using Genetic Algorithms (GA) and Teaching Learning Based Optimization (TLBO). The application of the Pareto evolutionary algorithm to the multi-objective design of isolated hybrid systems was presented by Jose L. et al. (Bernal, José, & Rodolfo, 2009). To discover the ideal combinations of components for the hybrid system and control method, they applied the approaches of a multi-objective evolutionary algorithm (MOEA) and a genetic algorithm (GA). M.S. Ismail et al. (Ismail, Moghavvemi, & Mahlia, 2014) suggested a method for sizing a hybrid system that includes photovoltaic (PV) panels, a backup source (microturbine or diesel), and a battery system. Their goal was to provide power to a tiny village in the Palestinian Territories. In order to optimize the objective function while covering the load demand with a set value for the loss of load probability, a Genetic Algorithm was applied to the system (LLP). Hongxing et al. (2008) proposed an optimal sizing technique for designing a hybrid solar–wind system with battery backup (Hongxing et al., 2008). The decision factors were PV module number, wind turbine number, battery number, PV module slope angle, and wind turbine installation height, and they used genetic algorithm (GA) Optimization approaches. It was possible to get the requisite loss of power supply probability (LPSP) with a minimum annualized cost of system (ACS). The authors of (Bao et al., 2013) used a Genetic algorithm to create a device that was commercially available. The design ensures that total system costs are reduced over a 20-year period, subject to the restriction that all load energy requirements are met. Katsigiannis et al. (Katsigiannis, Georgilakis, & Karapidakis, 2010) employed GA to improve hybrid energy systems for three remote Japanese islands. Based on the lowest total investment cost, the suggested technique identifies the optimal number of PV panels, WT, and battery banks. The authors of (Yang, Wei, & Chengzhi, 2009) advocated using GA to create an optimal design model for hybrid PV/wind/battery systems. B. Heyrman et al. (Heyrman & Dupré, 2013) provided a generic and efficient model for hybrid renewable-conventional electrical energy systems that minimizes the annualized cost of the systems while satisfying the custom necessary LOLP. The HOMER software was used to successfully validate the simulation model. In addition, two control systems for the distribution of electrical power are discussed. The authors employed a genetic algorithm technique to reduce the total cost of the system by optimizing the size of system components in the specified challenge. With the lowest total cost, improved system dependability, robustness, and efficiency were obtained.

In addition to these efforts, some authors combined genetic algorithms with other methodologies. When total solar radiation data is unavailable, Adel Mellit (Mellit, 2007) employed neural networks and genetic algorithms to forecast the appropriate size coefficient of Photovoltaic Supply (PVS) systems in remote places. By reducing the ideal cost (objective function) with the number of PV modules and the capacity of the batteries, a genetic algorithm (GA) is utilized to calculate the optimal coefficients for each site. Then, using simply geographical coordinates as input, a feed-forward neural network (NN) is used to predict the optimal coefficients in remote places. (Jemaa et al., 2013) used fuzzy adaptive Genetic Algorithms to optimize the setup of hybrid energy systems. Using historical data of hourly wind speed, solar irradiance, and load data, PV, wind generator, and load were stochastically modeled. The overall cost of the hybrid system is reduced thanks to the Hatata et al. paper's clonal selection technique (Hatata, Osmana & Aladla, 2018). By reducing the size of the wind turbine while increasing the size of the PV and Battery, the GA-based clonal algorithm provided improved outcomes.

Particle Swarm Optimization

The authors of (Mohamed et al., 2016) used Particle Swarm optimization to determine the optimal sizing of a hybrid renewable energy system employing a smart grid load management application based on available generation (PSO). One of the smart grid applications, load shifting, has been used to acquire the dispersed load profile, lower overall system costs, and reduce CO2 emissions. Motaz Amer and his co-authors PSO was utilized by (Motaz, Namaane, & M'sirdi, 2013) to calculate the power generated by a Hybrid Renewable Energy System to meet a typical load demand. PSO was successful in lowering the system's Levelized Cost of Energy (LCE) and optimizing it. They took into account both the total system power loss and the power stored in the standalone with battery storage and standalone with grid storage situations. Using Multi Objective Particle Swarm Optimization, (Hanieh et al., 2014) introduced a new way for optimizing micro-grid systems (MOPSO). This is accomplished using a hybrid wind/PV system with battery storage and a diesel generator. The load is subjected to the power management algorithm, and the MOPSO method is utilized to determine the best system configuration and component sizing. Using a comparative comparison of particle swarm optimization and ant colony optimization, Saeid Lotfi et al. (Lotfi, Farid, & Ghiamy, 2013) proposed a new method for optimal design of a stand-alone hybrid solar-wind-diesel power generating system. They employed a mathematical model of the hybrid system's numerous components. (Hameed et al., 2012) proposed an open space particle swarm optimization method for determining the optimal sizing of a hybrid renewable energy production system (HRES) with storage system (OSPSO). A reliable and cost-effective generation system is achieved by determining the optimal quantity of PV modules, wind turbines, and batteries. Maleki and Fathollah Pourfayaz optimized hybrid photovoltaic (PV)–wind turbine (WT) systems with battery storage for remote places (Maleki & Fathollah, 2015). Particle swarm optimization (PSO), tabu search (TS), and simulated annealing (SA) were used, as well as four variants: improved particle swarm optimization (IPSO), improved harmony search (IHS), improved harmony search-based simulated annealing (IHSBSA), and artificial bee swarm optimization (ABSO) (ABSO). The optimum sizing of a PV/ WT/battery hybrid system has been compared in terms of total annual cost (TAC). ABSO is more promising than the other algorithms, according to the results. A hybrid system incorporating wind, PV, and tidal energy with battery storage was proposed by (Bashir & Sadeh, 2012). They emphasized the advantages of tidal energy as being more predictable than wind and solar. For optimization, they used the PSO algorithm. The particle swarm optimization is used in (Kornelakis & Marinakis, 2010) to optimize the sizing of grid-connected PV installations. It is presented how to use Complex Mixed Integer Multi Objective Particle Swarm Optimization to obtain non-dominated solutions for optimal design, as well as a numerical example for a huge hotel (Lingfeng & Singh, 2007). PSO was used by (Boonbumroong et al., 2011) to reduce the life-cycle cost of a stand-alone PV/wind/diesel system to feed a certain load. The hourly energy demand has to be met by the amount of generated energy, according to the optimization constraint. Hakimi & Moghaddas-Tafreshi, 2009) PSO was utilized to reduce the total cost of a stand-alone hybrid energy system made up of wind turbines, electrolytes, a reformer, an anaerobic reactor, fuel cells, and hydrogen tanks while meeting demand. The stored energy in hydrogen tanks was the optimization restriction. As an example study, (Xiao Xu et al., 2020) constructed and investigated a hybrid PV/WT/hydropower/pump storage. To respect the maximum Loss of Power Supply Probability (LPSP) and least investment cost, they devised a techno-economic index. The problem is solved using MOPSO (Multi-Objective Particle Swarm Optimization). An intelligent energy management system for Hybrid Micro-Grid system at three stations in Iran is proposed in (Borhanazad et al., 2014). A multiobjective particle swarm optimization approach was used by the authors. In (Abedini, Moradi, & Hosseinian, 2016), the authors developed an energy management system for a stand-alone Hybrid Micro-Grid that was optimized utilizing a particle swarm method with Gaussian mutation.

Other Techniques

Sharmistha Sharma et al. (Sharma, Bhattacharjee, S., & Bhattacharya, A., 2016) improved the sizing of battery energy storage devices to reduce the operational planning of a micro-grid (MG) in terms of lowering energy expenses. Various restrictions were evaluated, including distributed generator (DG) power capacity, BES power and energy capacity, BES charge/discharge efficiency, operating reserve, and load demand satisfaction. To overcome the problem, they employed grey wolf optimization (GWO). In terms of solution quality and computing performance, GWO beats numerous optimization techniques such as genetic algorithm, particle swarm optimization, tabu search, and differential evolution. Ant Colony Optimization (ACO) was utilized by (Khorasaninejad & Fetanat, 2015) to size a hybrid photovoltaic (PV) and wind energy system. The findings are encouraging, demonstrating that the authors' proposed approach beats them in terms of finding the best solution and completing it quickly. The Bat algorithm is applied for size optimization of grid-connected PV systems by maximizing the specific yield in (Sulaiman et al., 2012; Othman et al., 2015). Erdinc et al. present a thorough examination of the various optimization strategies for sizing the design of hybrid renewable energy systems (Erdinc, & Uzunoglu, 2012). A brief overview of the advantages and disadvantages of various artificial intelligence approaches (GA, PSO, and SA) utilized for efficient and cost-effective hybrid system sizing is presented, as well as detailed intelligent techniques for hybrid system sizing. The authors developed a hybrid power system that included wind turbines (WT), photovoltaics (PV), and battery energy storage (BESS) technologies in (Hemeida et al., 2020). Load satisfaction is considered when evaluating the system's practicality. To determine the optimal solution, the linear TORSCHE optimization technique is utilized. The authors of (Alshammari & Asumadu, 2020) implemented and compared three algorithms to discover the best hybrid WT/PV/Biomass/BESS energy system design. When compared to the Jaya and PSO optimization methods, the Harmony Search Algorithm (HSA) was faster and more efficient at convergence. The teaching learning based optimization (TLBO) technique was applied to improve a hybrid renewable energy system in (Recioui & Dassa, 2017).

The artificial neural network is used in (Kolawole, 2014) to estimate the best sizing parameters for standalone PV systems. For the PV array size coefficient and the battery storage capacity coefficient, the ANN provided results. The generalized regression neural network is utilized in (Khatib & Elmenreich, 2014) to optimize the sizing coefficients and estimate the loss of load probability for freestanding PV systems. The ANFIS model is established in (Mellit, 2006) for optimizing the size coefficients of freestanding PV systems. Based on the costs of a solar panel, the ideal sizing parameters for these calculated sites were created. When compared to the site's known sizing parameters, the proposed adaptive neural fuzzy inference system model delivered the most accurate findings of the various network topologies.

Andrea G. Kraj et al. (Kraj, 2015) investigated the simulation of Multi-Renewable Energy Systems (MRESs) with the goal of making it easier to optimize multi-functional power systems for distant stand-alone power generation. They used an evolutionary algorithm technique to evaluate objectives within the simulated MRES configuration's geographical and economic constraints. For electrical energy generation, the simulated MRES uses real resource data for wind and solar energy, as well as model biomass resources, in combination with diesel backup and storage systems. The authors of (Barbaro & Castro,

2020) examined how to create an energy system with the highest renewable energy penetration. This system is made up of WT, PV, geothermal, diesel, and BESS, and the best system is found while keeping technological and cost limits in mind. Katsigiannis et al. (Katsigiannis, Georgilakis, and Karapidakis, 2010) employed multiobjective optimization approaches to size an A-HMG while minimizing LCOE and CO2 emissions at the same time.

Dedicated Software

To study and plan renewable energy systems, there are numerous software packages and tools available. HOMER (Hybrid optimization method for electric renewables), HOGA (Hybrid optimization by genetic algorithm), and HYBRD2 are the most commonly used software tools. NREL (National Renewable Energy Laboratory, USA) created the Hybrid Optimization Model for Electric Renewables (HOMER) in 1993. The software can be used to quickly perform prefeasibility, optimization, and sensitivity analysis in a variety of system configurations. Photovoltaic modules, wind turbines, biomass power, hydro power, generators, micro turbines, storage batteries, grid connections, fuel cells, and even electrolyzers may all be modelled using the software. It was created for both on-grid and off-grid applications. HOGA (Hybrid Optimization by Genetic Algorithm) is a hybrid system optimization tool developed by the University of Zaragoza's Electric Engineering Department (Spain). Genetic Algorithms are used to carry out the optimization, which can be single-objective or multi-objective. AC, DC, and/or hydrogen loads are all possibilities. The University of Massachusetts' Renewable Energy Research Laboratory (RERL) developed the hybrid system simulation program HYBRID2. The simulation is run at intervals ranging from 10 minutes to 1 hour.

In this context, Roy et al. (Roy, Majumder, & Chakraborty, 2010) calculated the ideal sizing for a hybrid solar-wind system for distributed power at Sugar, a remote off-grid island. Using HOMER software, they optimized the size of the generation units. The best configuration was subjected to a sensitivity analysis. A comparison of the hybrid system's various modes was also investigated. The solar PV-Wind Hybrid system was expected to have a lower cost per unit of power. When compared to solar PV DG, the capital investment cost was similarly lower when the system used wind DG. Nandi and Ghosh (Nandi & Ghosh, 2009) conducted research on the optimization of a hybrid PV/wind/battery system and its performance for a typical community load in Bangladesh, utilizing a hybrid optimization model for electric renewable energy (HOMER). In Inner Mongolia, Barley et al. (Barley, Lew, & Flowers, 1997) investigated a hybrid wind/PV system for generating electricity. They used the HYBRID2 software in conjunction with a reduced time-series model to simulate hybrid power systems. HYBRID2 can simulate hybrid systems with a surprising level of detail, but it does not optimize the size of the system components. The analysis of a stand-alone energy system using HRES was applied using HOMER software in Newfoundland, Canada (Khan & Iqbal, 2005). The authors of (Chaouachi et al., 2012) suggested a multiobjective optimization-based intelligent energy management of an HMG. To estimate RE generation and load demand, they used an artificial neural network. A approach for optimizing an HMG based on mixed integer linear programming is provided in (Ahmad et al., 2018). The HMG system's optimization step was carried out using HOMER software. HOMER software was used to optimize a hybrid system using solar and wind data in the Adrar region of Algeria (Chellali et al., 2011).

ENERGY MANAGEMENT IN MICROGRIDS WITH RENEWABLE ENERGIES

A microgrid is a medium- or small-scale distribution grid with distributed generation that includes renewable and conventional sources (hybrid systems) as well as storage units that provide electrical energy to end customers. The storage units that are used to compensate for the intermittency in PV and wind output power improve the microgrid's reliability (Thirugnanam et al., 2018; Yang et al., 2013; Atcitty et al., 2013). These microgrids have the communication systems that are required for real-time energy management (Lasseter, 2007). Microgrids can function independently (standalone) or connected to a grid. (Hatziargyriou et al., 2007).

Smart grids, which allow information interchange between consumers and dispersed generation, might be characterized as microgrids with renewable energies. An energy management system (EMS) is an information system that offers the essential functionality to ensure that energy is supplied at a low cost by generation, transmission, and distribution (Stanton, Giri & Bose, 2017). Microgrid energy management entails the use of control software that allows for the system's optimal operation (Su & Wang, 2012). This is accomplished by taking into account the cost as well as two microgrid operation options (isolated and inter-connected). When considering microgrids with renewable energy sources, the intermittence of resources such as solar irradiation and wind speed must be taken into account (Gildardo Gómez, 2016).

A comprehensive automated system is used to manage energy in a microgrid, with the primary goal of achieving optimal resource scheduling [Khan et al., 2016; Gamarra & Guerrero, 2015; Fathima & Palanisamy, 2015]. It is based on cutting-edge information technology and is capable of optimizing the administration of distributed energy sources and energy storage systems (Suchetha & Ramprabhakar, 2018).

The microgrid optimization challenge often seeks to achieve the following goals:

- Maximize generator output power at a specific moment;
- Minimize microgrid operating expenses;
- Maximize the lifetime of energy storage devices; and
- Minimize environmental costs.

A literature review of these approaches and solutions is presented in the following sub-sections. The information flow follows the same logic as the categorization and classification described in the section on renewable energy sizing review.

Energy Management Based on Classical Techniques

Ahmad et al. (Ahmad et al., 2018) proposed a mixed integer linear programming-based techno-economic optimization of an MG (MILP). The goal was to emphasize the benefits of programming distributed-source generation, regulating the intermittency and volatility of this sort of generation, and lessening load peaks. For MG energy management, Sukumar et al. (Sukumar et al., 2017) developed a mixed strategy. The utility grid and fuel cell power were combined to achieve this. The on/off states of the utility grid are solved using MILP, and the problem is solved using linear optimization methods. The ideal energy storage system size was determined using the particle swarm optimization (PSO) approach. Delgado and Dominguez-Navarro (Delgado & Dominguez-Navarro, 2014) proposed a linear programming-based approach for MG energy management to optimize the operation of generators or controllable/non-controllable loads. The optimal dispatch of generators (diesel) while satisfying the operational and economic restrictions imposed by the purchase and sale of energy corresponding to each grid component is the goal of the optimization issue. Correa et al. (Correa, Marulanda, & Garces, 2016) suggested a virtual power plant-based energy management system (VPP). Solar panels and storage systems are integrated on the MG. To reduce operating costs, they applied linear programming approaches. A model was created that was similar to the Colombian model in terms of energy. Umeozor and Trifkovic (Umeozor & Trifkovic, 2016) investigated how to manage energy in an MG based on MILP by parameterizing the uncertainty of solar and wind energy generation in the MG. The optimization was carried out in two stages: first, the parameterization scheme was chosen, and then operational decisions were made while market pricing and storage system disposition were taken into account. Tim and his co-authors (Paul et al., 2018) suggested a centralized energy management system for an interconnected MG. Quadratic programming was used to find the best economic dispatch. The technique was put to the test using a modified IEEE 33 node grid. (Xing et al., 2017) introduced a multi-time-scale energy management system. The optimization task was split into two parts: static programming and real-time dynamic compensation. The problem was solved using a mixed-integer quadratic programming method with optimal load flows and expected battery load states based on wind and solar radiation data. Cardoso et al. (Cardoso et al., 2018) studied a novel MG battery degradation model. The problem is tackled using stochastic mixed-integer linear programming, which takes into account a variety of parameters including loads and diverse energy generation sources, costs, limitations, grid topology, and local energy taxes. (Helal et al., 2017) investigated an energy management system for a hybrid MG in a remote hamlet using a photovoltaic desalination system. The mixed integer non-linear programming was used to solve the optimization problem. The objective function seeks to reduce daily operational costs as much as possible. (Rouholamini & Mohammadian, 2016) researched optimal energy management for a grid-connected hybrid generation system utilizing an interior search method. This system uses real-time electricity pricing based on simulation results to provide power to the local grid.

Energy Management Based on Metaheuristic Methods

(Dufo-López et al., 2007) developed an energy management system based on evolutionary algorithms for a hybrid system. The goal is to reduce operating expenses by charging the batteries or producing hydrogen in the electrolyzer with excess energy generated by renewable sources. (Das et al., 2018) looked into incorporating internal combustion engines and gas turbines into stand-alone hybrid MGs with solar modules. This system was optimized using a multi-objective genetic algorithm based on energy costs and overall efficiency. The load was tracked using both electric and thermal energy. In (Chalise et al., 2016), evolutionary algorithms were employed for energy supply alternatives via a diesel generator in a hybrid MG made of a diesel generator and photovoltaic system, with an economic dispatch with battery deterioration model. (Ogunjuyigbe, Ayodele, & Akinola, 2016) suggested a system for locating renewable generation and batteries in a stand-alone MG using a genetic algorithm. To optimize the MG, the optimization allows for differences in radiation and wind sources and collects data from a load profile. For residential and commercial MGs, (Aldaouab, Daniels, & Hallinan, 2017) used genetic algorithms. PV solar energy, microturbines, a diesel generator, and an energy storage system are all used in the MG. TRNSYS software was used to examine the performance of a hybrid system, and sizing was established using genetic algorithms in the HOGA software (formerly termed iHOGA), manual calculations, or

HOMER software (Behzadi & Niasati, 2015). For excess energy, three energy management options were tested: producing hydrogen, charging the battery, or both.

(Li et al., 2017) published a paper on MG optimization using the particle swarm technique, which may operate as a connected or isolated MG. To combat these oscillations, the strategy takes into account fluctuations in renewable energy sources and load needs, as well as appropriate 24-hour projections. Nivedha et al. (Nivedha, Singh, & Ongsakul, 2018) looked at an MG that included a wind turbine, fuel cells, a diesel generator, and an electrolyzer. Using the particle swarm optimization method, the fuel cell works to match the high load demand, resulting in cost-effective MG operation. Abedini et al. (Abedini, Moradi, & Hosseinian, 2016) provided an energy management system for a hybrid MG that used a particle swarm algorithm with Gaussian mutation to optimize capital and fuel expenses. (Wasilewski, 2018) used particle swarm algorithms and provide a metaheuristic optimization strategy for optimizing an MG. In a publication (Hossain et al., 2019), a particle swarm algorithm for energy management in a grid-connected MG is described. With one-hour time intervals, the proposed cost function cuts costs. (Azaza & Wallin, 2017) looked into how to manage energy in an MG with a hybrid system that included wind turbines, solar panels, a diesel generator, and battery storage. The probability of losing energy supply throughout 6 months during summer and winter is calculated using a multi-objective particle swarm optimization. (Nikmehr & Najafi-Ravadanegh, 2015) used an imperial competitive algorithm to determine the ideal generation for an MG. The goal is to resolve load unpredictability and scattered generators, as well as the cost-effective dispatching of generating units. (Marzband et al., 2017) used the artificial bee colony algorithm to produce an energy management system for an isolated MG (ABC). Neural networks and Markov chains are used to control non-dispatchable generation and load unpredictability. (Kuitaba et al., 2018) used fuzzy logic and the Grey Wolf algorithm to optimize an interconnected MG. The goal is to reduce both the costs of generating units and the amounts of emissions produced by fossil fuel sources. Energy management in an MG connected to a direct current electric grid was studied by (Papari et al., 2017). The crow search method is used for optimization (CSA). (Kumar & Saravanan, 2019) used the artificial fish swarm optimization method to develop an algorithm in a hybrid MG. The sources, load, and storage elements are all programmed using the algorithm.

Energy Management Using Hybrid and Other Artificial Intelligence Techniques

(Chaouachi et al., 2013) suggested a multi-objective, intelligent energy management system for an MG that uses multi-objective linear programming to reduce operating costs and environmental effect. Photovoltaic and wind power generation, as well as load demand, were predicted using an artificial neural network. An energy management model for an MG linked to the main power grid was proposed by (Venayagamoorthy et al., 2016). The MG makes the most of renewable energy sources while reducing carbon emissions. Two neural networks are used to model the management system, which is based on evolutionary adaptive dynamic programming and learning concepts. In a hybrid MG, (Motevasel & Seifi, 2014) introduced an expert system for energy management. Wind turbine generation is predicted using neural networks. The multi-objective problem is solved using an improved bacterial foraging-based fuzzy efficient algorithm.

(Prathyush & Jasmn, 2018) presented a fuzzy logic controller-based energy management system for an MG. The goal is to keep the battery state of charge while reducing grid power variation. An adaptive neural fuzzy inference system based on an echo state network was proposed by (Leonori et al., 2018). The goal was to make the most revenue possible from energy exchange with the grid. (Arcos-Avilés et al., 2017) provided an energy management method for a home grid-connected MG with distributed generation and batteries based on low-complexity fuzzy logic control. (Taha & Yasser, 2016) provided a method for an isolated MG based on a predictive control model. In the MG, this reduces the cost, energy consumption, and gas emissions caused by diesel generation. Luna et al. (2018) proposed a real-time energy management system that took into account perfect, poor, and exact forecasts. The optimization model was put to the test in both a connected and an isolated MG with large generation and load imbalances. (De Santis, Rizzi, & Sadeghian, 2017) proposed a fuzzy logic-based Mamdani algorithm-based energy management system for an interconnected MG. In the MG model, the system makes decisions on energy flow management tasks. A combination of fuzzy logic and generic algorithms were used to optimize the results. Energy management in an autonomous MG with optimal dispatch of the micro and utility grids to meet energy demand was studied by (Elseid et al., 2015). The authors employed an IBM-developed CPLEX algorithm.

Based on game theory, (Mondal et al., 2018) proposed a distributed energy management model for a smart MG. The MG maximizes its cost and makes efficient use of energy in this model. (Ma et al., 2016) investigated an energy management algorithm based on game theory and leaders and followers. The goal was to maximize the benefits to active customers while maintaining the Stackelberg balance to ensure that benefits were evenly distributed. For energy management in an MG, (Liu et al., 2017) developed a Stackelberg game strategy. Based on the MG operator profitability, a management system model for the fee for PV energy was introduced, as well as a utility model for PV consumers. (Nnamdi and Xiaohua, 2017) proposed an incentive-based demand response for grid-connected MG operations. Game theory was used to design the demand response program. For the energy management of an interconnected MG, (Jia et al., 2017) used an adaptive intelligence technique. The goal was to reduce load fluctuations caused by renewable energy generation uncertainty. Storage components and ultra-capacitors control the load profile.

REMARKS, TRENDS AND CONCLUSION

Because of the decreasing cost of PV and wind generators in recent years, hybrid renewable energy systems and microgrids are being more commonly used for remote and rural electrification. A review of power system optimization using hybrid renewable multi-source systems and microgrids has been provided in this chapter. Proper sizing and optimization of hybrid system components are essential to extract the required energy from renewable sources with an economically viable project over the life of the system. This chapter provides a concise but comprehensive overview of the techniques that academics have been utilizing for decades to optimize Hybrid Renewable Energy Systems in both independent and grid-connected configurations. The chapter's second section focused on two approaches to microgrid energy management using renewable energy sources: There are two approaches: centralized and decentralized. The centralized approach sends the ideal parameters to each microgrid partner via a computer center. The second method employs partial information optimization, in which each participant determines its own optimal parameters. Decentralized administration is widely used in multi-agent systems, while centralized management is typically used in metaheuristic approaches.

Despite the fact that a huge number of researchers have worked on improving Hybrid Energy Systems through the use of various computational intelligence techniques, there are still some gaps and obstacles in terms of their efficiency and optimal utilization. Innovative technology is required to harvest more useful power from renewable energy sources such as solar PV. Solar's low efficiency is a major impediment to its widespread adoption. Furthermore, the production cost of renewable energy sources must be reduced significantly, as this results in high capital costs and a longer payback period. In addition, research should focus on ensuring that power electrical equipment lose as little power as possible. To increase the life-cycle of storage technologies through innovative technologies, a lot of work needs to be done. Finally, standalone systems are less flexible to load fluctuations since big changes in load can cause the entire system to fail.

There has been very little effort reported on comparing simulation tools when it comes to optimizing hybrid renewable energy systems. In fact, there isn't a single technique that can solve the problem at hand. The time and effort spent applying two or more techniques to maximize the planned problem and comparing the results to determine why the applied strategy was favored or, alternatively, under what conditions the involved technique is important are significantly reduced. Additionally, efforts to develop hybrid methodologies for maximizing renewable energy systems of various resources to fulfill energy needs are restricted. So, in addition to determining the variables and parameters appropriate for a specific problem, research on these topics is recommended. To improve the quality of the solution and the complexity of computation, researchers must look into adopting hybrid nature-inspired technologies.

In terms of microgrids, the literature evaluation focused on energy management techniques. The selection of centralized or decentralized management guarantees that the microgrid designer and operator select the most appropriate management strategy for the microgrid. Various authors have tackled the issue and offered solutions utilizing approaches such as linear and nonlinear programming, heuristic methods, predictive control, dynamic programming, agent-based methods, and artificial intelligence. These solutions were adopted for the microgrid context because of their practicality, reliability, and resource availability. Data collection systems, supervisory control, human machine interface (HMI), and the monitoring and data analysis of meteorological factors are all part of a complete energy management model for a microgrid.

Lithium batteries have the potential to be a viable alternative to lead-acid batteries in microgrid storage systems. When used in microgrids, lead acid batteries are more cost-effective than lithium batteries, however research into them will result in lower acquisition costs, making them comparable with lead-acid batteries. As a result, further study into effective energy management of energy systems, as well as lithium battery technology and management, is necessary. In addition, more precise degradation models are required to reliably forecast battery lifetime in real-world settings.

ACKNOWLEDGMENT

This work has been accomplished at the Signals and Systems Research Laboratory, Institute of Electrical and Electronic engineering, University M'hamed Bougara of Boumerdes. This research received no specific grant from any funding agency in the public, commercial, or not-for-profit sectors.

REFERENCES

Abedini, M., Moradi, M. H., & Hosseinian, S. M. (2016). Optimal management of microgrids including renewable energy scources using GPSO-GM algorithm. *Renewable Energy, 90*, 430–439. doi:10.1016/j.renene.2016.01.014

Ahmad, J., Imran, M., Khalid, A., Iqbal, W., Ashraf, S. R., Adnan, M., Ali, S. F., & Khokhar, K. S. (2018). Techno economic analysis of a wind-photovoltaic-biomass hybrid renewable energy system for rural electrification: A case study of Kallar Kahar. *Energy, 148*, 208–234. doi:10.1016/j.energy.2018.01.133

Ai, B., Yang, H., Shen, H., & Liao, X. (2003). Computer-Aided Design of PV/Wind Hybrid System. *Renewable Energy, 28*(10), 1491–1512. doi:10.1016/S0960-1481(03)00011-9

Al-Shamma'a, A., & Addoweesh, K. (2014). Techno-economic optimization of hybrid power system using genetic algorithm. *International Journal of Energy Research, 38*, 1608-1623.

Al-Shamma'a, A. A., & Addoweesh, K. E. (2012). Optimum sizing of hybrid PV/wind/battery/diesel system considering wind turbine parameters using Genetic Algorithm. In *International Conference on Power and Energy*, (pp. 121-126). IEEE. 10.1109/PECon.2012.6450190

Aldaouab, I., Daniels, M., & Hallinan, K. (2017). Microgrid cost optimization for a mixed-use building. *Proceedings of the IEEE Texas Power and Energy Conference (TPEC)*. 10.1109/TPEC.2017.7868271

Alshammari, N., & Asumadu, J. (2020). Optimum Unit Sizing of Hybrid Renewable Energy System Utilizing Harmony Search, Jaya and Particle Swarm Optimization Algorithms. *Sustainable Cities and Society, 60*, 102255. doi:10.1016/j.scs.2020.102255

Ammari, C., Belatrache, D., Touhami, B., & Makhloufi, S. (2021). *Sizing, optimization, control and energy management of hybrid renewable energy system—A review*. Energy and Built Environment. doi:10.1016/j.enbenv.2021.04.002

Arcos-Aviles, D., Pascual, J., Guinjoan, F., Marroyo, L., Sanchis, P., & Marietta, M. P. (2017). Low complexity energy management strategy for grid profile smoothing of a residential grid-connected microgrid using generation and demand forecasting. *Applied Energy, 205*, 69–84. doi:10.1016/j.apenergy.2017.07.123

Atcitty, S., Neely, J., Ingersoll, D., Akhil, A., & Waldrip, K. (2013). *Battery Energy Storage System*. Green Energy Technol. doi:10.1007/978-1-4471-5104-3_9

Azaza, M., & Wallin, F. (2017). Multi objective particle swarm optimization of hybrid micro-grid system: A case study in Sweden. *Energy, 123*, 108–118. doi:10.1016/j.energy.2017.01.149

Bagul, A., Salameh, Z., & Borowy, B. (1996). Sizing of a Stand-Alone Hybrid Wind-Photovoltaic System using a Three-Event Probability Density Approximation. *Solar Energy, 56*(4), 323–335. doi:10.1016/0038-092X(95)00116-9

Baimel, D., Tapuchi, S., & Baimel, N. (2016). Smart grid communication technologies. *Journal of Power and Energy Engineering, 4*(8), 1–8. doi:10.4236/jpee.2016.48001

Banerji, A. (2013). Microgrid: A review. In *2013 IEEE Global Humanitarian Technology Conference: South Asia Satellite (GHTC-SAS)* (pp. 27-35). IEEE.

Bao, Y. (2013). Genetic algorithm based optimal capacity allocation for an independent wind/pv/diesel/battery power generation system. *Journal of Information and Computational Science*, *10*(14), 4581–4592. doi:10.12733/jics20102167

Barbaro, M., & Castro, R. (2020). Design optimisation for a hybrid renewable microgrid: Application to the case of Faial island, Azores archipelago. *Renewable Energy*, *151*, 434–445. doi:10.1016/j.renene.2019.11.034

Barley, C. D., Lew, D. J., & Flowers, L. T. (1997). Sizing wind/photovoltaic hybrids for Bashir, M., Sadeh. J. (2012). *Size optimization of new hybrid stand-alone renewable energy system considering a reliability index.* In *11th international conference on Environment and Electrical Engineering (EEEIC).* Academic Press.

Bayindir, R., Colak, I., Fulli, G., & Demirtas, K. (2016). Smart grid technologies and applications. *Renewable & Sustainable Energy Reviews*, *66*, 499–516. doi:10.1016/j.rser.2016.08.002

Behzadi, M. S., & Niasati, M. (2015). Comparative performance analysis of a hybrid PV/FC/battery stand-alone system using di_erent power management strategies and sizing approaches. *International Journal of Hydrogen Energy*, *40*(1), 538–548. doi:10.1016/j.ijhydene.2014.10.097

Belfkira, R. (2008). Modelling and optimal sizing of hybrid renewable energy system. *13th Power Electronics and Motion Control Conference, EPE PEMC 2008*, 1834-1839.

Benatiallah, Kadia, & Dakyob. (2010). Modelling and optimisation of wind energy systems. *JJMIE, 4*(1).

Bernal, A., José, L., & Rodolfo, D. L. (2009). Multi-objective design and control of hybrid systems minimizing costs and unmet load. *Electric Power Systems Research*, *79*(1), 170–180. doi:10.1016/j.epsr.2008.05.011

Bevrani, H., François, B., & Ise, T. (2017). *Microgrid dynamics and control.* John Wiley & Sons. doi:10.1002/9781119263739

Bhandari, R., & Stadler, I. (2011). Electrification using Solar Photovoltaic Systems in Nepal. *Applied Energy*, *88*(2), 458–465. doi:10.1016/j.apenergy.2009.11.029

Bhuiyan, M., & Ali, A. M. (2003). Sizing of a Stand-Alone Photovoltaic Power System at Dhaka. *Renewable Energy*, *28*(6), 929–938. doi:10.1016/S0960-1481(02)00154-4

Boonbumroong, U., Pratinthong, N., Thepa, S., Jivacate, C., & Pridasawas, W. (2011). Particle swarm optimization for AC-coupling stand-alone hybrid power systems. *Solar Energy*, *85*(3), 560–569. doi:10.1016/j.solener.2010.12.027

Borhanazad, H., Mekhilef, S., Gounder Ganapathy, V., Modiri-Delshad, M., & Mirtaheri, A. (2014). Optimization of micro-grid system using MOPSO. *Renewable Energy*, *71*, 295–306. doi:10.1016/j.renene.2014.05.006

Borowy, B. S., & Salameh, Z. M. (1996). Methodology for Optimally Sizing the Combination of a Battery Bank and PV Array in a Wind/PV Hybrid System. *IEEE Transactions on Energy Conversion, 11*(2), 367–375. doi:10.1109/60.507648

Calabrese, G. (1947). Generating Reserve Capacity Determined by the Probability Method. *Transactions of the American Institute of Electrical Engineers, 66*(1), 1439–1450. doi:10.1109/T-AIEE.1947.5059596

Cardoso, G., Brouhard, T., DeForest, N., Wang, D., Heleno, M., & Kotzur, L. (2018). Battery aging in multi-energy microgrid design using mixed integer linear programming. *Applied Energy, 231*, 1059–1069. doi:10.1016/j.apenergy.2018.09.185

Chalise, S., Sternhagen, J., Hansen, T. M., & Tonkoski, R. (2016). Energy management of remote microgrids considering battery lifetime. *The Electricity Journal, 29*(6), 1–10. doi:10.1016/j.tej.2016.07.003

Chaouachi, A., Kamel, R. M., Andoulsi, R., & Nagasaka, K. (2012). Multiobjective intelligent energy management for a microgrid. *IEEE Transactions on Industrial Electronics, 60*(4), 1688–1699. doi:10.1109/TIE.2012.2188873

Chauhan, A., & Saini, R. (2014). A review on integrated renewable energy system based power generation for stand-alone applications: Configurations, storage options, sizing methodologies and control. *Renewable and Sustainable Energy Reviews, 38*, 99-120.

Chedid, R., Karaki, S., & Rifai, A. (2005). A Multi-Objective Design Methodology for Hybrid Renewable Energy Systems. *Proc. of the IEEE on Power Tech*, 1-6. 10.1109/PTC.2005.4524339

Chedid, R., & Rahman, S. (1997). Unit Sizing and Control of Hybrid Wind-Solar Power Systems. *IEEE Transactions on Energy Conversion, 12*(1), 79–85. doi:10.1109/60.577284

Chellali, F., Khellaf, A., Belouchrani, A., & Recioui, A. (2011). A Contribution to the actualization of the wind map of Algeria. *Renewable & Sustainable Energy Reviews, 15*(2), 993–1002. doi:10.1016/j.rser.2010.11.025

Chellali, F., Recioui, A., Yaiche, M. R., & Bentarzi, H. (2014). A hybrid wind/solar/diesel stand alone system optimisation for remote areas in Algeria. *International Journal of Renewable Energy Technology, 5*(1), 12–24. doi:10.1504/IJRET.2014.059658

Correa, C. A., Marulanda, G., & Garces, A. (2016). Optimal microgrid management in the Colombian energy market with demand response and energy storage. In *Proceedings of the 2016 IEEE Power and Energy Society General Meeting (PESGM)* (pp. 1-5). IEEE. 10.1109/PESGM.2016.7741905

Das, B. K., Al-Abdeli, Y. M., & Kothapalli, G. (2018). Effect of load following strategies, hardware, and thermal load distribution on stand-alone hybrid CCHP systems. *Applied Energy, 220*, 735–753. doi:10.1016/j.apenergy.2018.03.068

De Santis, E., Rizzi, A., & Sadeghian, A. (2017). *Hierarchical genetic optimization of a fuzzy logic system for energy flows management in microgrids. Appl. Soft Comput. J.* doi:10.1016/j.asoc.2017.05.059

Delgado, C., & Dominguez-Navarro, J. A. (2014). Optimal design of a hybrid renewable energy system. *Proceedings of the Ninth International Conference on Ecological Vehicles and Renewable Energies (EVER)*. 10.1109/EVER.2014.6844008

Diaf, S., Notton, G., Belhamel, M., Haddadi, M., & Louche, A. (2008). Design and techno-economical optimization for hybrid PV/wind system under various meteorological conditions. *Applied Energy, 85*(10), 968–987. doi:10.1016/j.apenergy.2008.02.012

Ding, X., Guo, Q., Qiannan, T., & Jermsittiparsert, K. (2021). Economic and environmental assessment of multi-energy microgrids under a hybrid optimization technique. *Sustainable Cities and Society, 65*, 102630. doi:10.1016/j.scs.2020.102630

Dufo-López, R., Bernal-Agustín, J. L., & Contreras, J. (2007). Optimization of control strategies for stand-alone renewable energy systems with hydrogen storage. *Renewable Energy, 32*(7), 1102–1126. doi:10.1016/j.renene.2006.04.013

Ei-Bidairi, K. S., Nguyen, H. D., Jayasinghe, S. D. G., & Mahmoud, T. S. (2018). Multiobjective Intelligent Energy Management Optimization for Grid-Connected Microgrids. *Proceedings of the IEEE International Conference on Environment and Electrical Engineering and IEEE Industrial and Commercial Power Systems Europe (EEEIC / I&CPS Europe)*. 10.1109/EEEIC.2018.8493751

Ekanayaka, J. (2012). *The smart grid: Smart grid technology and applications* (1st ed.). Wiley. doi:10.1002/9781119968696

El-Khadimi, A., Bchir, L., & Zeroual, A. (2004). Dimensionnement et Optimisation Technico-Economique D'un Système D'Energie Hybride Photovoltaïque-Eolien Avec Système de Stockage. *Revue des Énergies Renouvelables, 7*, 73–83.

Elsied, M., Oukaour, A., Gualous, H., & Hassan, R. (2015). Energy management and optimization in microgrid system based ongreen energy. *Energy, 84*, 139–151. doi:10.1016/j.energy.2015.02.108

Erdinc, O., & Uzunoglu, M. (2012). Optimum design of hybrid renewable energy systems: Overview of different approaches. *Renewable & Sustainable Energy Reviews, 16*(3), 1412–1425. doi:10.1016/j.rser.2011.11.011

Fathima, A. H., & Palanisamy, K. (2015). Optimization in microgrids with hybrid energy systems—A review. *Renewable & Sustainable Energy Reviews, 45*, 431–446. doi:10.1016/j.rser.2015.01.059

Gamarra, C., & Guerrero, J. M. (2015). Computational optimization techniques applied to microgrids planning: A review. *Renewable & Sustainable Energy Reviews, 48*, 413–424. doi:10.1016/j.rser.2015.04.025

Gildardo Gómez, W. D. (2016). *Metodología para la Gestión Óptima de Energía en una Micro red Eléctrica Interconectada* (Unpublished Ph.D. Thesis). Universidad Nacional de Colombia, Medellín, Colombia.

Hakimi, S. M., & Moghaddas-Tafreshi, S. M. (2009). *Optimal sizing of a stand-alone hybrid power system via particle swarm optimization for Kahnouj area in South-East of Iran*. Academic Press.

Hameed, A. M. (2012). Optimum sizing of hybrid WT/PV systems via open-space particle swarm optimization. *Second Iranian Conference on Renewable Energy and Distributed Generation (ICREDG)*, 55-60. 10.1109/ICREDG.2012.6190468

Han, Y. (2014). *Microgrid Optimization, Operation, and Control* (Unpublished PhD thesis). Colorado State University.

Hanieh, B. (2014). Optimization of micro-grid system using MOPSO. *Renewable Energy*, *71*, 295–306. doi:10.1016/j.renene.2014.05.006

Hatata, A. Y., Osmana, G., & Aladla, M. M. (2018). An optimization method for sizing a solar/wind/battery hybrid power system based on the artificial immune system. *Sustainable Energy Technologies and Assessments*, *27*, 83–93. doi:10.1016/j.seta.2018.03.002

Hatziargyriou, N., Asano, H., Iravani, R., & Marnay, C. (2007). Microgrids: An Overview of Ongoing Research, Development, and Demonstration Projects. *IEEE Power & Energy Magazine*, *5*(4), 78–94. doi:10.1109/MPAE.2007.376583

Helal, S. A., Najee, R. J., Hanna, M. O., Shaaban, M. F., Osman, A. H., & Hassan, M. S. (2017). An energy management system for hybrid microgrids in remote communities. *Can. Conf. Electr. Comput. Eng.*

Hemeida, A., El-Ahmar, M., El-Sayed, A., Hasanien, H. M., Alkhalaf, S., Esmail, M., & Senjyu, T. (2020). Optimum design of hybrid wind/PV energy system for remote area. *Ain Shams Engineering Journal*, *11*(1), 11–23. doi:10.1016/j.asej.2019.08.005

Heyrman, B., & Dupré, L. (2013). Efficient Modeling, Control and Optimization of Hybrid Renewable-Conventional Energy Systems. *International Journal of Renewable Energy Research*, *3*(4), 781–788.

Hocaoğlu, F. O., Gerek, Ö. N., & Kurban, M. (2009). A novel hybrid (wind-photovoltaic) system sizing procedure. *Solar Energy*, *83*(11), 2019–2028. doi:10.1016/j.solener.2009.07.010

Hongxing, Y. (2008). Optimal sizing method for stand-alone hybrid solar–wind system with LPSP technology by using genetic algorithm. *Solar Energy*, *82*(4), 354–367. doi:10.1016/j.solener.2007.08.005

Hossain, M. A., Pota, H. R., Squartini, S., & Abdou, A. F. (2019). *Modified PSO algorithm for real-time energy management in grid-connected microgrids. Renew. Energy.*

Ismail, M. S., Moghavvemi, M., & Mahlia, T. M. I. (2014). Genetic algorithm based optimization on modelling and design of hybrid renewable energy systems. *Energy Conversion and Management*, *85*, 120–130.

Jemaa, A. B. (2013). Optimum Sizing of Hybrid PV/Wind/Battery System Using Fuzzy-Adaptive Genetic Algorithm. *Proceedings of the 3rd International Conference on Systems and Control.*

Jia, K., Chen, Y., Bi, T., Lin, Y., Thomas, D., & Sumner, M. (2017). Historical-Data-Based Energy Management in a Microgrid with a Hybrid Energy Storage System. *IEEE Trans. Ind. Inform.*

Jiang, X., He, C., & Jermsittiparsert, K. (2020). Online Optimal Stationary Reference Frame Controller for Inverter Interfaced Distributed Generation in a Microgrid System. *Energy Reports*, *6*, 134–145.

Kaabeche, A., Belhamel, M., & Ibtiouen, R. (2011). Sizing optimization of grid-independent hybrid photovoltaic/wind power generation system. *Energy*, *36*(2), 1214–1222.

Karaki, S., Chedid, R., & Ramadan, R. (1999). Probabilistic Performance Assessment of Autonomous Solar-Wind Energy Conversion Systems. *IEEE Transactions on Energy Conversion*, *14*(3), 766–772.

Katsigiannis, Y. A., Georgilakis, P. S., & Karapidakis, E. S. (2010). *Genetic algorithm solution to optimal sizing problem of small autonomous hybrid power systems*. Lecture.

Katsigiannis, Y. A., Georgilakis, P. S., & Karapidakis, E. S. (2010). Multiobjective genetic algorithm solution to the optimum economic and environmental performance problem of small autonomous hybrid power systems with renewables. *IET Renewable Power Generation*, *4*(5), 404–419.

Kellogg, W. (1996). Optimal unit Sizing for a Hybrid Wind/Photovoltaic Generating System. *Electric Power Systems Research*, *39*(1), 35–38.

Kellogg, W. (1998). Generation Unit Sizing and Cost Analysis for Stand-Alone Wind, Photovoltaic, and Hybrid Wind/PV Systems. *IEEE Transactions on Energy Conversion*, *13*(1), 70–75.

Khan, A. A., Naeem, M., Iqbal, M., Qaisar, S., & Anpalagan, A. (2016). A compendium of optimization objectives, constraints, tools and algorithms for energy management in microgrids. *Renewable & Sustainable Energy Reviews*.

Khan, M., & Iqbal, M. (2005). Pre-feasibility study of stand-alone hybrid energy systems for applications in Newfoundland. *Renewable Energy*, *30*, 835–854.

Khatib, T., Azah, M., & Kamaruzzaman, S. (2012). A software tool for optimal sizing of PV systems in Malaysia. *Modelling and Simulation in Engineering*, 10.

Khatib, T., & Elmenreich, W. (2014). An Improved Method for Sizing Standalone Photovoltaic Systems Using Generalized Regression Neural Network. *International Journal of Photoenergy*, 1–8.

Khatod, D. K., Pant, V., & Sharma, J. (2010). *Analytical approach for well-being assessment*. Academic Press.

Khorasaninejad, E., & Fetanat, A. (2015). Size optimization for hybrid photovoltaic–wind energy system using ant colony optimization for continuous domains based integer programming. *Applied Soft Computing*, *31*, 196–209.

Kingsley, A., Shongwe, T., & Joseph, M. K. (2018). Renewable Energy Integration in Ghana: The Role of Smart Grid Technology. In *International Conference on Advances in Big Data, Computing and Data Communication Systems* (pp. 1-7). IEEE.

Ko, M. J. (2015). Multi-objective optimization design for a hybrid energy system using the genetic algorithm. *Energies*, *8*(4), 2924–2949.

Kolawole, S. (2014). Application of Neural Networks For Predicitng The Optimal Sizing Parameters Of Stand-Alone Photovoltaic Systems. *SOP Trans. Appl. Phys.,* 12–16.

Kornelakis, A., & Marinakis, Y. (2010). Contribution for optimal sizing of grid-connected PV-systems using PSO. *Renewable Energy*, *35*, 1333–1341.

Koutroulis, E. (2006). Methodology for optimal sizing of stand-alone photovoltaic/wind-generator systems using genetic algorithms. *Solar Energy*, *80*(9), 1072–1088.

Koutroulis, E., Dionissia, K. A. P., & Kostas, K. (2006). Methodology for optimal sizing of stand-alone photovoltaic/wind-generator systems using genetic algorithms. *Solar Energy*, *80*(9), 1072–1088.

Kraj, A. (2015). *Intelligent computational infrastructures for optimized autonomous distributed energy generation in remote communities*. Brazil Wind Power.

Kumar, K. P., & Saravanan, B. (2019). *Day ahead scheduling of generation and storage in a microgrid considering demand Side management*. J. Energy Storage.

Kusakana, K., & Vermaak, H. J. (2013). Hybrid Renewable Power Systems for Mobile Telephony Base Stations in developing Countries. *Renewable Energy*, *51*, 419–425.

Lasseter, R. H. (2007). CERTS Microgrid. *Proceedings of the 2007 IEEE International Conference on System of Systems Engineering*.

Leonori, S., Rizzi, A., Paschero, M., & Mascioli, F. M. F. (2018). Microgrid Energy Management by ANFIS Supported by an ESN Based Prediction Algorithm. *Proceedings of the International Joint Conference on Neural Networks (IJCNN)*.

Li, H., Eseye, A. T., Zhang, J., & Zheng, D. (2017). *Optimal energy management for industrial microgrids with high-penetration renewables*. Prot. Control Mod. Power Syst.

Li, J., Wei, W., & Xiang, J. (2012). A Simple Sizing Algorithm for Stand-Alone PV/Wind/Battery Hybrid Microgrids. *Energies*, *5*(12), 5307–5323.

Lingfeng, W., & Singh, C. (2007). Compromise between cost and reliability in optimum design of an autonomous hybrid power system using mixed-integer PSO algorithm. *International conference on clean electric power*.

Liu, N., Yu, X., Wang, C., Wang, J. (2017). Energy Sharing Management for Microgrids with PV Prosumers: A Stackelberg Game Approach. *IEEE Trans. Ind. Inform.*

Lotfi, S., Farid, L. T., & Ghiamy, M. (2013). Optimal design of a hybrid solar-wind-diesel power system for rural electrification using imperialist competitive algorithm. *International Journal of Renewable Energy Research*, *3*(2), 403–411.

Luna, A. C., Meng, L., Diaz, N. L., Graells, M., Vasquez, J. C., Guerrero, J. M. (2018). Online Energy Management Systems for Microgrids: Experimental Validation and Assessment Framework. *IEEE Trans. Power Electron.*

Ma, L., Liu, N., Zhang, J., Tushar, W., Yuen, C. (2016). Energy Management for Joint Operation of CHP and PV Prosumers Inside a Grid-Connected Microgrid: A Game Theoretic Approach. *IEEE Trans. Ind. Inform.*

Mahesh, A., & Sandhu, K. S. (2015). Hybrid wind/photovoltaic energy system developments: Critical review and findings. *Renewable and Sustainable Energy Reviews*, *52*, 1135-1147.

Maleki, A., & Askarzadeh, A. (2014). Comparative study of artifcial intelligence techniques for sizing of a hydrogen-based stand-alone photovoltaic/wind hybrid system. *International Journal of Hydrogen Energy*, *39*(19), 9973–9984.

Maleki, A., & Fathollah, P. (2015). Optimal sizing of autonomous hybrid photovoltaic/wind/battery power system with LPSP technology by using evolutionary algorithms. *Solar Energy*, *115*, 471–483.

Maleki, A., & Pourfayaz, F. (2015). Sizing of stand-alone photovoltaic/wind/diesel system with battery and fuel cell storage devices by harmony search algorithm. *Journal of Energy Storage*, *2*, 30–42.

Markvart, T. (1996). Sizing of hybrid photovoltaic-wind energy systems. *Solar Energy*, *57*(4), 277–281.

Markvart, T., Fragaki, A., & Ross, J. (2006). PV System Sizing using Observed Time Series of Solar Radiation. *Solar Energy*, *80*(1), 46–50.

Marzband, M., Azarinejadian, F., Savaghebi, M., & Guerrero, J. M. (2017). An optimal energy management system for islanded microgrids based on multiperiod artificial bee colony combined with markov chain. *IEEE Systems Journal*.

Mellit, A. (2006). Artificial intelligence based-modeling for sizing of a Stand-Alone Photovoltaic Power System: Proposition for a New Model using Neuro-Fuzzy System (ANFIS). *Proceedings of the 3rd International IEEE Conference Intelligent Systems*, 606–611.

Mellit, A. (2007). Sizing of a stand-alone photovoltaic system based on neural networks and genetic algorithms: Application for remote areas. *Istanbul Univ. J. Electr. Electron. Eng.*, *7*, 459–469.

Mohamed, M. A. (2016). PSO-based smart grid application for sizing and optimization of hybrid renewable energy systems. *PLoS One*, *11*(8).

Momoh, J. (2012). Smart Grid Fundamentals of Design and Analysis. Academic Press.

Mondal, A., Misra, S., Patel, L. S., Pal, S. K., & Obaidat, M. S. (2018). DEMANDS: Distributed energy management using noncooperative scheduling in smart grid. *IEEE Systems Journal*.

Motaz, A., Namaane, A., & M'sirdi, N. K. (2013). Optimization of hybrid renewable energy systems (HRES) using PSO for cost reduction. *Energy Procedia*, *42*, 318–327.

Motevasel, M., & Seifi, A. R. (2014). Expert energy management of a micro-grid considering wind energy uncertainty. *Energy Conversion and Management*.

Nafeh, A. S. A. (2011). Optimal economical sizing of a PV-wind hybrid energy system using genetic algorithm. *International Journal of Green Energy*, *8*(1), 25–43.

Nandi, S. K., & Ghosh, H. R. (2009). A wind-PV-battery hybrid power system at Sitakunda in Bangladesh. *Energy Policy*, *37*(9), 3659–3664.

Nikmehr, N., & Najafi-Ravadanegh, S. (2015). Optimal operation of distributed generations in micro-grids under uncertainties in load and renewable power generation using heuristic algorithm. *IET Renewable Power Generation*.

Nivedha, R. R., Singh, J. G., & Ongsakul, W. (2018). PSO based economic dispatch of a hybrid microgrid system. *Proceedings of the 4th International Conference on Power, Signals, Control and Computation (EPSCICON 2018)*.

Nojavan, S., & Jermsittiparsert, K. (2020). Risk-Based Performance of Combined Heat and Power Based Microgrid Using Information Gap Decision Theory. *IEEE Access: Practical Innovations, Open Solutions*, *8*(1), 93123–93132.

Nwulu, N. I., & Xia, X. (2017). Optimal dispatch for a microgrid incorporating renewables and demand response. *Renewable Energy*.

Ogunjuyigbe, A. S. O., Ayodele, T. R., & Akinola, O. A. (2016). Optimal allocation and sizing of of small autonomous power systems with solar and wind energy sources. *IEEE Transactions on Energy Conversion, 25*(2), 535-545.

Othman, Z., Sulaiman, S. I., Musirin, I., & Mohamad, K. (2015). Bat inspired algorithm for sizing optimization of grid-connected photovoltaic system. *Proceedings of the 2015 SAI Intelligent Systems Conference (IntelliSys)*, 195–200.

Papari, B., Edrington, C. S., Vu, T. V., & Diaz-Franco, F. (2017). A heuristic method for optimal energy management of DC microgrid. *Proceedings of the IEEE Second International Conference on DC Microgrids (ICDCM).*

Paul, T. G., Hossain, S. J., Ghosh, S., Mandal, P., Kamalasadan, S. (2018). A Quadratic Programming Based Optimal Power and Battery Dispatch for Grid-Connected Microgrid. *IEEE Trans. Ind. Appl.*

Pillai, N. V. (2008). *Loss of Load Probability of a Power System.* Munich Personal Repec Archive, Paper No. 6953.

Prathyush, M., & Jasmin, E. A. (2018). Fuzzy Logic Based Energy Management System Design for AC Microgrid. *Proceedings of the International Conference on Inventive Communication and Computational Technologies (ICICCT).*

Rajendra, P. A., & Natarajan, E. (2006). Optimization of integrated photovoltaic–wind power generation systems with battery storage. *Energy, 31*(12), 1943–1954.

Ramesh Babu, N. (2017). *Smart Grid System Modeling and Control.* Apple Academic Press.

Ramoji, K. B. S. K. (2014). Optimal economical sizing of a PV-wind hybrid energy system using genetic algorithm and teaching learning based optimization., *International Journal of Advanced Research in Electrical, Electronics and Instrumentation Engineering, 3*(3).

Ramoji, S. K., Bibhuti, B. R., & Kumar, V. D. (2014). Optimization of hybrid PV/wind energy system using genetic algorithm (GA). *Journal of Engineering Research and Applications, 4*, 29–37.

Recioui, A., & Bentarzi, H. (2021). *Optimizing and Measuring Smart Grid Operation and Control.* IGI Global.

Recioui, A., & Dassa, K. (2017). Design of Standalone Micro-Grid Systems Using Teaching Learning Based Optimization. *Algerian Journal of Signals and Systems, 2*(2), 75–85.

Rouholamini, M., & Mohammadian, M. (2016). Heuristic-based power management of a grid-connected hybrid energy system combined with hydrogen storage. *Renewable Energy.*

Roy, P. C., Majumder, A., & Chakraborty, N. (2010). Optimization of a stand-alone Solar PV-Wind-DG Hybrid System for Distributed Power Generation at Sagar Island. *AIP Conference Proceedings, 1298*(1), 260–265.

Russell, S. J., & Norvig, P. (2003). Artificial Intelligence: A Modern Approach (2nd ed.). Academic Press.

Shaahid, S. M., & Elhadidy, M. A. (2008). Economic analysis of hybrid photovoltaic–diesel–battery power systems for residential loads in hot regions—A step to clean future. *Renewable & Sustainable Energy Reviews*, *12*(2), 488–503.

Sharma, S., Bhattacharjee, S., & Bhattacharya, A. (2016). Grey wolf optimisation for optimal sizing of battery energy storage device to minimise operation cost of microgrid. *IET Generation, Transmission & Distribution*, *10*(3), 625–637.

Sopian, K. (2008). Optimal operational strategy for hybrid renewable energy system using genetic algorithms. *WSEAS Transactions on Mathematics*, *7*(4), 130–140.

Speer, B. (2015). Role of smart grids in integrating renewable energy. National Renewable Energy Lab. (NREL).

Sriyakul, T., & Jermsittiparsert, K. (2020). Economic scheduling of a smart microgrid utilizing the benefits of plug-in electric vehicles contracts with a comprehensive model of information-gap decision theory. *Journal of Energy Storage*, *32*, 102010.

Stanton, K. N., Giri, J. C., & Bose, A. (2017). *Energy management*. Syst. Control Embed. Syst. Energy Mach.

Su, W., & Wang, J. (2012). Energy Management Systems in Microgrid Operations. *The Electricity Journal*.

Suchetha, C., & Ramprabhakar, J. (2018). *Optimization techniques for operation and control of microgrids-Review*. J. Green Eng.

Sukumar, S. (2017). Mix-mode energy management strategy and battery sizing for economic operation of grid-tied microgrid. *Energy*, *118*, 1322–1333.

Sukumar, S., Mokhlis, H., Mekhilef, S., Naidu, K., & Karimi, M. (2017). Mix-mode energy management strategy and battery sizing for economic operation of grid-tied microgrid. *Energy*.

Sulaiman, S. I., Rahman, T. K. A., Musirin, I., Shaari, S., & Sopian, K. (2012). An intelligent method for sizing optimization in grid-connected photovoltaic system. *Solar Energy*, *86*, 2067–2082.

Taha, M. S., & Mohamed, Y. A. R. I. (2016). Robust MPC-based energy management system of a hybrid energy source for remote communities. *Proceedings of the IEEE Electrical Power and Energy Conference (EPEC)*.

Thirugnanam, K., Kerk, S.K., Yuen, C., Liu, N., Zhang, M. (2018). Energy Management for Renewable Microgrid in Reducing Diesel Generators Usage with Multiple Types of Battery. *IEEE Trans. Ind. Electron.*

Tina, G., Gagliano, S., & Raiti, S. (2006). Hybrid solar/wind power system probabilistic modelling for long-term performance assessment. *Solar Energy*, *80*(5), 578-588.

Ton, D. T., Wang, W. M., & Wang, W. T. P. (2011). *Smart Grid R&D by the US Department of Energy to optimize distribution grid operations. In 2011 IEEE Power and Energy Society General Meeting*. IEEE.

Umeozor, E. C., & Trifkovic, M. (2016). Energy management of a microgrid via parametric programming. *IFAC-PapersOnLine*.

Venayagamoorthy, G. K., Sharma, R. K., Gautam, P. K., Ahmadi, A. (2016). Dynamic Energy Management System for a Smart Microgrid. *IEEE Trans. Neural Netw. Learn. Syst.*

Wasilewski, J. (2018). Optimisation of multicarrier microgrid layout using selected metaheuristics. *International Journal of Electrical Power & Energy Systems*.

Xing, X., Meng, H., Xie, L., Li, P., Toledo, S., Zhang, Y., & Guerrero, J. M. (2017). Multi-time-scales energy management for grid-on multi-layer microgrids cluster. *Proceedings of the IEEE Southern Power Electronics Conference (SPEC)*.

Xu, X., Hu, W., Cao, D., Huang, Q., Chen, C., & Chen, Z. (2020). Optimized sizing of a standalone PV-wind-hydropower station with pumped-storage installation hybrid energy system. *Renewable Energy*, *147*, 1418–1431.

Yang, H., Lu, L., & Burnett, J. (2003). Weather Data and Probability Analysis of Hybrid Photovoltaic-Wind Power Generation Systems in Hong Kong. *Renewable Energy*, *28*(11), 1813–1824.

Yang, H., Lu, L., & Zhou, W. (2007). A Novel Optimization Sizing Model for Hybrid Solar-Wind Power Generation System. *Solar Energy*, *81*(1), 76–84.

Yang, H., Wei, Z., & Chengzhi, L. (2009). Optimal design and techno-economic analysis of a hybrid solar-wind power generation system. *Applied Energy*, *86*(2), 163–169.

Yang, N., Paire, D., Gao, F., & Miraoui, A. (2013). Power management strategies for microgrid—A short review. *Proceedings of the 2013 IEEE Industry Applications Society Annual Meeting*.

KEY TERMS AND DEFINITIONS

Evolutionary Optimization: An evolutionary algorithm is a population-based optimization algorithm that uses mechanisms inspired by biological evolution, such as reproduction, mutation, recombination, and selection. Evolution of the population then takes place after the repeated application of the previous mechanisms.

Metaheuristics: A metaheuristic is a higher-level procedure to find, generate, or select a sufficiently good solution to an optimization problem. Metaheuristics require a few assumptions about the optimization problem being solved and so may be usable for a variety of problems.

Microgrid: A microgrid is a decentralized group of electricity sources and loads that is able to disconnect from the interconnected grid and to function autonomously. Microgrids improve the security of supply within the microgrid cell, and can supply emergency power, changing between autonomous and connected modes.

Optimization: Optimization is the process of choosing the best element from a set of available alternatives under some constraints. This process amounts to minimizing or maximizing the objective or cost function of the problem.

Renewable Energy Sources: Renewable energy sources are natural resources which will replace the portion depleted by usage and consumption. Common sources of renewable energy include solar, geothermal and wind power.

Smart Grid: A smart grid is an electricity network that uses digital and other advanced technologies to monitor and manage the transport of electricity from all generation sources to meet the varying electricity demands of end-users.

Chapter 6
Artificial Neural Network:
A New Tool for the Prediction of Hydrate Formation Conditions

Anupama Kumari
Indian Institute of Technology, Roorkee, India

Mukund Madhaw
DHC Trading India Private Limited, Delhi, India

C. B. Majumder
Indian Institute of Technology, Roorkee, India

Amit Arora
Shaheed Bhagat Singh State University, Ferozepur, India

ABSTRACT

The analysis and collection of data is an integral part of all research fields of the modern world. There is a need to perform forward mathematical modeling to improve the operations and calculations with modern technologies. Artificial neural network signifies the structure of the human brain. They can provide reasonable solutions quickly for the problems that classical programming cannot solve. An in-depth systematic study is presented in this chapter related to artificial neural network applications (ANN) for predicting the equilibrium conditions for gas hydrate formation, which can assist in designing future dissociation technology for gas hydrate so that this white gold can make world energy free for the future generation. This chapter can also help to develop a novel inhibitor for gas hydrate formation and save millions of dollars for the oil and gas industry.

INTRODUCTION

Gas hydrates are solid crystalline compounds formed by the physical interaction between water and gas molecules (Kumari et al., 2020). The gas molecules are entrapped inside the cages formed by the

DOI: 10.4018/978-1-7998-8561-0.ch006

hydrogen-bonded water molecules. The gas hydrates can be formed as methane (CH_4), ethane (C_2H_6), propane (C_3H_8), and carbon dioxide (CO_2), as well as nitrogen (N_2), hydrogen sulfide (H_2S), and natural gas (Kumari et al., 2020). At low temperatures and high pressures, the gas molecules, usually methane, react with water molecules and form the gas hydrates. Hydrates can be found within deep-water permafrost and oceanic regions. Each gas hydrate volume can contain 184 volumes of gas at STP; hence, hydrates can be considered a potential unconventional energy source. The volume of gas recovered from hydrate reservoirs is about ten times the volume of all known retrievable gas in the whole world. Hence, this makes gas hydrate reservoirs a future energy source to fulfill the world's future energy requirements. Gas hydrate can be formed at approximately 300-800 m water depth depending upon the local temperature of bottom water (E. Dendy Sloan & Koh, 1998; Max et al., 2005; Kumari et al., 2021). Most of the Gas hydrates are formed with methane gas; hence it is termed methane hydrates. The methane gas present in the onshore and offshore gas hydrate reservoir is 3000 times larger than the methane gas present in the atmosphere. Hence, the rapid release of methane gas from gas hydrate could significantly impact the atmosphere's composition and can affect the global climate. Methane gas could release slowly in the atmosphere to oxidize carbon dioxide by the chemical and microbial processes. The methane gas released in the atmosphere should react with the hydroxyl radicals in ten years approximately. Hence, the role of methane gas hydrate in global climate change depends on the release rate of methane gas from gas hydrates, which is primarily unknown. An example of the slow release of methane gas from oceanic sediments is in the Gulf of Mexico, and the rapid release of methane gas is in the Blake Ridge (Kvenvolden, 1999).

The natural gas hydrate can form three types of structure: cubic structure I and II (sI, sII), and hexagonal structure H (sH). sI structured hydrate is formed with the guest molecules (methane, ethane, carbon dioxide) of diameters between 4.2 and 6 Å. The guest molecules of diameters less than 4.2 Å (nitrogen and hydrogen) and diameters between 6 to 7 Å (Propane or isobutane) can form sII structures. When the guest molecules of larger diameter between 7 to 9 Å (iso-pentane or neohexene) are mixed with methane, hydrogen sulfide, or nitrogen, then they will form sH structures (Koh et al., 2009). Based on the geologic and reservoir conditions, the natural reservoirs of gas hydrates are classified into three classes. Class 1 type contains a hydrate layer followed by a two-phase zone of mobile gas and water. Class 2 type consists of a hydrate layer followed by one phase zone of mobile water. Class 3 type consists of a hydrate layer with the absence of underlying zones of mobile fluids (Moridis et al., 2009; Xu & Li, 2015). Three main methods are available for the dissociation of gas hydrates: thermal stimulation, depressurization and chemical injection. In the depressurization method, hydrate is dissociating after decreasing the pressure and this method needs a mobile or permeable fluid zone to produce the gas from the gas hydrates. In the thermal stimulation method, the additional equipment and costs for the injection of hot water or steam are required for the gas hydrate dissociation. Chemical injection requires the insertion of inhibitors such as salts and alcohols and leads to a rapid rate of dissociation and rupturing of the reservoirs (S. H. Khan et al., 2020). The production efficiency of these dissociation methods is affected because of some drawbacks. The depressurization method is favorable for the widespread gas hydrate deposits in a closure. The efficiency of the thermal stimulation method is low due to the high heat loss and high energy consumption. The chemical injection method is not suitable for the ocean area and it is very expensive and creates environmental pollution. Due to these drawbacks, there is a requirement to develop new dissociation techniques for gas hydrates that will be safe and economical (Amit Arora, 2015; Arora et al., 2015).

The gas hydrates can partially or entirely chock the pipelines and lead to severe damage to equipment and the environment. The formation of gas hydrates is a severe flow assurance problem in natural gas production, transport, and processing. There are three methods to prevent hydrate formation: (1) injection of thermodynamic inhibitors, (2) natural gas dehydration, and (3) maintenance of the operating conditions of pipelines separated from the stability zone of hydrate (Najibi et al., 2009). But these above methods are not economically viable, and the inhibitor recovery is complex and detrimental to the environment. The cost arises by offshore gas and oil transport process to prevent hydrate and can be approximately one million dollars per mile (V. T. John, K. D. Papadopoulos, 1985; Gudmundsson & Borrehaug, 1996; Jassim et al., 2008). An accurate estimation of the phase equilibrium conditions for the formation of single and mixture gas hydrate is essential for performing the field and experimental studies (Kumari et al., 2020).

PHASE EQUILIBRIUM CONDITIONS OF GAS HYDRATES

The phase equilibria of gas hydrates are critical parameters for the thermodynamic and kinetic study of the formation of gas hydrates. The phase equilibrium data are an essential parameter to stop the hydrate plug agglomeration in subsea gas and oil flow lines. This is also an important factor to understand the applications of gas hydrates as a source of energy, in the separation of gas and the desalination of seawater. Gas hydrates are formed after the inclusion of small gas molecules (methane) inside the cages of water molecules at low temperature and high pressure. The availability of water and light hydrocarbons at high pressure and low temperature provide the necessary thermodynamic conditions for the gas hydrates formation inside the subsea gas and oil flowlines. The formation of gas hydrates inside the pipelines lead to the blockage of flow line. The severe economic and safety risks because of the formation of the gas hydrate in oil and gas flowline is the major research area currently (Sloan Jr & Koh, 2007; Zerpa et al., 2012). The estimated amount of natural gas is more than 10^{15} m^3 available in permafrost and oceanic deposits of gas hydrates (Sloan Jr & Koh, 2007; Collett et al., 2014). These guest (gas) molecules and the cages of host water are not bonded with each other chemically. They are connected by the weak Vander Waals forces between the molecules of guest and host. Hence, the kinetics of gas hydrate formation and dissociation can be fast (M. N. Khan et al., 2016). This can make gas hydrates as an attractive source for the storage of energy (Collett et al., 2014), transportation of natural gas (E Dendy Sloan, 2003), separation of gases (Aaron & Tsouris, 2005), desalination of seawater (Park et al., 2011) and sequestration of carbon dioxide gas (Park et.al., 2006). The prediction of stability conditions of gas hydrates with good accuracy is difficult for these applications. The available models for the prediction of the phase equilibrium conditions of gas hydrate does not give results with good accuracy. These models are not favorable for the ultra-high pressure conditions of gas hydrates due to the very low availability of experimental data at very high pressure (M. N. Khan et al., 2016). Figure 1 shows the phase diagram of gas hydrates.

Figure 1. Phase diagram of gas hydrates

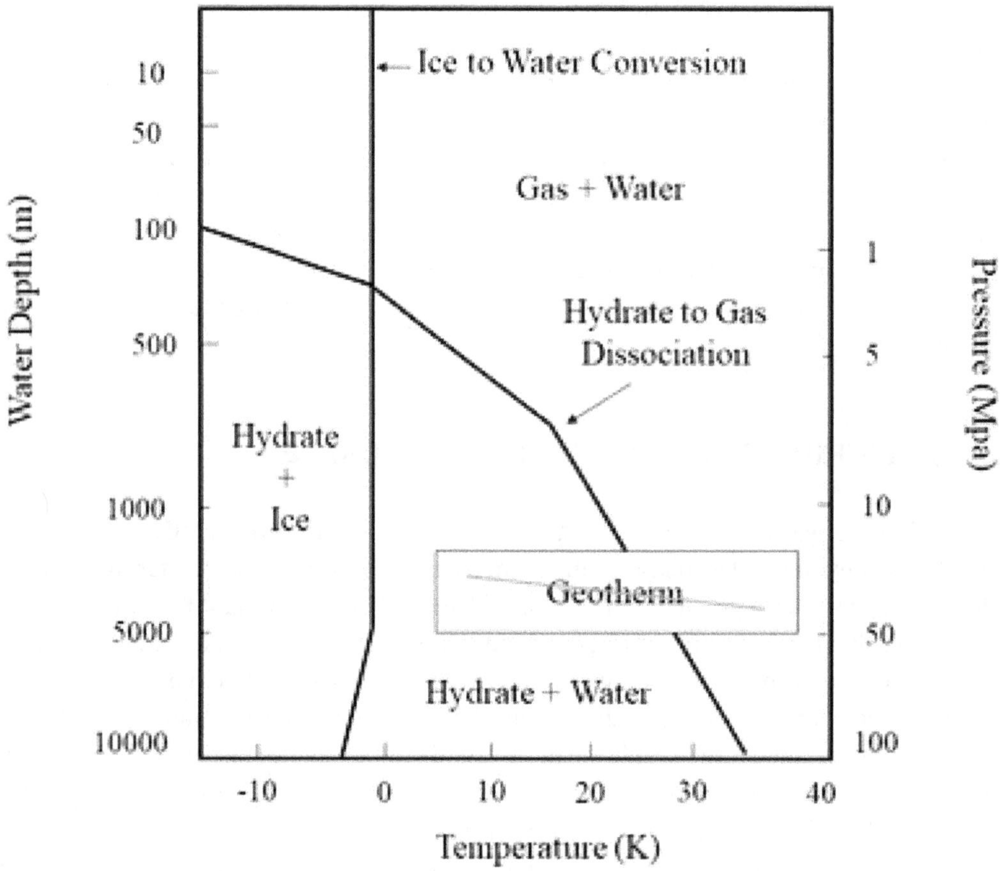

METHODS FOR THE ESTIMATION OF PHASE EQUILIBRIUM OF GAS HYDRATES

There are three methods available for the estimation of the phase equilibrium conditions of gas hydrates: experimental method, dynamic method and static method.

Experimental Method: The phase equilibrium of hydrates can be explained by Gibbs phase rule (Gibbs, 1928) and this rule can be applied to predict the number of the independent variable needed for the determination of equilibrium point. The Gibbs phase rule gave a relation between the number of degrees of freedom or independent variables (F), Number of components (C) and Number of phases (P).

$$F=C-P+2 \quad (1)$$

The independent variables for the experimental method are the pressure and temperature of the system and the initial composition of gas hydrates (M. N. Khan et al., 2016).

Dynamic Methods: The most common approach for the dynamic methods is phase recirculation, in which the fluid mixture is maintained at the target pressure and temperature in an equilibrium vessel and liquid and/or vapor phases are recirculated continuously. A different approach for these methods is a single-pass method in which a stream of the gaseous component is circulated by a stationary liquid phase continuously in the equilibrium vessel. Different streams of liquid and vapor are mixed and sent to the equilibrium vessel. After this the liquid and vapor phases exit from the equilibrium vessel separately (Sloan Jr & Koh, 2007; M. N. Khan et al., 2016).

Static Methods: In this method, the components do not flow continuously. The components are fed in the equilibrium vessel and then the pressure and temperature of the vessel are managed. The volume of the equilibrium vessel is maintained constant frequently. The contents of the equilibrium cell are often agitated to decrease the time of equilibration. In this method, the samples of liquid and vapor phases are frequently strained and examined by gas chromatography for the determination of phase compositions (Raal & Mühlbauer, 1994). A different approach of this method is called the static total pressure method in which known quantities of components are fed to the equilibrium vessel and then the total equilibrium pressure of the process is observed for a desired temperature. The compositions of vapor and liquid phases are then calculated iteratively (M. N. Khan et al., 2016).

ANN model can be applied for the prediction of formation of pressure of gas hydrates with and without thermodynamic inhibitors. (Elgibaly & Elkamel, 1998) applied the ANN model for the estimation of formation pressure of the gas hydrate system and observed a 19% average error. They also observed the quantity of required thermodynamic inhibitors for different types of gas hydrate systems. (Chapoy et al., 2007) applied the ANN model to predict the formation conditions of natural gas hydrate system in the absence and presence of inhibitors. (Mohammadi, Belandria, et al., 2010; Mohammadi & Richon, 2010) predicted the formation conditions of methane and hydrogen hydrates with the promoters by applying the ANN models. (Mohammadi, Martínez-López, et al., 2010) applied the ANN model for the prediction of hydrate formation conditions of methane, nitrogen or carbon dioxide hydrate (Mohammadi, Martínez-López, et al., 2010), hydrogen and tetra-n-butyl ammonium bromide hydrate (Mohammadi et al., 2010). (Rebai et al., 2019) applied the ANN model for the estimation of thermodynamic properties such as viscosity (Rai et al., 2005; Rebai et al., 2019), density, compressibility factor (Bouchard & Granjean, 1995), vapor pressure (Laugier & Richon, 2003), heat transfer coefficient (Potukuchi & Wexler, 1997) and vapor liquid equilibrium (Petersen et al., 1994; Sharma et al., 1999; Sablani et al., 2002; Ganguly, 2003). Heydari et al., 2006 and Zahedi et al., 2009 applied the ANN model to predict the temperature for the hydrate formation. (Mehrizadeh, 2020) estimated the formation conditions of hydrate in the presence of thermodynamic inhibitors by applying the ANN model. Souroush and colleagues applied ANNs for the estimation of the formation temperature of natural gas hydrate and observed the total mean square error of 0.349 only (Soroush et al., 2015).

DISSOCIATION METHODS OF GAS HYDRATES

The gas hydrate dissociation is an important factor for the production of gas from natural gas hydrate reservoir and in the prevention of lug formation in gas pipelines. The dissociation of gas hydrates is an endothermic reaction in which heat must be provided externally. This supplied heat breaks the hydrogen bonds between the molecules of water and weaken the Vander Waals interaction forces between the molecules of guest and water. The gas hydrates will be produced water and gas after the dissociation. There are different methods are available for the dissociation of gas hydrate plug or core such as depressurization, thermal stimulation, CO_2- CH_4 replacement, injection of thermodynamic inhibitor or a combination of the above dissociation methods (Kumari et al., 2020).

Thermal Stimulation: In the thermal stimulation method, the production of gas from the natural deposits of gas hydrate is performed by increasing the temperature of the local natural gas hydrate deposits by the application of heat, injection of water or steam, microwave radiation and electrical heating etc. Thermal stimulation method can be undertaken by three different methods: hot water circulation, hot water huff and puff, and wellbore heating. The gas can be produced by two coaxial cylindrical wells in the natural gas hydrate deposits. The hot water is transported through the inner wall and the gas and water is produced through the outer well. Thermal stimulation can be performed by wellbore heating method by using the electric or other heating method. The heat loss due to transmission of heat through the wells can be neglected as compared to the hot water injection. The main issue of the thermal stimulation through the injection of water or steam is the diffusion and effective permeability of the water/steam in the natural gas hydrate reservoirs (E. Dendy Sloan & Koh, 1998; Amit Arora, 2015; Arora et al., 2015).

Depressurization: In depressurization method, the production of gas from the natural deposits of gas hydrate is performed by reducing the pressure below the natural gas hydrates equilibrium pressure at the local temperature. During the dissociation process, there is continuous drop is observed in the local temperature. At a specific pressure, the dissociation of gas hydrates might stop when the local temperature decreases to the equilibrium temperature. Hence, a sufficient supply of heat or energy is required. The sustainability of production of gas by depressurization method depends upon the diffusion of pressure, effective permeability and saturation of gas hydrates in the natural gas hydrate deposits (E. Dendy Sloan & Koh, 1998).

Chemical Injection: The inhibitor injection method is performed by shifting the phase equilibrium curve of the natural gas hydrate to higher pressure and lower temperature. The natural gas hydrate will become unstable in the local pressure and temperature after the injection of inhibitor. Chemical inhibitors are of two types such as thermodynamic and kinetic inhibitors. Thermodynamic inhibitor disturbs the equilibrium conditions of natural gas hydrates while kinetic inhibitors decrease the rate of formation of gas hydrates. The most frequently used thermodynamic inhibitors are ethanol and methanol glycol. The main issue of the chemical injection method is the effective permeability and diffusion of the chemicals in the natural reservoirs of gas hydrates. The rate of dissociation of gas hydrates through chemical injection method is depends upon the concentration of inhibitors, injection rate of inhibitors, interfacial area of hydrate and inhibitor along with temperature and pressure (E. Dendy Sloan & Koh, 1998).

CO_2-CH_4 replacement: In this method, the CO_2 gas can be sequestrated directly for the recovery of CH_4 gas from natural gas hydrate reservoirs. Through this method CH_4 gas is replaced by CO_2 gas by forming CO_2 hydrates. The CO_2 hydrate has more stability than the CH_4 hydrate under the same pressure and temperature conditions because the enthalpy of formation of CO_2 hydrate is lower than the enthalpy of formation of CH_4 hydrate. In this method, the CH_4 hydrate will dissociate first and then the produced water again forms the CO_2 hydrate. The molecules of CO_2 gas replace the molecules of CH_4 gas directly. The cavities of hydrate remain unchanged during the replacement process (E. Dendy Sloan & Koh, 1998; Park et.al., 2006). Figure 2 shows the dissociation methods of gas hydrates.

Figure 2. Dissociation methods of gas hydrates

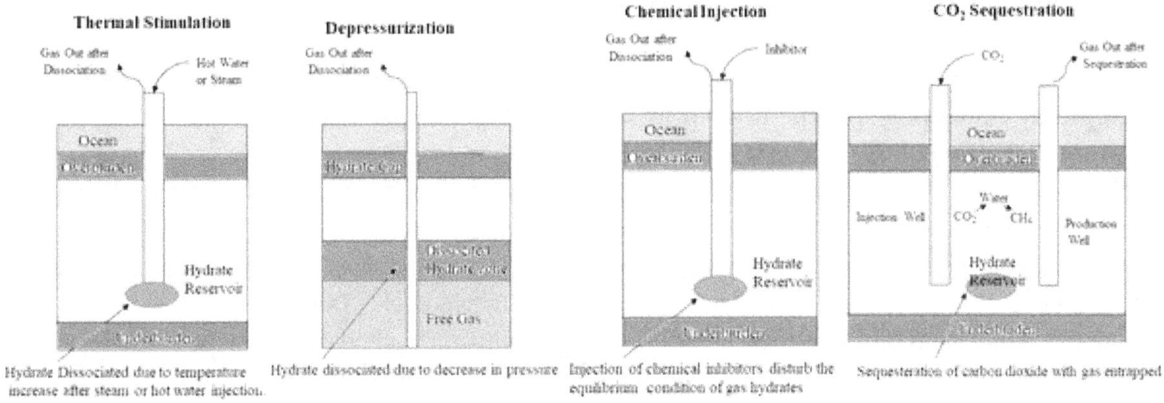

ARTIFICIAL INTELLIGENCE

Artificial intelligence consists of several number of neural networks just like a human brain through which a computer program is developed to achieve some information (Kutsurelis, 1998). The first computer was invented in 1940 and after this, artificial intelligence was established in the world (Bilge, 2007). ANNs are used for the development of prediction models and it is a subsegment of artificial intelligence. These networks are made up from processing devices as similar to the structure of human brain and its data processing (Blackard & Dean, 1999). McCulloch and Pitts performed the first study in 1956 based on artificial intelligence by a calculation model related to the logic modeling. They used physiology, Turing's calculation notion and artificial nerve cells for the logical modeling. Hebb (1949) created the ANNs for the system by the application of a rule hypothesis and changed the power of connection between the nerve cell. McCarthy developed a first computer (SNARC) based on the ANNs. It came in use in 1980 for the development of professional systems for the various engineering fields, business management and economics (Akın, 1997; Staub et al., 2015).

TECHNIQUES OF ARTIFICIAL INTELLIGENCE

There are some techniques which are available for the development of artificial intelligence such as expert systems, fuzzy systems, ANNs, multi-layered detectors and learning algorithms and genetic algorithms.

Expert Systems: Expert systems are those types of computer programs that can solve problems in a non-procedural manner by applying the knowledge from human experts for the simulation of human reasoning. These systems use knowledge for their processing rather than information. The expert systems search for the solutions dependent upon the expertise area of people. The main areas of the application of these systems are medicine and biomedical (Detore, 1989).

Fuzzy Systems: These systems deal with incomplete and imprecise sets of data. These systems are advantageous in complicated structural control systems. In the fuzzy systems, the membership is indicated by a number between 0 and 1. Fuzzy sets have clearly defined boundaries. The fuzzy logic uses the combination of learning and decision-making properties of ANNs (Zadeh, 1996; Tanyildizi et al.,

2009). It is one of the best tools for predicting time sequence and networks using the back propagation approach (Ghiassi et al., 2005; Canan & Yildirim., 2008).

Artificial Neural Networks: The structure of ANNs work resembles the neurons of the human brain. The information on systems is collected by the artificial neurons in ANNs. These network structure are developed by connecting artificial neurons by various connection geometries (Hopfield, 1988; Hammerstrom, 1993; Emel & Taşkın, 2002).

Multi-Layered Detectors and Learning Algorithms Model: In the multi-layered detectors and learning algorithms model, the processors in a layer are attached to the processors in another layer. Several layers are available in this model containing an input and one or more outputs. Several teaching algorithms can be applied during the training process of this network (Chen et al., 2008; Emel & Taşkın, 2002).

Genetic Algorithm (GA): Genetic algorithm is a probabilistic method related to the natural genetics and the selection mechanism utilizing the details of purpose function and functions with the help of the coding clusters of parameters (Shapiro, 2001; Emel & Taşkın, 2002).

Case Based Reasoning (CBR): A problem is solved by recalling the similar types of previous problems and assuming that both have similar solutions. Various previous problems are required to adjust the methods or solutions according to the new problem. CBR acknowledges these problems and becomes easier to solve by the repeated efforts (Aamodt & Plaza, 1994).

Rule Based Systems (RBS): In these systems, the problems are solved by the rules obtained by the expert knowledge. Those rules contain the action and condition parts. These systems contain a pattern matcher (working memory) and a rule applier. The working memory decides which rules are applicable and then the rule applier selects the rule to apply (Hayes-Roth, 1985).

Cellular Automata (CA): These are dynamic models which discrete in the state, time, and space. These systems contain a well-organized lattice of cells that can interact with the neighbors. These cell states are synchronously improved in time as per the local rules and determine the new state of that cell at time t+1 by applying the previous state and neighboring cells at time t (Codd, 1968).

ARTIFICIAL NEURAL NETWORKS (ANN)

The structure of ANNs and the structure of the human brain are remarkably similar. In both structures, the information is obtained by networks and the information is stored in the connections between artificial neurons. ANNs can become functional when the artificial network used as processor to store the information (Gelir, 1994). It formed by the integration of constant non-linear functions and the establishment of neural networks shows the extent of neural networks (Kröse et al., 1993; Chenoweth, Obradovic & Stephen, 1996). These networks are similar to the biological neural networks and give results with good accuracy in spite of the presence of the superficial connections between the ANNs (Emir, 2013; Staub et al., 2015).

There are three layers in the ANNs: the input, hidden and output layers. The groups of input data are passed through the input layer to the network. The parameters for the input layers must be chosen before the analysis (Blackard & Dean, 1999). The number of input data and neurons are same in an input layer. Each input neuron is transferred to the hidden layer and the hidden layer is the basic function in the network. The number of neurons in the hidden layer should be between the input and output layer's size of the input and output layer, or 2/3 of the combined size, or less than the twice of the size of the input layer (Heaton, 2008). Output data obtained through the input layer is processed properly in the hidden layer. Output data obtained through hidden layer is then transferred to the output layer (Dag, 2012). The output layer is the last layer of the network, which processes the output data obtained after the hidden layer and gives the final output. The number of neurons in the network is same as the number of outputs obtained by the network. The data obtained from the output layer is the output for the given problem (Haykin, 1994; Abdi, 2003; Dag, 2012; Staub et al., 2015).

ANN is a very efficient technique to obtain the results for the complex non-linear problems. The fault tolerance of these networks is sufficiently high because the information is dispersed in a regular way around the system. These networks can be rearranged to improve their own values and can modify themselves for the required result of the given problem. ANNs have the capability to generate the desired results throughout the training process by generalization (Staub et al., 2015).

ANNs comprise conventional network compounds including linear functions and feed-forward connections (Kramer, 1991). The two drawbacks of neural networks are slow learning speed and the local minimum convergence (Castillo et al., 2006). The ANNs can be classified as feedback networks, feed forward networks, radial and memory-based networks and module ANNs based on the architectural structures. The most used structure is feedback networks and feed forward networks. The most crucial characteristic of feed-forward network is that it can recognize missing or fake data before the termination of the process. In these networks, data processed from the input units to the output units (Peterson et al., 1994; Bishop, 1995; Ripley, 1996; Bennell & Sutcliffe, 2004; Staub et al., 2015).

For the development of multi-layered perceptron neural networks, an activation function is required in the middle layer and the output layer and a linear function representing the input layer. An activation function can approximate the arbitrary accuracy of each function. The standard error backpropagation method can be applied to train the neural network. The Scaled Conjugate Gradient (SCG), Levenberg–Marquardt (LM) and Polak-Ribiere Conjugate Gradient (CGP) can be utilized as the learning algorithms (Haykin, 1994; Vakil-Baghmisheh, 2002). Early stopping method can be applied to test and train data driven models and to stop overfitting (Møller, 1993). For the development of ANN model, all data sets are randomly divided in three groups such as training, cross-fitting and testing (Coulibaly et al., 2000). The optimal designing for these network structures can be carried out by choosing the optimal number of hidden layer and hidden layer neurons by trial and error method (Denai et al., 2004). The best learning functions and the most applicable number of hidden layer neurons can be achieved from various learning functions applied in error backpropagation training algorithm throughout the training phase. The criteria for the selection of optimum network in the training phase is the calculation of minimum root mean square error (RMSE), maximum coefficient of determination (R^2) and coefficient of variation (COV) (Mehrizadeh, 2020).

An artificial neural collects the processing elements (neurons) and arranged them in layers and then transforms the input set (i) to the desired output set. From every layer, the neurons are linked to all different neurons emerging in the next layer (j). These neurons are joined with different types of synaptic weights of neurons (w_{ij}). A threshold value (b_j) is added as the bias term after the summation of the weighted input values (scalar (x_i)) at each neuron. At last, a non-linear transfer function ($f(.)$) is used as the activation function. This activation function is applied to this linear combination of inputs and then output of the neuron is produced through the linear combiner (o_j). The bias term is added to the weights depending on the situation. A model for the artificial neuron is represented in Figure 3.

$$o_j = f\left(\sum_{i=1}^{n} w_{j-1,i}x_{j-1,i} + b_j\right) \quad (2)$$

Figure 3. A model of an artificial neuron j in the hidden layer

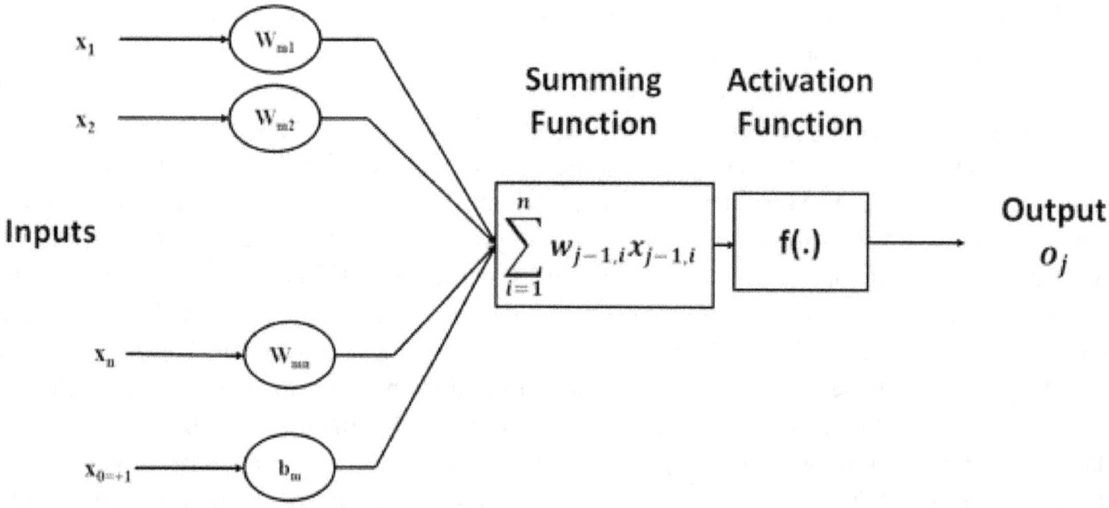

The output from one neuron will be the input for the neuron of the next layer. The neural network utilizes a training algorithm such as backpropagation for the adjustment of its weights on the demonstration of a training data set and then minimizes the performance function of the network. The default function used to estimate the performance of the neural network is mean square error (MSE). It is the average squared error obtained for the output of the network and the desired values for the given inputs. MSE is given by following equation:

$$MSE = \frac{1}{n}\sum_{i=1}^{n}(t_i - o_i)^2 \quad (3)$$

In the above equation, n represents the number of training data, the target data and calculated output is represented by t_i and o_i respectively. When the training phase of the network model has been completed successfully then the validation of the performance of the trained model has been performed by an independent testing set.

Figure 4 represents the architecture of an artificial neural network with input branching nodes (*I*), hidden neurons (*H*) and output neurons (*O*). The number of hidden neurons *H* is defined by MSE. The value of MSE is the average squared variation between the output and the target. The minimum value of MSE denotes the better performance of the neural network (Naderpour et al., 2018). The number of hidden neurons can be determined by cross-validating the model with different configurations and observing the MSE value by plotting the graph between the MSE value and the number of hidden neurons. The cross validation of a model can be performed by dividing all datasets into training and testing data sets and then train the model using the training data set. Now, the model has to be validated on the test data set. After the repetition of the above steps, the MSE value can be minimized (Paneiro & Rafael, 2021).

Figure 4. Architecture of Artificial Neural Network

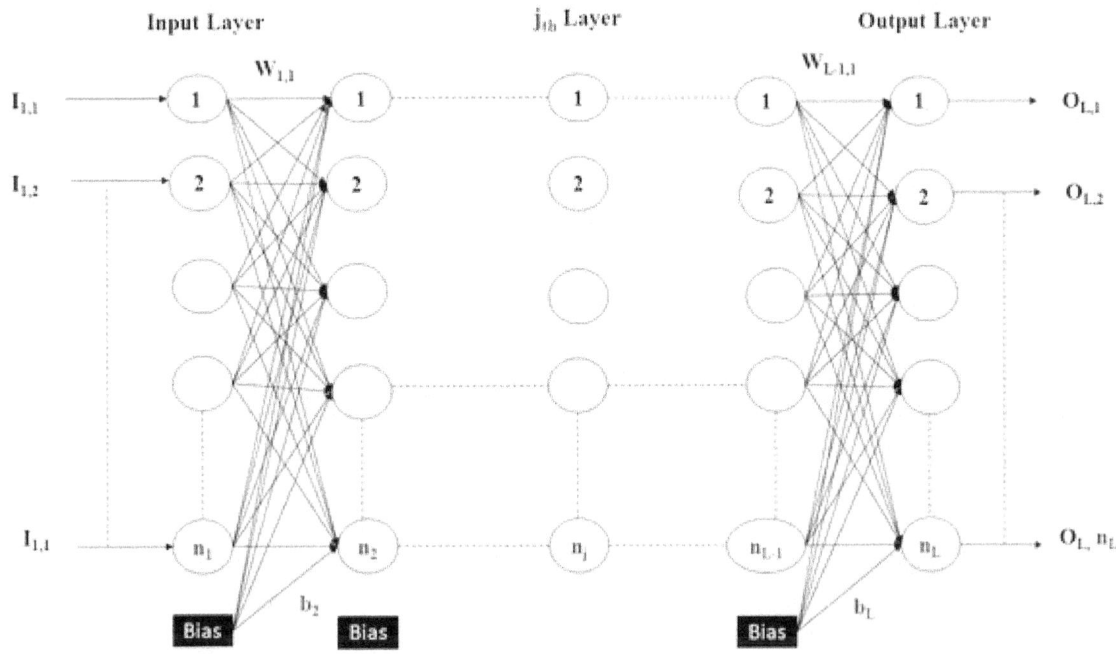

MODEL EVALUATION CRITERIA

Various statistic parameters such as Classification Accuracy, Logarithmic Loss, Confusion Matrix, Area under curve, F1 Score, Mean Squared Error (MSE), COV, RMSE and R^2 can be used to check the performance of the function used in the prediction models. These parameters can be applied to obtain the best structure of the ANN model. Various other factors such as mean absolute error (MAE) and mean absolute percentage of error (MAPE) can be applied to compare the optimum efficiency of the model (Mehrizadeh, 2020).

MAIN FOCUS OF THE CHAPTER

This chapter deals with the uses of ANN for the prediction of phase equilibrium conditions of Gas Hydrates. The phase equilibrium condition of gas hydrate is an important factor for the gas hydrate formation and dissociation. The gas hydrate dissociation depends upon the disturbance of phase equilibrium conditions of gas hydrates.

APPLICATION OF ARTIFICIAL NEURAL NETWORK FOR GAS HYDRATES

There are various other methods are available for prediction of phase equilibrium conditions of gas hydrates such as thermodynamic modeling, mixing rule modification and interaction coefficient approximation between molecules (Berecz and Balla-Achs, 1983; Holder et al., 1988; Sloan, 1990; Sloan and Koh, 1998; Ballard, 2002). The main drawbacks of these models are that they can be applied for particular systems with the restricted constituents. These methods are dependent on the equality of chemical potentials of different components in distinct phases (Mohammadi & Tohidi, 2005). These methods are dependent on the limited data and their limited applications. But the modeling with a sustainable uncertainty of temperature-pressure by these available thermodynamic models needs the calculation of various parameters. A tedious process can adjust these parameters, and it is not certain that the obtained set is the best. The ANNs can constitute within the experimental uncertainties and calculate the phase boundaries of gas hydrates accurately (Chapoy et al., 2007).

The wide applicability of ANNs in pharmaceutical and chemical areas has been expanded due of their ability and flexibility to model linear and non-linear systems without the previous understanding of an empirical model (Jouyban et al., 2004). ANN removes the limitations of the classical approaches by extracting the desired information using the input data. The ANNs can be applied to a system that requires enough input and output data rather than a mathematical equation. The ANN can be trained by utilizing the input and output data to adjust to the system and it can be applied to solve the problems with the inaccurate and incomplete input data (Akcayol & Cinar, 2005; Heydari et al., 2006).

Baghban et al. use the Katz Chart Data Points in the Least Square Vector Machine (LSSVM) model to predict the hydrate formation temperature. The LSSVM model applied the chemical properties of gases and the structure type of hydrates (Baghban et al., 2016).

The SVMs use the kernel functions, and ANNs use the hidden layers to transform the data and send these data into a higher-dimensional space. The one advantage of the ANNs is that their size is fixed, and they are the parametric models, but the SVMs are the non-parametric models. The ANNs may have several outputs, but the SVMs have only one. Hence, ANN train the models in one chance, and SVM must train one by one.

ANN is based on the intrinsic relationships between independent and dependent variables (Mehrizadeh, 2020). The ANN can reproduce any type of a function if adequate examples of that function are given. These examples are selected on the basis of behavior of the reproducing function (Rebai et al., 2019). The ANN models are of data driven models which are very recent models which can be applied to estimate the formation conditions of gas hydrates. The data-driven models include direct training of model related to the data in which basic assumptions is not needed. Also, the advantages of these models are that there is no need of the preliminary understanding of the connection between parameters and the capability to set up a connection between a set of input and outputs for the prediction of the output (Torrecilla et al., 2004; Kaul et al., 2005; Singh et al., 2007). These types of models can represent any nonlinear continuous function and add flexibility to face any potential error (Haykin & Network, 2004; Azadeh et al., 2007).

ANNs can be applied for the prediction of formation pressure or temperature of the single, binary and natural gas hydrate system in presence of inhibitors and promoters also. The input parameters for the different gas hydrate system depend upon the type of the gas hydrate system. For example, for natural gas hydrate system the input parameter is specific gravity of the natural gas but it will be changed to concentration of inhibitors or promoters for gas hydrate system with inhibitors or promoters. This model can also be applied for the prediction of thermodynamic properties based on hydrate formation such as viscosity (Rai et al., 2005; Rebai et al., 2019), density, compressibility factor (Bouchard & Granjean, 1995), vapor pressure (Laugier & Richon, 2003), heat transfer coefficient (Potukuchi & Wexler, 1997) and vapor liquid equilibrium (Petersen et al., 1994; Sharma et al., 1999; Sablani et al., 2002; Azadeh et al., 2007).

The neural network will be constructed using the experimental data of the formation process of the gas hydrates. The available experimental data are divided into three random subsets such as training, validation, and testing sets. The input parameters for the development of ANN depend upon the structure of the gas hydrates. The input parameters are mole % of CH_4 or C_2H_6 for gas hydrates of sI structure, total mole % of C_3H_8, C_4H_{10} and N_2 for gas hydrates of sII structure and mole % of gases which cannot form gas hydrates. The composition of CO_2 and H_2S gases are taken as the separate parameter because they are soluble in water. Specific gravity of gases, pressure and acid gas loading are the remaining parameters for the neural network. The output parameter for the developed neural networks of the gas hydrates system is the formation temperature of gas hydrates (Heydari et al., 2006; Soroush et al., 2015; Hesami et al., 2017).

METHODOLOGY

The gas hydrates can be formed in the different types of gas systems such as gas systems with hydrocarbon and non-hydrocarbon materials or their mixtures. Gas system can be formed in the absence or presence of inhibitors. ANN models are data-driven models which estimate a desired variable dependent upon input and output conditions of system or the initial and final conditions of the system. For the development of data-driven models, the interrelations between effective parameters in hydrate formation conditions were considered. These models consider the dependency of formation pressure of hydrate on the characteristics of the gas hydrate systems. The components which can disturb the formation process of hydrate are pressure and temperature of hydrate formation, composition or mole fraction of gas components, density of gas and the weight percentage of inhibitors. To develop an ANN model for the prediction of hydrate formation pressure or temperature, the hydrate formation temperature or pressure is assumed to be a function of the remaining variables. ANN models can be developed as three different patterns based upon the input variable of the data driven models (Mehrizadeh, 2020). In pattern A, Pressure (P) depends upon the Temperature (T) and the density (γ) of gas such as CH_4, C_2H_6, C_3H_8, i-C_4H_{10} and their mixture. In pattern B, Pressure (P) depends upon the temperature (T) and the percentage of gas components and their mixtures. Pattern C is similar to the pattern B including the non-hydrocarbon components of hydrate such as CO_2, N_2 and H_2S.

The input neurons transmit temperature and specific gravity or concentration of inhibitors or promoters to the neurons of the hidden layer. To determine the inputs of the hidden layer, a weighted summation of outputs of input layer has been performed. The output from the hidden layer now becomes the input of the output layer. This output layer is determined by shifting its input after adding a non-linear transfer function. At last, pressure is determined by shifting the input from the output layer by adding the linear transfer function.

The ANN model modifies the values of synaptic weights between computing cells according to the differences in the output and target values to achieve the targets. An appropriate learning algorithm is needed for the training of the network.

Figure 5. General block structure of ANN

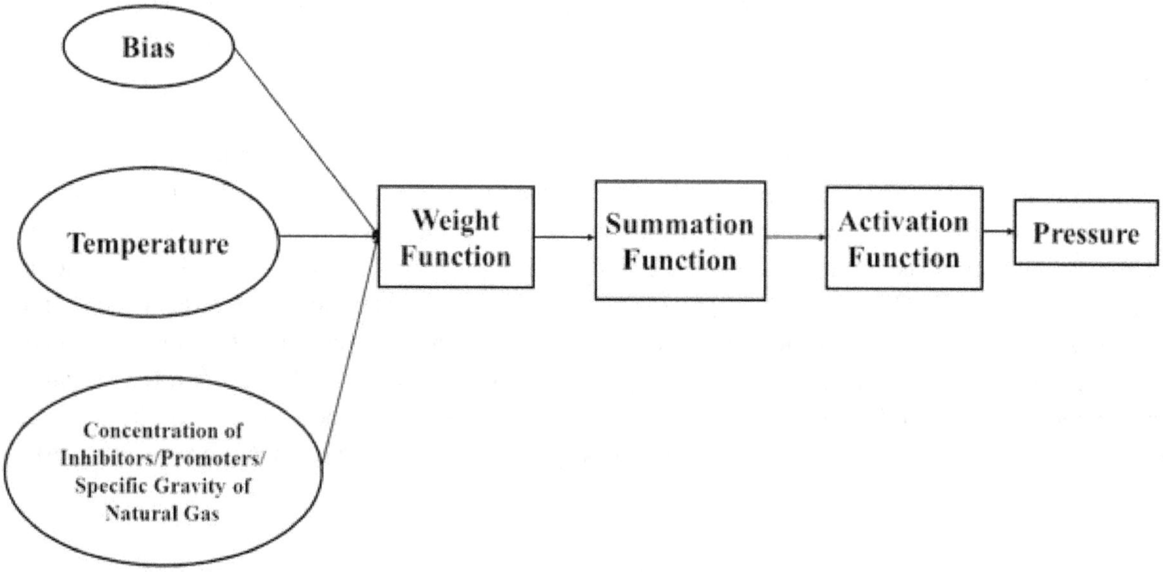

SOLUTIONS AND RECOMMENDATIONS

Application of ANN in the field of gas hydrates is a new and novel idea which can attempt many burning issues of the current century like energy securities and carbon capture. It is recommended to apply ANN for CO_2 sequestration of methane hydrates so that methane comes out of hydrates economically and hydrate of CO_2 gets formed resulting into giving energy at the cost of reducing global warming making each cleaner and greener and giving energy in addition. ANN can be applied for determination of phase equilibrium conditions of gas hydrates with and without inhibitors or promoters. There is need to develop several methods for the prediction of phase equilibrium conditions of gas hydrates formed in natural sediments and porous materials. For this system, the pore size of the porous materials can be added as the input parameter of the ANN model.

FUTURE RESEARCH DIRECTIONS

The application of an ANN model is being extended for the gas hydrate system with different parameters such as gas hydrate formation in porous media, sand and different types of inhibitors and promoters. It can be extended for gas hydrates of different compositions based upon the composition of gases.

CONCLUSION

The application of ANN to estimate the phase equilibrium conditions for formation of gas hydrates can assist in designing an economic viable technology for the gas hydrate dissociation and can make the world energy-independent for many centuries. Moreover, ANN can also help in developing a future economically viable method for the inhibition of gas hydrates in pipelines for which antifreeze proteins currently used are very costly. For getting both these benefits to mankind it is necessary to look it into various aspects of ANN in detail so that along with above two benefits gas hydrate becomes a technology for carbon capture, desalination, gas separation as well as hydrogen storage.

ABBREVIATIONS AND NOTATIONS

ANN- Artificial Neural Network
Genetic Algorithm- GA
Case Based Reasoning- CBR
Rule Bases Systems- RBS
Cellular Automata- CA
Levenberg–Marquardt- LM
Scaled Conjugate Gradient- SCG
Polak-Ribiere Conjugate Gradient- CGP
Root Mean Square Error- RMSE
Mean Squared Error- MSE
Maximum Coefficient of Determination- R^2
Coefficient of Variation- COV
Input Ser- i
Next Layer- j
Weights of neurons- w_{ij}
Threshold Value- b_j
Scalar- x_i
Non-Linear Transfer Function- $f(.)$
Linear Combiner- o_j
Target Data- t_i
Calculated Output- o_i
Input Branching Nodes- I
Hidden Neurons- H
Output Neurons- O
Independent Variables- F
Number of Components- C
Number of Phases- P

ACKNOWLEDGMENT

This study is supported by Gas Hydrate Research & Technology Centre (*GHRTC*), Oil and Natural Gas (ONGC), PANVEL and Ministry of Human Resource Development (MHRD), India.

REFERENCES

Aamodt, A., & Plaza, E. (1994). Case-based reasoning: Foundational issues, methodological variations, and system approaches. *AI Communications*, *7*(1), 39–59. doi:10.3233/AIC-1994-7104

Aaron, D., & Tsouris, C. (2005). Separation of CO2 from flue gas: A review. *Separation Science and Technology*, *40*(1–3), 321–348. doi:10.1081/SS-200042244

Akcayol, M. A., & Cinar, C. (2005). Artificial neural network based modeling of heated catalytic converter performance. *Applied Thermal Engineering*, *25*(14–15), 2341–2350. doi:10.1016/j.applthermaleng.2004.12.014

Akın, H. L. (1997). *The body and brain problem in artificial intelligence. Computer and brain*. Pomegranate Publications.

Amit Arora, S. S. (2015). Techniques for Exploitation of Gas Hydrate (Clathrates) an Untapped Resource of Methane Gas. *Journal of Microbial & Biochemical Technology*, *07*(02), 108–111. doi:10.4172/1948-5948.1000190

Arora, A., Chandrajit, B., & Cameotra, S. S. (2015). Natural Gas Hydrate (Clathrates) as an Untapped Resource of Natural Gas. *Journal of Petroleum & Environmental Biotechnology*, *6*(04).

Azadeh, A., Ghaderi, S. F., & Sohrabkhani, S. (2007). Forecasting electrical consumption by integration of neural network, time series and ANOVA. *Applied Mathematics and Computation*, *186*(2), 1753–1761. doi:10.1016/j.amc.2006.08.094

Baghban, A., Namvarrechi, S., Phung, L. T. K., Lee, M., Bahadori, A., & Kashiwao, T. (2016). Phase equilibrium modelling of natural gas hydrate formation conditions using LSSVM approach. *Petroleum Science and Technology*, *34*(16), 1431–1438. doi:10.1080/10916466.2016.1202966

Ballard, A. L. (2002). *Non-ideal hydrate solid solution model for a multi-phase equilibria program, A*. Colorado School of Mines. Arthur Lakes Library.

Bennell, J., & Sutcliffe, C. (2004). Black–Scholes versus artificial neural networks in pricing FTSE 100 options. *Intelligent Systems in Accounting, Finance & Management*, *12*(4), 243–260.

Berecz, E., & Balla-Achs, M. (1983). *Gas hydrates*. Academic Press.

Bilge, U. (2007). Artificial Intelligence and Expert Systems in Medicine. *Informatics Association of Turkey Congress*, 113–118.

Bishop, C. M. (1995). *Neural networks for pattern recognition*. Oxford University Press.

Blackard, J. A., & Dean, D. J. (1999). Comparative accuracies of artificial neural networks and discriminant analysis in predicting forest cover types from cartographic variables. *Computers and Electronics in Agriculture*, *24*(3), 131–151. doi:10.1016/S0168-1699(99)00046-0

Bouchard, C., & Granjean, B. P. A. (1995). A neural network correlation for the variation of viscosity of sucrose aqueous solutions with temperature and concentration. *Lebensmittel-Wissenschaft + Technologie*, *28*(1), 157–159. doi:10.1016/S0023-6438(95)80029-8

Castillo, E., Guijarro-Berdinas, B., Fontenla-Romero, O., Alonso-Betanzos, A., & Bengio, Y. (2006). A Very Fast Learning Method for Neural Networks Based on Sensitivity Analysis. *Journal of Machine Learning Research*, *7*(7).

Chapoy, A., Mohammadi, A. H., & Richon, D. (2007). Predicting the hydrate stability zones of natural gases using artificial neural networks. *Oil & Gas Science and Technology-Revue de l'IFP*, *62*(5), 701–706. doi:10.2516/ogst:2007048

Chen, S. H., Jakeman, A. J., & Norton, J. P. (2008). Artificial Intelligence techniques: An introduction to their use for modelling environmental systems. *Mathematics and Computers in Simulation*, *78*(2–3), 379–400. doi:10.1016/j.matcom.2008.01.028

Codd, E. F. (1968). *Cellular automata*. Academic Press.

Collett, T. S., Boswell, R., Cochran, J. R., Kumar, P., Lall, M., Mazumdar, A., Ramana, M. V., Ramprasad, T., Riedel, M., Sain, K., Sathe, A. V., & Vishwanath, K. (2014). Geologic implications of gas hydrates in the offshore of India: Results of the National Gas Hydrate Program Expedition 01. In Marine and Petroleum Geology (Vol. 58, Issue PA, pp. 3–28). doi:10.1016/j.marpetgeo.2014.07.021

Coulibaly, P., Anctil, F., & Bobée, B. (2000). Daily reservoir inflow forecasting using artificial neural networks with stopped training approach. *Journal of Hydrology (Amsterdam)*, *230*(3–4), 244–257. doi:10.1016/S0022-1694(00)00214-6

Denai, M. A., Palis, F., & Zeghbib, A. (2004). ANFIS based modelling and control of non-linear systems: a tutorial. *2004 IEEE International Conference on Systems, Man and Cybernetics (IEEE Cat. No. 04CH37583)*, *4*, 3433–3438. 10.1109/ICSMC.2004.1400873

Detore, A. W. (1989). *An Introduction to Expert Systems*. Academic Press.

Elgibaly, A. A., & Elkamel, A. M. (1998). A new correlation for predicting hydrate formation conditions for various gas mixtures and inhibitors. *Fluid Phase Equilibria*, *152*(1), 23–42. doi:10.1016/S0378-3812(98)00368-9

Emel, G. G., & Taşkın, Ç. (2002). Genetic Algorithms and Their Application Areas. *Journal of Uludağ University Faculty of Economics and Administrative Sciences*, *21*(1), 129–152.

Emir, Ş. (2013). *Comparison of the classification performance of artificial neural networks and support vector machine methods: an application on the prediction of stock market index direction*. Istanbul University.

Ganguly, S. (2003). Prediction of VLE data using radial basis function network. *Computers & Chemical Engineering*, *27*(10), 1445–1454. doi:10.1016/S0098-1354(03)00068-1

Ghiassi, M., Saidane, H., & Zimbra, D. K. (2005). A dynamic artificial neural network model for forecasting time series events. *International Journal of Forecasting*, *21*(2), 341–362. doi:10.1016/j.ijforecast.2004.10.008

Gibbs, J. W. (1928). *The collected works of J. Willard Gibbs*. Yale Univ. Press.

Gudmundsson, J., & Borrehaug, A. (1996). Frozen hydrate for transport of natural gas. *NGH 96: 2nd International Conference on Natural Gaz Hydrates*, 415–422.

Hammerstrom, D. (1993). Working with neural networks. *IEEE Spectrum, 30*(7), 46–53. doi:10.1109/6.222230

Hayes-Roth, F. (1985). Rule-based systems. *Communications of the ACM, 28*(9), 921–932. doi:10.1145/4284.4286

Haykin, S. (1994). *Neural networks: A comprehensive foundation*. Mc Millan.

Haykin, S., & Network, N. (2004). A comprehensive foundation. *Neural Networks, 2*, 41.

Heaton, J. (2008). *Introduction to neural networks with Java*. Heaton Research, Inc.

Hebb, D. O. (1949). The organization of behavior: A neuropsycholocigal theory. *A Wiley Book in Clinical Psychology, 62*, 78.

Hesami, S. M., Dehghani, M., Kamali, Z., & Ejraei Bakyani, A. (2017). Developing a simple-to-use predictive model for prediction of hydrate formation temperature. *International Journal of Ambient Energy, 38*(4), 380–388. doi:10.1080/01430750.2015.1100678

Heydari, A., Shayesteh, K., & Kamalzadeh, L. (2006). Prediction of hydrate formation temperature for natural gas using artificial neural network. *Oil and Gas Business*.

Holder, G. D., Pradhan, N., Equilibria, P., & Model, T. (1988). *Phase Behaviour In Systems Containing Clathrate Hydrates. A Review Contents Structure of Gas Hydrates*. Academic Press.

Hopfield, J. J. (1988). Artificial neural networks. *IEEE Circuits and Devices Magazine, 4*(5), 3–10. doi:10.1109/101.8118

Jassim, E. I., Abdi, M. A., & Muzychka, Y. (2008). A CFD-based model to locate flow-restriction induced hydrate deposition in pipelines. *Offshore Technology Conference*. 10.4043/19190-MS

John, V. T., & Papadopoulos, K. D. G. D. H. (1985). John V. T.; Papadopoulos K. D.; Holder G. D. A Generalized Model for Predicting Equilibrium Conditions for Gas Hydrates. *AIChE Journal, 31*(2), 252–259. doi:10.1002/aic.690310212

Jouyban, A., Majidi, M.-R., Jabbaribar, F., & Asadpour-Zeynali, K. (2004). Solubility prediction of anthracene in binary and ternary solvents by artificial neural networks (ANNs). *Fluid Phase Equilibria, 225*, 133–139. doi:10.1016/j.fluid.2004.08.031

Kaul, M., Hill, R. L., & Walthall, C. (2005). Artificial neural networks for corn and soybean yield prediction. *Agricultural Systems, 85*(1), 1–18. doi:10.1016/j.agsy.2004.07.009

Khan, M. N., Warrier, P., Peters, C. J., & Koh, C. A. (2016). Review of vapor-liquid equilibria of gas hydrate formers and phase equilibria of hydrates. *Journal of Natural Gas Science and Engineering, 35*, 1388–1404. doi:10.1016/j.jngse.2016.06.043

Khan, S. H., Kumari, A., Chandrajit, G. D., & Amit, B. M. (2020). Thermodynamic modeling and correlations of CH 4, C 2 H 6, CO 2, H 2 S, and N 2 hydrates with cage occupancies. *Journal of Petroleum Exploration and Production Technology*, (8), 3689–3709. Advance online publication. doi:10.100713202-020-00998-y

Koh, C. A., Sum, A. K., & Sloan, E. D. (2009). Gas hydrates: Unlocking the energy from icy cages. *Journal of Applied Physics*, *106*(6), 061101. Advance online publication. doi:10.1063/1.3216463

Kramer, M. A. (1991). Nonlinear principal component analysis using autoassociative neural networks. *AIChE Journal. American Institute of Chemical Engineers*, *37*(2), 233–243. doi:10.1002/aic.690370209

Kröse, B., Krose, B., van der Smagt, P., & Smagt, P. (1993). *An introduction to neural networks*. Academic Press.

Kumari, A., Hasan, S., Majumder, K. C. B., Arora, A., & Dixit, G. (2021). Physio - chemical and mineralogical analysis of gas hydrate bearing sediments of Andaman Basin. *Marine Geophysical Researches*, *42*(1), 1–12. doi:10.100711001-020-09423-9

Kumari, A., Khan, S. H., Misra, A. K., Majumder, C. B., & Arora, A. (2020). Hydrates of Binary Guest Mixtures: Fugacity Model Development and Experimental Validation. Journal of Non-Equilibrium Thermodynamics, 45(1), 39–58. doi:10.1515/jnet-2019-0062

Kutsurelis, J. E. (1998). *Forecasting financial markets using neural networks: An analysis of methods and accuracy*. Naval Postgraduate School.

Kvenvolden, K. A. (1999). Potential effects of gas hydrate on human welfare. *Proceedings of the National Academy of Sciences of the United States of America*, *96*(7), 3420–3426. doi:10.1073/pnas.96.7.3420 PMID:10097052

Laugier, S., & Richon, D. (2003). Use of artificial neural networks for calculating derived thermodynamic quantities from volumetric property data. *Fluid Phase Equilibria*, *210*(2), 247–255.

Max, M. D., Johnson, A. H., & Dillon, W. P. (2005). *Economic geology of natural gas hydrate* (Vol. 9). Springer Science & Business Media.

Mehrizadeh, M. (2020). Prediction of gas hydrate formation using empirical equations and data-driven models. *Materials Today: Proceedings*. doi:10.1016/j.matpr.2020.06.058

Mohammadi, A. H., Belandria, V., & Richon, D. (2010). Use of an artificial neural network algorithm to predict hydrate dissociation conditions for hydrogen+ water and hydrogen+ tetra-n-butyl ammonium bromide+ water systems. *Chemical Engineering Science*, *65*(14), 4302–4305.

Mohammadi, A. H., Martínez-López, J. F., & Richon, D. (2010). Determining phase diagrams of tetrahydrofuran+ methane, carbon dioxide or nitrogen clathrate hydrates using an artificial neural network algorithm. *Chemical Engineering Science*, *65*(22), 6059–6063.

Mohammadi, A. H., & Richon, D. (2010). Hydrate phase equilibria for hydrogen+ water and hydrogen+ tetrahydrofuran+ water systems: Predictions of dissociation conditions using an artificial neural network algorithm. *Chemical Engineering Science*, *65*(10), 3352–3355.

Mohammadi, A. H., & Tohidi, B. (2005). A novel predictive technique for estimating the hydrate inhibition effects of single and mixed thermodynamic inhibitors. *Canadian Journal of Chemical Engineering*, *83*(6), 951–961.

Møller, M. F. (1993). A scaled conjugate gradient algorithm for fast supervised learning. *Neural Networks*, *6*(4), 525–533.

Moridis, G. J., Collett, T. S., Boswell, R., Kurihara, M., Reagan, M. T., Koh, C., & Sloan, E. D. (2009). Toward production from gas hydrates: Current status, assessment of resources, and simulation-based evaluation of technology and potential. *SPE Reservoir Evaluation & Engineering*, *12*(5), 745–771. https://doi.org/10.2118/114163-PA

Naderpour, H., Poursaeidi, O., & Ahmadi, M. (2018). Shear resistance prediction of concrete beams reinforced by FRP bars using artificial neural networks. *Measurement: Journal of the International Measurement Confederation*, *126*(June), 299–308. doi:10.1016/j.measurement.2018.05.051

Najibi, H., Chapoy, A., Haghighi, H., & Tohidi, B. (2009). Experimental determination and prediction of methane hydrate stabilityin alcohols and electrolyte solutions. *Fluid Phase Equilibria*, *275*, 127–131.

Paneiro, G., & Rafael, M. (2021). Artificial neural network with a cross-validation approach to blast-induced ground vibration propagation modeling. *Underground Space (China)*, *6*(3), 281–289. doi:10.1016/j.undsp.2020.03.002

Park. (2006). Sequestering carbon dioxide into complex structures of naturally occurring gas hydrates. *Proceedings of the National Academy of Sciences of the United States of America*, *103*(34), 12690–12694. https://doi.org/10.1073/pnas.0602251103

Park, K., Hong, S. Y., Lee, J. W., Kang, K. C., Lee, Y. C., Ha, M.-G., & Lee, J. D. (2011). A new apparatus for seawater desalination by gas hydrate process and removalcharacteristics of dissolved minerals (Na+,Mg2+,Ca2+,K+,B3+). *Desalination*, *274*, 91–96.

Petersen, R., Fredenslund, A., & Rasmussen, P. (1994). Artificial neural networks as a predictive tool for vapor-liquid equilibrium. *Computers & Chemical Engineering*, *18*, S63–S67.

Peterson, C., Rgnvaldsson, T., & Lnnbladb, L. (1994). *JETNET 3.0 - A versatile artificial neural network package*. Academic Press.

Potukuchi, S., & Wexler, A. S. (1997). Predicting vapor pressures using neural networks. *Atmospheric Environment*, *31*(5), 741–753.

Raal, J. D., & Mühlbauer, A. L. (1994). The Measurement of High Pressure Vapour-Liquid-Equilibria: Part I: Dynamic Methods. *Developments in Chemical Engineering and Mineral Processing*, *2*(2-3), 69–87.

Rai, P., Majumdar, G. C., DasGupta, S., & De, S. (2005). Prediction of the viscosity of clarified fruit juice using artificial neural network: A combined effect of concentration and temperature. *Journal of Food Engineering*, *68*(4), 527–533.

Rebai, N., Hadjadj, A., Benmounah, A., Berrouk, A. S., & Boualleg, S. M. (2019). Prediction of natural gas hydrates formation using a combination of thermodynamic and neural network modeling. *Journal of Petroleum Science Engineering*, *182*(March), 106270. https://doi.org/10.1016/j.petrol.2019.106270

Ripley, B. D. (1996). *Pattern Recognition and Neural Networks*. Cambridge University Press.

Sablani, S. S., Baik, O.-D., & Marcotte, M. (2002). Neural networks for predicting thermal conductivity of bakery products. *Journal of Food Engineering*, *52*(3), 299–304.

Shapiro, J. (2001). Genetic Algorithms in Machine Learning BT - Machine Learning and Its Applications: Advanced Lectures. Springer Berlin Heidelberg. https://doi.org/10.1007/3-540-44673-7_7.

Sharma, R., Singhal, D., Ghosh, R., & Dwivedi, A. (1999). Potential applications of artificial neural networks to thermodynamics: Vapor–liquid equilibrium predictions. *Computers & Chemical Engineering*, *23*(3), 385–390.

Singh, T. N., Sinha, S., & Singh, V. K. (2007). Prediction of thermal conductivity of rock through physico-mechanical properties. *Building and Environment*, *42*(1), 146–155.

Sloan. (2003). Fundamental principles and applications of natural gas hydrates. *Nature*, *426*(6964), 353–359.

Sloan, E. D. (1990). Natural gas hydrate phase equilibria and kinetics: Understanding the state-of-the-art. *Revue de l'Institut Français du Pétrole*, *45*(2), 245–266.

Sloan, E. D. Jr, & Koh, C. A. (2007). *Clathrate hydrates of natural gases*. CRC Press.

Soroush, E., Mesbah, M., Shokrollahi, A., Rozyn, J., Lee, M., Kashiwao, T., & Bahadori, A. (2015). Evolving a robust modeling tool for prediction of natural gas hydrate formation conditions _ Elsevier Enhanced Reader.pdf. *Journal of Unconventional Oil and Gas Resources*, *12*, 45–55.

Staub, S., Karaman, E., Kaya, S., Karapınar, H., & Güven, E. (2015). Artificial Neural Network and Agility. *Procedia: Social and Behavioral Sciences*, *195*, 1477–1485. https://doi.org/10.1016/j.sbspro.2015.06.448

Tanyildizi, H., Özcan, F., Atis, C. D., Karahan, O., & Uncuog, E. (2009). *Advances in Engineering Software Comparison of artificial neural network and fuzzy logic models for prediction of long-term compressive strength of silica fume concrete*. doi:10.1016/j.advengsoft.2009.01.005

Torrecilla, J. S., Otero, L., & Sanz, P. D. (2004). A neural network approach for thermal/pressure food processing. *Journal of Food Engineering*, *62*(1), 89–95.

Vakil-Baghmisheh, M.-T. (2002). Farsi character recognition using artificial neural networks. In *Faculty of Electrical Engineering*. University of Ljubljana.

Xu, C.-G., & Li, X.-S. (2015). Research progress on methane production from natural gas hydrates. *RSC Advances*, *5*(67), 54672–54699.

Zadeh, L. A. (1996). Fuzzy sets. In *Fuzzy sets, fuzzy logic, and fuzzy systems: selected papers by Lotfi A Zadeh* (pp. 394–432). World Scientific.

Zahedi, G., Karami, Z., & Yaghoobi, H. (2009). Prediction of hydrate formation temperature by both statistical models and artificial neural network approaches. *Energy Conversion and Management*, *50*(8), 2052–2059.

Zerpa, L. E., Aman, Z. M., Joshi, S., Rao, I., Sloan, E. D., Koh, C., & Sum, A. (2012). Predicting hydrate blockages in oil, gas and water-dominated systems. *Offshore Technology Conference*.

Chapter 7
DNA Computing:
Future of Renewable Smart Computation Systems

Mandrita Mondal
Indian Statistical Institute, Kolkata, India

ABSTRACT

The modern era of classical silicon-based computing is at the edge of a number of technological challenges which include huge energy consumption, requirement of massive memory space, and generation of e-waste. The proposed alternative to this pitfall is nanocomputing, which was first exemplified in the form of DNA computing. Recently, DNA computing is gaining acceptance in the field of eco-friendly, unconventional, nature-inspired computation. The future of computing depends on making it renewable, as this can cause a drastic improvement in energy consumption. Thus, to save the natural resources and to stop the growing toxicity of the planet, reversibility is being imposed on DNA computing so that it can replace the traditional form of computation. This chapter reflects the foundation of DNA computing and renewability of this multidisciplinary domain that can be produced optimally and run from available natural resources.

1. INTRODUCTION

The history of modern silicon-based computer technology has evolved through a long sequence of changes and is heading towards faster analysis and calculation. Literally classical computing is facing several technological challenges, like, consumption of large amount of energy, requirement of huge memory space and generation of vast e-waste. The massive growth in industrial activities is gradually leading us towards the global climate change which is of real concern. To overcome these drawbacks several modes of unconventional, nature-inspired computing have been proposed where a handful of atoms are being used to perform computation. Here comes the term nano-computing. The concept of nano-computing was proposed by Richard P. Feynman.

DOI: 10.4018/978-1-7998-8561-0.ch007

On 9th December 1959, Feynman delivered lecture on "There's Plenty of Room at the Bottom" at the annual meeting of the American Physical Society at Caltech (Feynman, 1960). In this seminal talk, Feynman accentuated to control and manipulate things on nano-scale to solve computing problems. He stated that nano particles like quantum molecules, DNA molecules are capable to perform computation.

Natural biochemical nano computers exist in all living organisms which store, process and retrieve all information in form of chemical structures and interactions. But these nano-computers of nature are basically uncontrollable by humans. The concept of developing a biochemical nano computer was established by Leonard Adleman of University of Southern California in 1994 (Adleman, 1994). He first exemplified DNA computing by solving the seven-point Hamiltonian Path Problem, a NP complete problem, using DNA strands. This piece of work first explored the idea that computing is possible by directly controlling molecules in nano-scale level.

Besides performing computations (Adleman, 1994; Winfree et al., 1998; Benenson et al., 2001; Chang et al., 2003; Green et al., 2006; Akerkar and Sajja, 2009), various algorithms have been developed to resolve reasoning and classification problems (Yeung and Tsang, 1997; Ray and Mondal, 2011a; Ray and Mondal, 2011b; Ray and Mondal, 2016) using DNA strands. Computation using DNA sequences requires some operations to handle the DNA strands *viz.* synthesis, merging, melting and annealing, amplification, separation, extraction, cutting, ligation, substituting, marking, destroying, detection and reading. By manipulating synthetic DNA strands the mathematical and logical aspects of computation have been replaced by unique DNA chemistry.

Different innovative and novel research works are being performed globally to develop renewable smart computation systems, where the intelligent eco-friendly molecular arrangement can be reused leading to less energy consumption. Recently DNA computing is gaining acceptance in the field of eco-friendly unconventional computation. The reversible DNA computing models (Garg et al., 2018; Eshra et al., 2019) recycles complex structures constructed by single-stranded (ssDNA), partially double-stranded and complete double-stranded (dsDNA) DNA sequences forming DNA gate structures and hairpin complexes.

Section 2 of this chapter review the sphere of DNA computing which includes definition, advantages, innovative DNA models to perform computation. The renewable or reversible aspect of DNA computation is highlighted in section 3 where the restorable models are demonstrated to perform computation and construct DNA logic gates. Section 4 contains the conclusion of the chapter with future scope of research.

2. DNA COMPUTING

Before delving deeper into the DNA computing, a recapitulation of basic concepts around DNA is a mandate. The next subsection illustrates some of those basic premises of DNA in molecular biology.

DNA in Molecular Biology

Deoxyribonucleic Acid or DNA can be defined as genetic material that carries instructions throughout the generations of all the cellular organisms and most of the viruses. This genetic blueprint determines all the characteristic of each organism. Generally, DNA molecules are arranged in the chromosomes placed in the cell nucleus (Watson and Crick, 1953). The molecular structure of DNA was revealed by James D. Watson and Francis Crick in 1953.

Some unique structural and chemical features of the DNA molecule distinguish it from other biomolecules. The dsDNA molecules are actually in a form of a double-helix where two polynucleotide chains are twisted around a common axis (Figure 1). *Deoxyribonucleotides (or nucleotides)* are the building blocks of the long polymeric chain of DNA molecule. The components of this individual monomer i.e., nucleotide, are: 2'-deoxyribose (sugar), nitrogen base and phosphate group.

Two structural families of nitrogen bases are, (i) purine (double-ringed structure), (ii) pyrimidine (single-ringed structure). The nitrogen bases of DNA, adenine (A) and guanine (G) are in the purine group and cytosine (C) and thymine (T) are in the pyrimidine group.

Figure 1. Structure of DNA (A) DNA helix (B) Complementary base-pairing

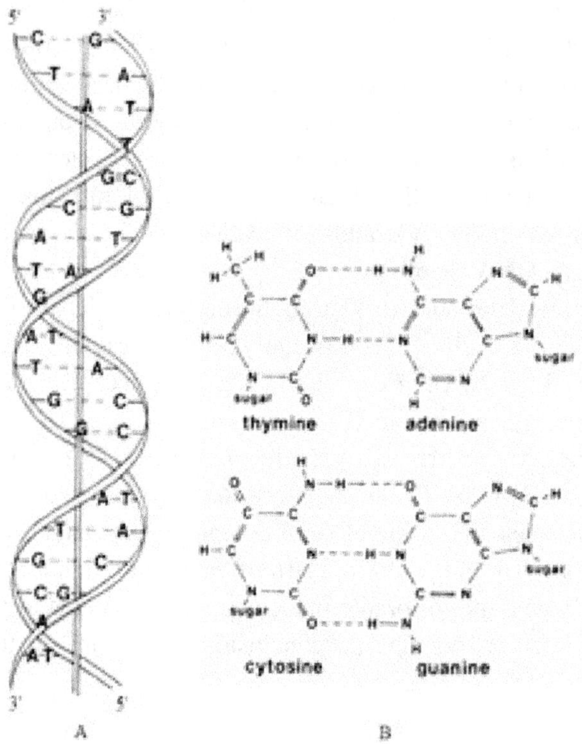

The two double-helical polynucleotide chains of DNA molecule are held together in antiparallel orientation by weak and non-covalent hydrogen-bonds (Figure 1). The adenine base of a strand forms two hydrogen bonds only with thymine on the other strand (A = T) and the guanine of a DNA strand forms three hydrogen bonds with cytosine on the other strand (G ≡ C). This pairing between two polynucleotide strands is termed Watson-Crick pairing or complementary base-pairing, which leads to the self-coding characteristics of DNA molecule. The base pairing holds the strands with opposite polarity, i.e., base at 5' end of one strand is paired with the base at 3' end of the complementary strand. For example, if a DNA strand contains sequence 5'-AGATCTA-3', the opposite strand i.e., complementary strand must have the complementary sequence 3'-TCTAGAT-5'. The weak hydrogen bonds between complementary strands of DNA molecules controls the thermodynamic stability of the helix and induce the fundamental property of denaturation and hybridization of the DNA molecule.

The complementary pairing can be represented as follow;

If σi denotes a single base of a DNA sequence, then $\bar{\sigma}_i$ ($0 < i < m$-1, where m is the length of the DNA sequence in base-pair) can be denoted as the complementary base of σ_i. Let, alphabet $\Sigma = \{ \sigma_0, \sigma_1,$ ...., $\sigma_{m-1}, \bar{\sigma}_0, \bar{\sigma}_1, \ldots, \bar{\sigma}_{m-1} \}$. A strand is represented by one or more symbols in Σ. dsDNA sequence can be formed only if the two ssDNA sequences are complementary to each other. The complete double-strand is symbolized as $\begin{bmatrix} \sigma_i \\ \bar{\sigma}_i \end{bmatrix}$.

Computation using DNA sequences requires some operations to manipulate the DNA strings (Adleman, 1994; Adleman, 1996; Kari, 1997). Usually, these operations are performed by molecular biologists in a wet-laboratory.

1. *Synthesis:* For computation with the DNA, oligonucleotides are required. These are short polynucleotide chains with definite chemical structure. Oligonucleotides can be synthesized and purified in wet-laboratory environment. One of the popular methods of synthesis is the solid phase method.
2. *Merging:* This is the union of two test tubes containing DNA solution into one. For example, if B_1 and B_2 are two test tubes with different DNA solutions, a third test tube B_3 can be generated where $merge(B_1, B_2) = B_3 = B_1 \cup B_2$.
3. *Melting and Annealing:* Melting or denaturation of dsDNA molecules occurs by heating which leads to the generation of two single-stranded complementary strands. If the strands are cooled again these anneal or hybridize to form dsDNA sequence. If input

$$B = \{\sigma_0\sigma_1, \bar{\sigma}_1\bar{\sigma}_2\}, \text{ then } annealing(B) = \begin{bmatrix} \sigma_1\sigma_2 \\ \overline{\sigma_1\sigma_2} \end{bmatrix} \text{ and } denaturation(B) = \{\sigma_0\sigma_1, \overline{\sigma_1\sigma_2}\}.$$

4. *Amplification: Polymerase chain reaction* or *PCR* is revolutionary method for rapid production of huge number of copies of a DNA sequence or a specific portion it. DNA sample, primers, ddNTPs and DNA polymerase are needed for amplification. The process of amplification can be illustrated as, if input

$$B = \begin{bmatrix} \alpha_0\sigma_1\sigma_2\beta_0 \\ \overline{\alpha_0\sigma_1\sigma_2\beta_0} \end{bmatrix}, amplify\ (B, \{\sigma_0\sigma_1, \overline{\sigma_1\sigma_2}\}) = \begin{bmatrix} \frac{\alpha_0\sigma_1\sigma_2\beta_0}{\alpha_0\sigma_1\sigma_2\beta_0}, \frac{\sigma_1\sigma_2}{\sigma_1\sigma_2}, \frac{\sigma_1\sigma_2}{\sigma_1\sigma_2}, \frac{\sigma_1\sigma_2}{\sigma_1\sigma_2}, \frac{\sigma_1\sigma_2}{\sigma_1\sigma_2}, \frac{\sigma_1\sigma_2}{\sigma_1\sigma_2}, \ldots\ldots \end{bmatrix}.$$

5. *Separation:* The *gel electrophoresis* method is used to separate the DNA sequences according to their size i.e., the number of bases. The function *separation* (B_1, I, B_2) sorts the DNA sequences of test tube B_1 and generates a new test tube B_2 containing only the sequences having length I (where I is an integer).

6. *Extraction:* The DNA sequences having desired pattern can be extracted from a solution using *affinity purification* technique. Another methodology of extraction is the *magnetic bead separation*. *Extract*(B_1, P, B_2) isolates all DNA strands having pattern P in the solution of test tube B_1 and generates a new test tube B_2 with the extracted sequences.

Input: $B_1 = \{\alpha_0\sigma_1\sigma_2\beta_0, \gamma_0\sigma_2\alpha_0\beta_0, \beta_0\gamma_0\sigma_2\}$.

Extract $(B_1, \{\sigma_1\sigma_2\}, B_2)$ generate test tube B_2, where $B_2 = \{\gamma_0\sigma_2\alpha_0\beta_0, \beta_0\gamma_0\sigma_2\}$.

7. *Cutting:* Restriction enzyme (or restriction endonuclease) cleaves ssDNA or dsDNA sequence at or near a specific site, termed the restriction site. For example, *cut*$(B, \sigma_1\sigma_2)$ cleaves each dsDNA sequence in test tube B containing the specific segment $\left[\frac{\sigma_1\sigma_2}{\sigma_1\sigma_2}\right]$ and generates two double-strands as,

$$\left[\frac{\alpha_0\sigma_1\sigma_2\beta_0}{\alpha_0\sigma_1\sigma_2\beta_0}\right] \rightarrow \left[\frac{\alpha_0\sigma_1}{\alpha_0\sigma_1}\right], \left[\frac{\sigma_2\beta_0}{\sigma_2\beta_0}\right].$$

8. *Ligation:* Ligase enzyme can catalyze the ligation of dsDNA molecules with compatible sticky ends or even blunt ends of ssDNA molecules producing new chemical bonds. *Ligate*(T) ligates the sticky ends of dsDNA molecules;

$$\left[\frac{\alpha_0\sigma_1\sigma_2}{\alpha_0\sigma_1}\right], \left[\frac{\beta_0}{\sigma_2\beta_0}\right] = \left[\frac{\alpha_0\sigma_1\sigma_2\beta_0}{\alpha_0\sigma_1\sigma_2\beta_0}\right].$$

9. *Substituting:* PCR site-directed mutagenesis is an *in vitro* technique by which certain bases can be inserted, deleted or substituted from a DNA sequence.
10. *Marking:* Sequences can be marked by making it dsDNA duplex again unmarked by denaturation.
11. *Destroying:* If the marked DNA sequences are treated with restriction enzyme, it can be destroyed by generating shorter fragments. The unaffected strands can be separated by performing gel electrophoresis.
12. *Detection and Reading:* Sequencing method can detect the order of nucleotides in a DNA sequence.

Depending on the backbone of molecular biology and the wet-lab operations on DNA sequences, the concept of DNA computing has been developed. The next subsection focuses solely on the domain of DNA computation.

DNA Computing and its Advantages

DNA computing is the amalgamation of biological science with computer science where the biomolecule DNA is used to perform computation. The advent of DNA computation is leading the modern world towards the paradigm shift, from silicon to carbon, where, it is expected, DNA computing will be capable of replacing classical silicon-based computation in the coming decade. The unique properties of the genetic molecule DNA, which stimulate the notion of DNA computing, are the complementary base-pairing property and capability of storing, processing and retrieving information. In comparison with the traditional computing technology, the precedence of DNA computing is discussed below;

1. *Energy efficiency:* The energy consumption of the DNA operations required to perform computation is trivial compared to a modern computer. DNA computers can be 10^9 times more energy-efficient.
2. *Massive parallelism:* One of the prominent advantages of DNA computing is its massive parallelism. Though the operations performed in DNA computation are comparatively slow, the speed is faster because of the huge parallelism. About 10^{18} processors can perform in parallel in an in v*itro assay*.
3. *Potential for information storage:* DNA requires a trillion times less space than existing silicon-based storage media due to its durability and density. DNA is said to be approximately 300 times more durable and 1000 times more dense than modern storage media.
4. *Speed:* As a result of the vast parallelism of DNA computing, in some cases it is 10^5 times faster than the most recent super-computer as certain DNA computing models can perform 330 trillion operations per second.
5. *Non-toxicity:* Traditional microprocessors are constructed using toxic materials; but DNA biochips and other components used in molecular computers are completely biodegradable; thus, massive production of e-waste can be avoided.

In the following subsection, the basic models of DNA computing are thoroughly discussed. The first highlighted model is by the pioneer of DNA computing, Professor Leonard Adleman. He solved the directed Hamiltonian path problem having seven vertices by using DNA molecules in 1994. The next discussed model is the path breaking DNA computing model by Richard J. Lipton. He solved a satisfiability (SAT) problem using DNA strands.

Adleman's Experiment

In mathematical field of graph theory, Hamiltonian path problem is a NP-complete problem in which computational complexity combinatorically increases with the linear increase in problem size. The problem enquires whether a route can be found in a directed graph from the initial vertex to final vertex, visiting each vertex exactly once.

Leonard Adleman solved the seven-vertex instance of Hamiltonian path problem by using DNA strands (Adleman, 1994). He employed the directed graph G (Figure 2), where the number of vertices is seven. The objective is to find Hamiltonian path in G starting from an initial vertex v_0 (i.e., v_{in}) and ending at a final vertex v_6 (i.e., v_{out}) entering every vertex only once. Adleman determined the path in form of DNA sequence which encodes compatible one-way edges.

The nondeterministic steps followed by Adleman to solve the problem are;

Step 1: Generation of random paths through the graph.

Figure 2. Directed graph

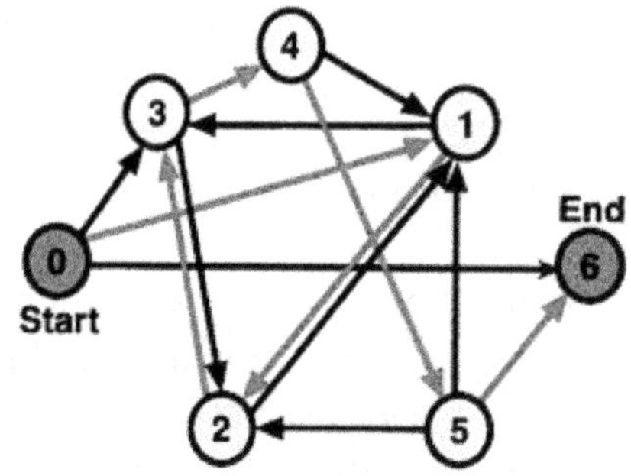

He involved randomly chosen ssDNA sequences to encode the vertices and edges of graph G. Each vertex i in G was encoded by 20-mer (i.e., 20 bases long) sequence denoted by v_i. To encode each edge $i \rightarrow j$ (i.e., $e_{i \rightarrow j}$) in G, the ssDNA string was selected using complement of the 3'-10-mer of v_i followed by the complement of the 5'-10-mer of v_j.

Let v_1 is encoded by the sequence,
5' TATCGGATCGGTATATCCGA 3'

and v_2 is encoded by the sequence,
5' GCTATTCGAGCTTAAAGCTA 3'

then, the DNA sequence representing directed edge $e_{1 \rightarrow 2}$ is,
3' CATATAGGCTCGATAAGCT 3'

All the ssDNA sequences representing the vertices and directed edges were mixed in a solution and allowed to be ligated. The vertices are hybridized with the compatible edges which leads to the generation of partially double-stranded DNA strings coding random routes through G.

For example, edge $e_{1 \rightarrow 2}$ connects two vertices v_1 and v_2 respectively. The path formed due to hybridization and ligation reaction is shown in Figure 3.

Figure 3. Directed path connecting two vertices

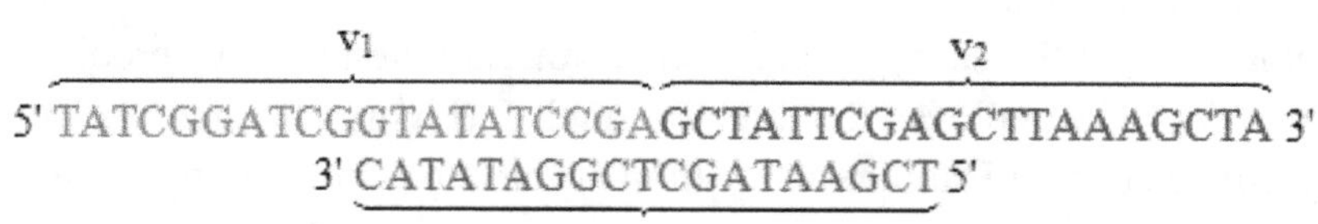

Step 2: Only the paths those initiate with v_{in} and terminate with v_{out} are kept.

The resultant sequences of step 1 is amplified in large scale by PCR using the primers v_0 and complementary sequence of v_6. This step results the amplification only the DNA sequences those encode the routes initiating with vertex 0 and terminating with vertex 6.

Step 3: The routes those enter exactly n vertices are kept (where n = number of vertices in G).

Gel electrophoresis by agarose gel is performed with the product of step 2. DNA sequences from 140 base pairs (bp) band of the gel are extracted. As the graph *G* has seven vertices (n=7) and each vertex is encoded by 20-mer DNA sequence, the 140 bp (7 × 20 bp = 140 bp) long resultant sequences of step 2 represent routes which enter exactly seven vertices. These sequences are again amplified by PCR using appropriate primers and purified for several times by gel electrophoresis to enhance purity.

Step 4: The routes those enter all of the vertices at least once are retained.

Affinity purification is conducted with the resultant sequences of step 3. To perform this process, first the sequences are denatured into ssDNA and then incubated with magnetic beads which has complementary sequence of v_1 attached to it. As a result, only ssDNA molecules, those contain the sequence encoding v_1 are retained. This process is repeated successively with complementary oligonucleotides representing all vertices v_i ($2 \leq i \leq 5$) of *G*.

Step 5: If any route remains, return "Yes"; otherwise, "No".

The consequential sequences of step 4 are amplified by PCR and gel electrophoresis is performed. The resultant sequence encodes the Hamiltonian path in *G* (Figure 4).

Figure 4. Hamiltonian path in Adleman's graph

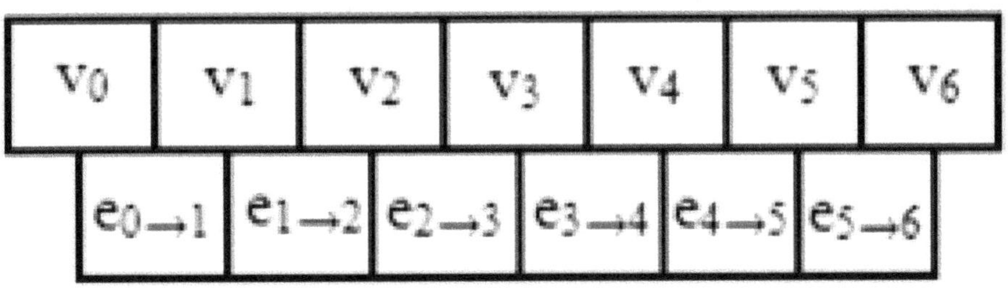

The DNA computing model proposed by Adleman was semi-autonomous in nature as it require human interference. This experiment initiates with a single test-tube and terminates with one or more final tubes.

Lipton's Experiment

Now we are going to demonstrate the contribution of Richard J. Lipton in the domain of DNA computing (Lipton, 1995). He proposed DNA model to resolve satisfiability problem (SAT). By using massive parallelism, unusual information density and incomparable energy efficiency of DNA chemistry, Lipton solved the first known example of an NP-complete problem i.e., SAT problem.

In computer science, the SAT problem determines whether the variables of a Boolean formula can be assigned in such a way that the formula evaluates to true. If yes, truth-value assignment of the variables has to be determined. If no such assignment exists, the formula is said to be identically false for all possible variable assignments, hence unsatisfiable. Otherwise, it is satisfiable.

An instance of SAT problem is a Boolean formula expressed using only Boolean variables, (logical) negation (NOT; ~), (logical) conjunction (AND; ^), (logical) disjunction (OR; ˅) and parentheses.

Now, formula *F* is considered, expressed using the propositional variables, $x_1, x_2, \ldots\ldots$, and connectives ~, ^, ˅ (i.e., negation, conjunction, disjunction). Thus,

$$F = (x_1 \vee x_2 \vee {\sim}x_3 \vee x_4) \wedge ({\sim}x_2 \vee {\sim}x_3) \wedge (x_3 \vee x_4) \wedge ({\sim}\, x_4)$$

As the variables x_1, x_2, x_3, x_4 are Boolean, these occur into the set {0, 1} or {false, true}. There are four clauses in the given formula *F*. The general form of the SAT problem involves Boolean formula of the form $C_1 \wedge C_2 \wedge \ldots\ldots\ldots \wedge C_m$, where C_i is a clause ($i = 1, 2, \ldots.., m$). If the association of truth-value to each variable can be done in such a way that *F* evaluate to 1, then, *F* is called to be satisfiable. We have to find the set of truth-values for which *F* is true. On other word, it can be said that the variables that make each clause having the value 1, has to be determined.

The fourth clause of *F*, i.e. $({\sim}\, x_4) = 1$ implies that x_4 should be 0. If x_4 is 0, then to make the third clause $(x_3 \vee x_4)$ true, x_3 must be equal to 1. Likewise, we can assign that $x_1 = 1$ and $x_2 = 0$. Thus, it is clear that this set of truth-value assignments $\{x_1 = 1, x_2 = 0, x_3 = 1, x_4 = 0\}$ makes *F* satisfiable.

However, in the general case an exhaustive search method is followed; i.e., if a formula with *n* variables is given, then all possible *n* truth-value assignments have to be searched; thus, the satisfiability problem is known to be an NP-complete problem. As the size of instances i.e., the number of variables increases, the running time of exhaustive search becomes forbiddingly large.

To solve SAT problem Lipton used DNA sequences and performed some basic operations on the sequences. He solved a propositional formula containing two variables, *x* and *y*. The formula is,

$$E = (x \vee y) \wedge ({\sim}x \vee {\sim}y)$$

Figure 5. Graph G_2 encoding two-bit number

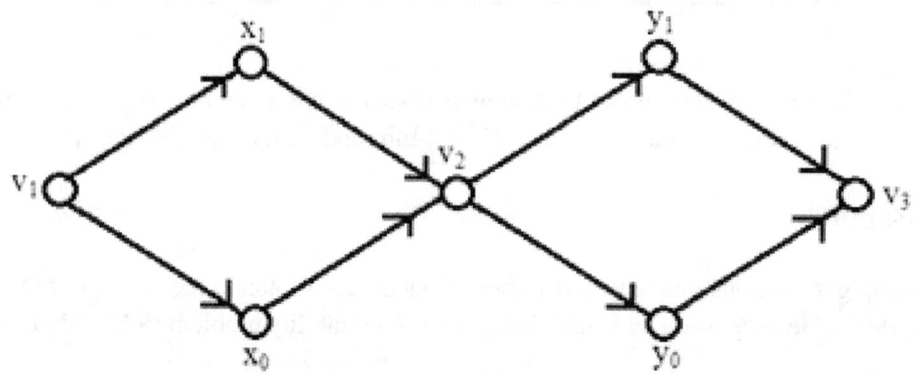

To solve the problem, he followed a simple graph G_2 (Figure 5). The variable and the possible truth-value assignments of the above said formula were represented by the vertices and the edges of the above said graph G_2.

The vertices of the graph are v_1, x_1, x_0, v_2, y_1, y_0, v_3. It contains edges from v_1 to both x_1 and x_0 and from both x_1, x_0 to v_2; again, from v_2 to both y_1 and y_0 and from both y_1, y_0 to v_3. All the paths of G_2 that start at v_1 and end at v_3 encode a 2-bit binary number. For the vertices v_1 and v_2 the paths have two options; if it chooses the vertex with suffix 1 (i.e., paths $v_1x_1v_2$ and $v_2y_1v_3$) it will encode 1; if it chooses the vertex with suffix 0 (i.e. paths $v_1x_0v_2$ and $v_2y_0v_3$) it will encode 0. Thus, the path $v_1x_1v_2y_0v_3$ encodes the binary values 10.

Lipton represented all the vertices and edges of the graph G_2 in the form of DNA sequences. Each vertex of the graph was encoded by 20 bases (l) long random DNA sequence. The first half of each encoded sequence is denoted by p_i and the second half is represented by q_i. Therefore, p_iq_i is the symbol associated with i^{th} vertex. Each directed edge from i^{th} vertex to j^{th} vertex (i.e., $i \rightarrow j$) is encoded by DNA sequence in 3' to 5' direction of the form $\hat{q}_i \hat{p}_j$ (where, $\hat{x}$ denotes the Watson-Crick complement to the oligonucleotide x). All the strands representing the vertices and the edges are taken in a test tube which is the initial test tube. This starting test tube consists of the following strands:

i. Many copies of the DNA sequences representing the vertices. These DNA strands are in 5' to 3' direction.
ii. Many copies of the edges are also taken which are in 3' to 5' direction.
iii. DNA sequence of length 10 bases ($l/2$) in 3' to 5' direction which is complementary to the first half of the initial vertex (i.e. v_1) is added. That means $\hat{p}_1$ is added.
iv. DNA sequence of length 10 bases ($l/2$) in 3' to 5' direction which is complementary to the last half of the final vertex (i.e. v_3) is added. That means $\hat{q}_3$ is added.

After adding, hybridizations between the appropriate complementary sequences occur in the test tube. As each edge contains the complementary sequence of the associated vertices, the newly formed hybridized double-stranded DNA sequence represents a path through the graph which is in "vertex, edge, vertex, edge,......." form. Therefore, after hybridization, the DNA sequence representing the vertices and the edge of the graph G_2 (see Figure 5) forms the path shown in Figure 6.

Figure 6. Path formed through the graph G_2

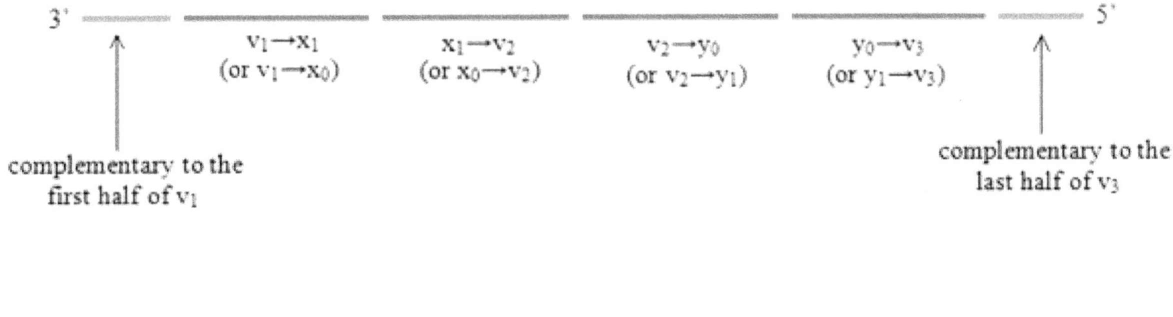

The top 5' → 3' strand of the newly formed dsDNA sequence (Figure 6) comprises of successions of vertices and the bottom 3' → 5' strand comprises of successions of edges. From the vertex v_1 or v_2 the path can go either left or right. If it goes through the left vertex "1" is encoded, otherwise, it will encode "0". Each of the possible four paths through G_2 specifies one of the four truth-value assignments for the Boolean variables x and y.

To solve SAT problem of the propositional formula $E = (x \vee y) \wedge (\sim x \vee \sim y)$, the following steps are to be followed;

Step 1: The initial test tube T_0 is constructed in such a way that it contains DNA strands encoding each path and, thus, represents one of the possible truth-value assignments.
Step 2: From T_0 strands with truth-value assignment of $x = 1$ are separated and kept in another test tube T_1. This operation is denoted by $S(T_0, 1, 1)$. This means, separate the strands from T_0 which encode 1 in the first position.
Step 3: After separation of the selected strands from T_0, let the remainder be in the test tube T_1'. This means, T_1' contains the strands $S^-(T_0, 1, 1)$.
Step 4: From T_1' separate those strands for which $y = 1$ i.e. $S(T_1', 2, 1)$. The separated DNA strands are kept in the test tube T_2.
Step 5: The contents of T_1 and T_2 are merged to form the test tube T_3.
Step 6: From T_3 separate those strands which encode 0 in the first position i.e. $x = 0$ and keep in the test tube T_4.
Step 7: After separation of the selected strands from T_3, let the remainder be in the test tube T_4'.
Step 8: From T_4' separate those strands for which $y = 0$ and form the test tube T_5.
Step 9: The contents of T_4 and T_5 are merged to form the test tube T_6.
Step 10: The presence of any DNA strand is detected in the final test tube T_6. If yes, it can be concluded that the given formula is satisfiable. The satisfying assignments are encoded by the DNA sequence present in T_6.

The algorithm to solve the SAT problem of given formula, $E = (x \vee y) \wedge (\sim x \vee \sim y)$ is summarized in Table 1.

Now, we will demonstrate the generalized version of the SAT problem resolved by Lipton by DNA computing model. The general form the given formula in a SAT problem is;

$$Z = C_1 \wedge C_2 \wedge \ldots\ldots\ldots\ldots\ldots \wedge C_m$$

Let us assume, this problem has n number of variables and m clauses. Thus, the corresponding directed graph (G_n) is shown in Figure 7.

The generalized form of SAT problem having n variables and m clauses can be solved by at most order m extract steps and one detect step. Starting from initial test tube (T_0) with n-bit sequences, going through all the clauses of Z and retaining desired strands, finally those strands only remain that encode truth-value assignments satisfying Z.

The next section illustrates comparatively new and important aspect of DNA computing i.e., renewable or reversable DNA computing. This technological advent has taken us nearer towards the paradigm shift, from silicon to carbon.

Table 1. Algorithm to solve the SAT problem of given formula $E = (x \vee y) \wedge (\sim x \vee \sim y)$

Step	Operation	Set of truth-value assignment present in the corresponding test tube
1	Input(T_0)	{00, 01, 10, 11}
2	$T_1 = S(T_0, 1, 1)$	{10, 11}
3	$T_1' = S^-(T_0, 1, 1)$	{00, 01}
4	$T_2 = S(T_1', 2, 1)$	{01}
5	$T_3 = \text{merge}(T_1, T_2)$	{10, 11, 01}
6	$T_4 = S(T_3, 1, 0)$	{01}
7	$T_4' = S^-(T_3, 1, 0)$	{10, 11}
8	$T_5 = \text{S}(T_4', 2, 0)$	{10}
9	$T_6 = \text{merge}(T_4, T_5)$	{01, 10}
10	detect(T_6)	{01, 10}

Figure 7. The graph G_n encoding n-bit number

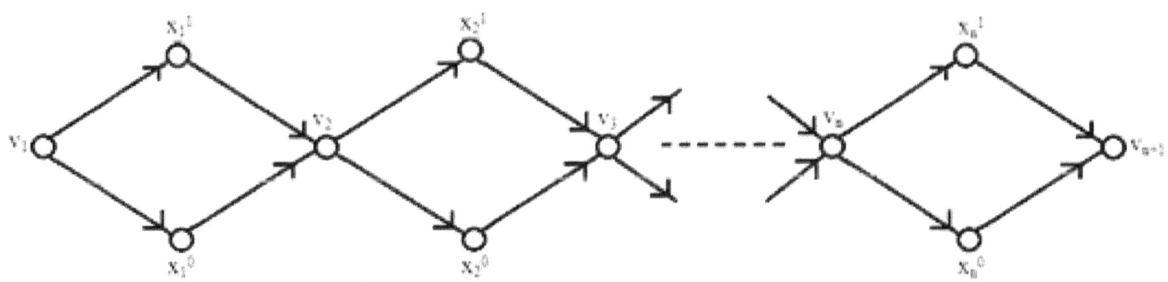

3. RENEWABLE DNA COMPUTING

Renewable or reversible DNA computing is a smart computation system, where the intelligent eco-friendly molecular arrangement can be reused resulting in less energy consumption. The reversible models recycle complex DNA structures which can be constructed of single-stranded, double-stranded or partially double-stranded DNA sequences. The renewability of DNA computing leads to energetically efficient methodology to perform computation. It can be mentioned as the seed of renewable smart computation system with less energy consumption. *Reif et al.* proposed the DNA models of reversible computing using double-stranded gate structures (Garg et al., 2018) and hairpin complexes (Eshra et al., 2019) which are highlighted in the subsections.

Renewable DNA Computing Presenting Time-Responsive Circuit

Researchers have designed time-responsive asynchronous DNA circuits using renewable theory of DNA computing (Garg et al., 2018) based on the toehold mediated strand displacement mechanism (Cardelli, 2013; Zhang & Winfree 2009; Green & Tibbetts, 1981). The uniqueness of the proposed double-stranded gate structured model is that, after running experiment, the model can be restored to its initial state by manual addition of DNA strands. The time-responsive property allows the DNA circuit to change its computed output depending upon the change in input over time.

The designed DNA logic gates (Figures 8 and 10) are based on Boolean logic circuits which can receive either single input (i.e., NOT) or double inputs (i.e., AND, NOT, OR, NAND etc.). The input and output variables of the of the logic gate are represented by DNA strands consisting five toehold domains. At the initial state the output strands are fully hybridized with the gate structure. When the input strand is added to the gate structure, it eventually displaces the output strand by toehold mediated DNA strand displacement mechanism. The released strand acts as the input to the downstream gate structure. Only the three toehold domains at the center of the input strand hybridizes with the gate structure to form partially double-stranded intermediate structure as explained in Figures 8 and 10. The output strands representing Boolean values 0 and 1 are denoted as Z_0 and Z_1 respectively.

Single-Input Single-Output Time-Responsive DNA Gate

Figure 8 gives the pictorial representation of single-input single-output time-responsive NOT DNA-gate structure. The input strands X_0 (representing Boolean value 0) and X_1 (representing Boolean value 1) having five toehold domains are shown in Figure 8(a). The gates GZ_0 and GZ_1 are initially hybridized with the output strands Z_1 and Z_0 forming partially double-stranded structures GZ_0:Z_1 and GZ_1:Z_0 respectively. When the input strands are added to the gate structures, these displaces the pre-hybridized output strands by toehold mediated branch migration and DNA strand displacement. The output strands are realized and intermediate structures are formed with gate and input strands, pictorially presented as GZ_0:X_0 and GZ_0:X_1 in Figure 8(c). The possible output strands are shown in Figure 8(d).

Figure 8. Single-input single-output time-responsive DNA gate: (a) possible input strands, (b) gate structure attached with output strands, (c) time-responsive intermediate structure, (d) possible output strands

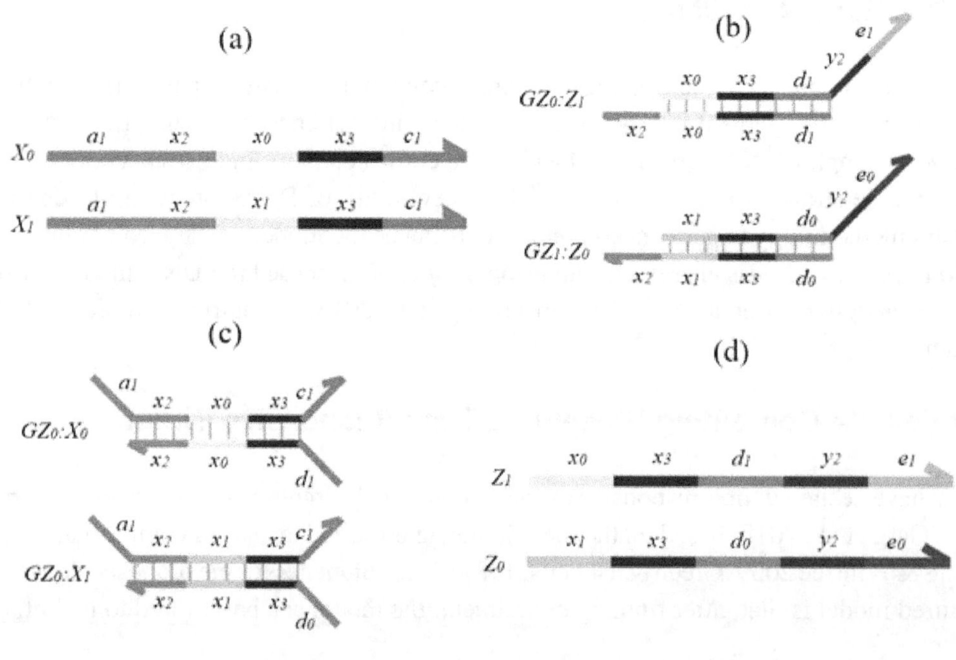

Figure 9. Renewability of single-input single-output DNA gate

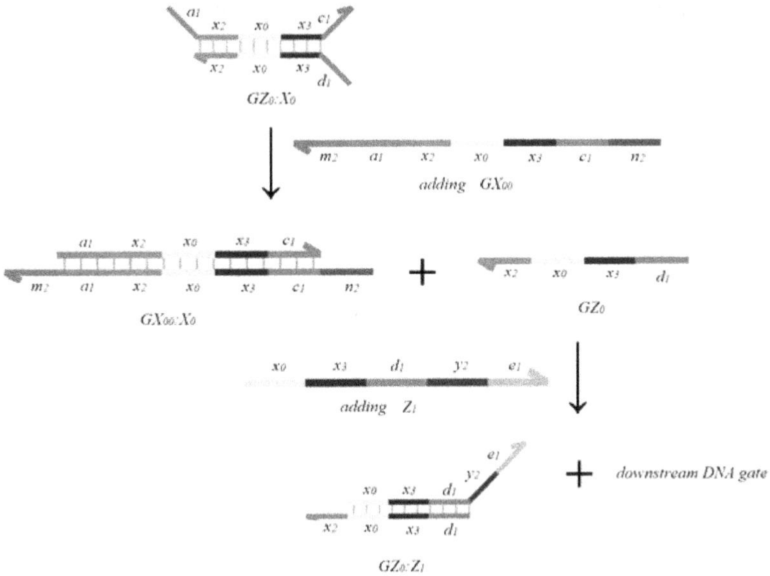

Figure 10. Double-input single-output time-responsive DNA gate: (a) possible input strands, (b) gate structure attached with output strands, (c) time-responsive intermediate structure, (d) possible output strands

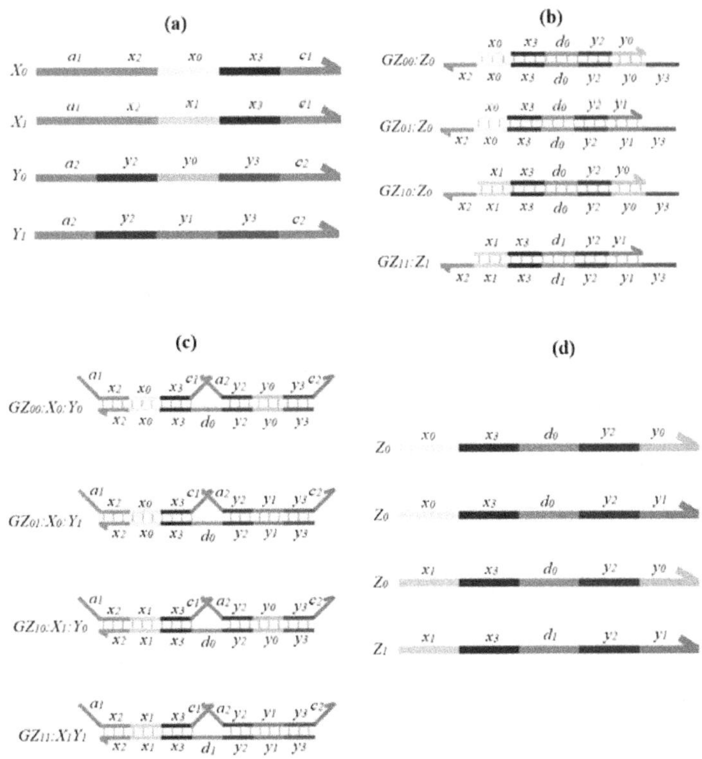

The renewability of the initial gate structure GZ_0:Z_1 is illustrated in Figure 9. The DNA circuit model can be restored through cascade of reactions by adding of ssDNA and reused again to generate the output strand corresponding to the given input DNA strand.

Double-Input Single-Output Time-Responsive DNA Gate

Likewise, the double-input single-output time-responsive OR DNA-gate structure is demonstrated in Figure 10. Figure 11 illustrates the renewability of one of the four gate conformations of the proposed DNA circuit model, GZ_{00}:Z_0 through cascade of reaction by addition of two ssDNA sequences, GX_{00} and Z_0.

So far, we have focused on renewable energy efficient DNA model using gate conformations which successfully represents logic circuit in the domain of DNA computing. The next subsection demonstrates the mechanism of energy efficient, renewable hairpin DNA structures used for computation.

Figure 11. Renewability of double-input single-output DNA gate

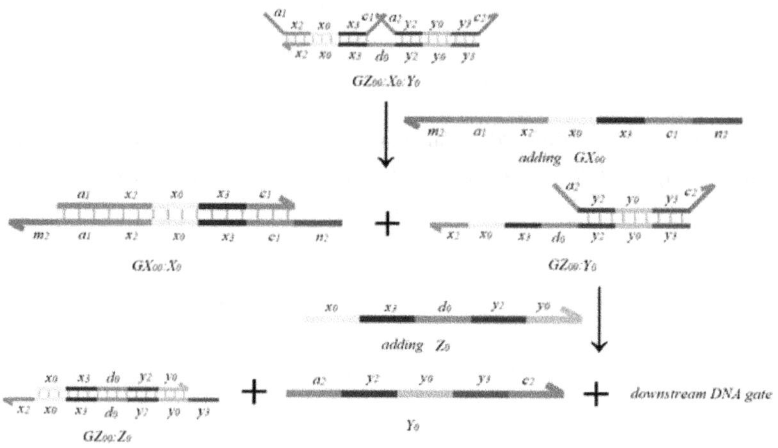

Renewable DNA Hairpin Structures Performing Computation

An intelligent DNA system has been developed using renewable hairpin motifs (Eshra et al., 2019) which can efficiently perform computation and generate output. The DNA hairpin or stem-loop structure is comparatively stable secondary structure which occurs when two distant domains of the same ssDNA are complementary to each other in opposite directions. These two domains hybridize to each other forming the stem region, and the unhybridized single-stranded domain linking the ends of the stem region is termed a loop. The proposed stem-loop structure unlocks to perform computation, again reverts back to its stable initial form which can be reused for performing further computation and producing more output sequences. Thus, the open configuration of the stem-loop structure indicates its ON mode, which is competent to perform computation and conversely the OFF mode is specified by its closed configuration. The mechanism of the renewable DNA hairpin structures has three steps listed below:

- Addition of input strand and booster strand

- Generation of detectable output
- Renewability of the DNA hairpin structure and reporter strand

Addition of Input Strand and Booster Strand

The gate structure is constructed by hairpin motif H having an additional toehold domain T_1*; T_1* denote the Watson-Crick complement of T_1. Figure 12 represents the interaction of the renewable gate structure H with the input strand I_n and the booster strand B_o. The input strand hybridizes with H through the toehold domain T_1 and extends the hybridization by the branch migration mechanism. Finally it opens the stem (domain D_2) to generate $H.I_n$ complex. The single-stranded loop of H having two toehold domains, T_1* and T_2, is now open to perform further computation. The booster strand is designed to have the domain D_2, T_1 and an additional domain D_b. If B_o is further added to the $H.I_n$ complex, it replaces the pre-hybridized I_n strand by toehold mediated strand displacement and form the complex $H.B_o$. The released single-stranded I_n is now available to interact with new gate structure and generate output. Thus, the addition of booster strand leads to the renewability of the input strand.

Figure 12. Interaction of hairpin motif with input strand and booster strand

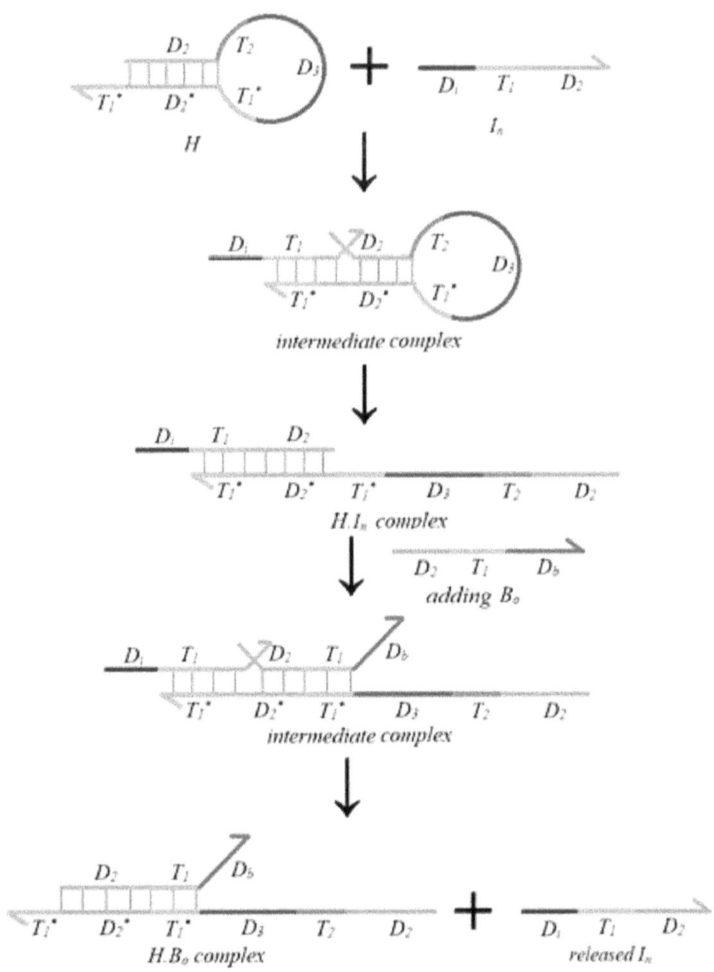

Generation of Detectable Output

The addition of the partially double-stranded reporter complex with the resultant complexes of the previous step leads to the release of detectable output molecule which is single-stranded fluorescent DNA molecule. The reporter complex R_e is designed to have two complementary domains corresponding to those of single-stranded loop region of H i.e., T_2* and D_3*. R_e is tagged by fluorophore in shorter strand and quencher molecule in longer strand as shown in Figure 13. Quencher decreases the fluorescent intensity of the fluorophore, thus, R_e emits no light. As R_e hybridizes with $H.I_n$ and $H.B_o$, new complexes are formed and a single-stranded DNA output molecule is released by the mechanism of strand displacement using the toehold domain T_2. As the effect of quenching molecule is inversely proportional to the distance between fluorophore and quencher, fluorescent emission occurs in this step because of the dissociation of these two molecules. The fluorescence indicates the completion of reaction and the release of an output strand.

Figure 13. Generation of detectable output: (a) Interaction of reporter strand with $H.I_n$ complex (b) Interaction of reporter strand with $H.B_o$ complex

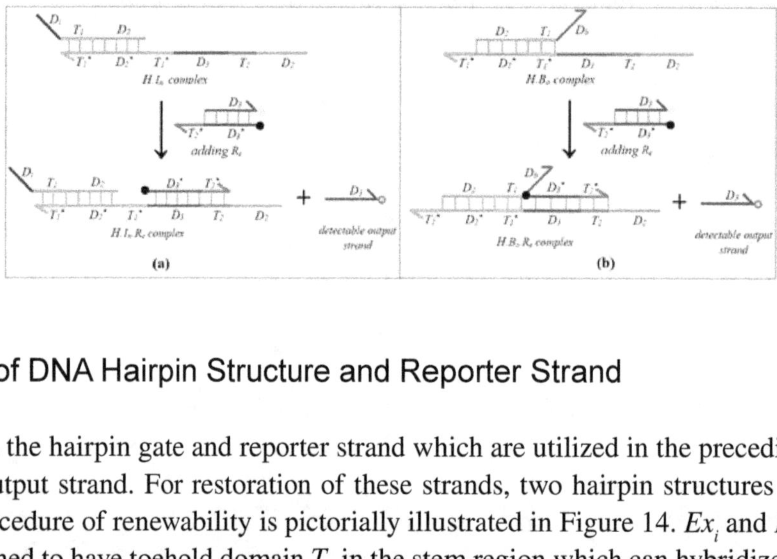

Renewability of DNA Hairpin Structure and Reporter Strand

This step renews the hairpin gate and reporter strand which are utilized in the preceding step to generate detectable output strand. For restoration of these strands, two hairpin structures are used *viz.* Ex_i and Ex_b. The procedure of renewability is pictorially illustrated in Figure 14. Ex_i and Ex_b are termed as extractors, designed to have toehold domain T_1 in the stem region which can hybridize with $H.I_n.R_e$ and $H.B_o.R_e$ complexes respectively. Figure 14(a) and (b) demonstrate the extraction of the input strand and booster strand by the extractor and formation of $H.R_e$ complex which is further employed for restoration of the initial stem-loop motif H. The extracted input and booster strands turn out to be the waste as these cannot be reused further. The development of $H.R_e$ leads to the hybridization starting from the edges of the complex and formation of double-stranded stem structure as shown in Figure 14 (c). The intensity of the stem hybridizations increases which results into opening up the toehold domain (T_1) of the pre-hybridized reporter strand R_e. If the output strand is further added, it hybridizes with R_e because of their complementarity. This reaction ends up in extraction of the complete reporter strand and formation of the initial hairpin motif H which can be reused again in further production of output sequence.

Figure 14. Renewability of the DNA hairpin structure: (a) Formation of $H.R_e$ complex from $H.I_n.R_e$ complex, (b) Formation of $H.R_e$ complex from $H.B_o.R_e$ complex, (c) Restoration hairpin structure and reporter strand

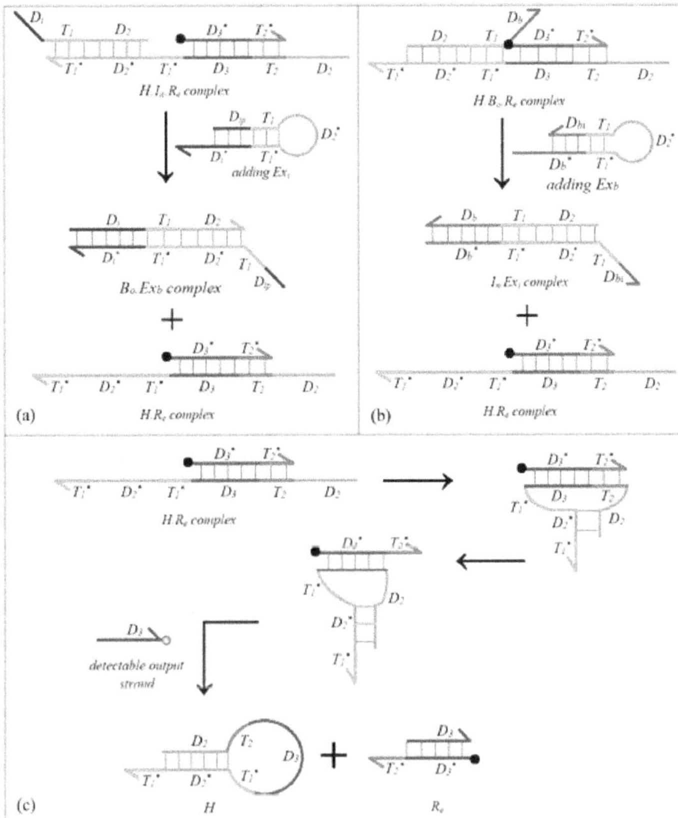

The researchers have experimentally verified the efficacy of reversible DNA hairpin model and successfully applied it to mimic the Boolean logic gates used in classical computation. These research works are considerable stages to start a journey towards the future of reversible, energy-efficient, eco-friendly, minimal-energy computation.

4. CONCLUSION

The contemporary technology of molecular computing is still too inefficient to surpass classical silicon-based technology. The domain of nanotechnology was first exemplified in the form of DNA computing which amalgamates biological science with the computer science. If DNA molecular devices can be handled more effectively, this interdisciplinary domain would be able to overcome all drawbacks of modern microelectromechanical devices.

The future of computing depends on making it renewable, as this can cause a drastic improvement in its energy efficiency. Though reversible computing is not a new domain, but it didn't get considerable attention in the era of silicon-based technology. Today, to save the natural resources and to stop the growing toxicity of the planet, reversibility should be imposed more successfully into every aspect of DNA computing so that it can replace the traditional form of computation. DNA computing is the novel smart-computing technology that can be optimally produced and run from available natural resources. Along with all the advantages over classical technology, DNA computing ensures environmental cognizance and better management system.

ACKNOWLEDGMENT

This research is supported by Council of Scientific & Industrial Research: Human Resource Development Group (CSIR: HRDG), Government of India (grant number 09/093(0180)/2018 EMR-I).

REFERENCES

Adleman, L. (1994). Molecular computation of solutions to combinatorial problems. *Science*, *266*(5187), 1021–1024. doi:10.1126cience.7973651 PMID:7973651

Adleman, L.M. (1995). On constructing a molecular computer. *DNA Based Computers*, *27*, 1-21.

Akerkar, R., & Sajja, P. S. (2009). Bio-inspired computing: Constituents and challenges. *International Journal of Bio-inspired Computation*, *1*(3), 135–150. doi:10.1504/IJBIC.2009.023810

Benenson, Y., Paz-Elizur, T., Adar, R., Keinan, E., Livneh, Z., & Shapiro, E. (2001). Programmable and autonomous computing machine made of biomolecules. *Nature*, *414*(6862), 430–434. doi:10.1038/35106533 PMID:11719800

Chang, W. L., Ho, M., & Guo, M. (2004). Molecular Solutions for the Subset-sum Problem on DNA-based Supercomputing. *Bio Systems*, *73*(2), 117–130. doi:10.1016/j.biosystems.2003.11.001 PMID:15013224

Eshra, A., Shah, S., Song, T., & Reif, J. (2019). Renewable DNA hairpin-based logic circuits. *IEEE Transactions on Nanotechnology*, *18*, 252–259. doi:10.1109/TNANO.2019.2896189

Feynman, R. P. (1960). There's Plenty of Room at the Bottom. *Engineering and Science*, *23*(5), 22–36.

Garg, S., Shah, S., Bui, H., Song, T., Mokhtar, R., & Reif, J. (2018). Renewable Time-Responsive DNA Circuits. *Small*, *14*(33), 1801470. doi:10.1002mll.201801470 PMID:30022600

Green, C., & Tibbetts, C. (1981). Reassociation rate limited displacement of DNA strands by branch migration. *Nucleic Acids Research*, *9*(8), 1905–1918. doi:10.1093/nar/9.8.1905 PMID:6264399

Green, S. J., Lubrich, D., & Turberfield, A. J. (2006). DNA hairpins: Fuel for autonomous DNA devices. *Biophysical Journal*, *91*(8), 2966–2975. doi:10.1529/biophysj.106.084681 PMID:16861269

Kari, L. (1997). DNA computing: Arrival of biological mathematics. *The Mathematical Intelligencer*, *19*(2), 9–22. doi:10.1007/BF03024425

Lipton, R. (1995). DNA solution of hard computational problems. *Science*, *268*(5210), 542–545. doi:10.1126cience.7725098 PMID:7725098

Ray, K. S., & Mondal, M. (2011). Similarity-based Fuzzy Reasoning by DNA Computing. *International Journal of Bio-inspired Computation, 3*(2), 112–122. doi:10.1504/IJBIC.2011.039910

Ray, K. S., & Mondal, M. (2011b). Classification of SODAR data by DNA computing. *New Mathematics and Natural Computation, 7*(3), 413–432. doi:10.1142/S1793005711002074

Ray, K. S., & Mondal, M. (2016). Logical inference by DNA strand algebra. *New Mathematics and Natural Computation, 12*(1), 29–44. doi:10.1142/S1793005716500046

Watson, J. D., & Crick, F. H. C. (1953). A Structure for Deoxyribose Nucleic Acid. *Nature, 171*(4356), 737–738. doi:10.1038/171737a0 PMID:13054692

Winfree, E., Liu, F., Wenzler, L. A., & Seeman, N. C. (1998). Design and self-assembly of two-dimensional DNA crystals. *Nature, 394*(6693), 539–544. doi:10.1038/28998 PMID:9707114

Yeung, D. S., & Tsang, E. C. C. (1997). A comparative study on Similarity based fuzzy reasoning method. *IEEE Transactions on Systems, Man, and Cybernetics. Part B, Cybernetics, 27*(2), 216–227. doi:10.1109/3477.558802 PMID:18255859

Zhang, D. Y., & Winfree, E. (2009). Control of DNA strand displacement kinetics using toehold exchange. *Journal of the American Chemical Society, 131*(47), 17303–17314. doi:10.1021/ja906987s PMID:19894722

Chapter 8
Hybrid Optimization Methods Application on Sizing and Solving the Economic Dispatch Problems of Hybrid Renewable Power Systems

Muhammet Tahir Guneser
https://orcid.org/0000-0003-3502-2034
Karabuk University, Turkey

Abdurazaq Elbaz
Karabuk University, Turkey

Cihat Seker
Karabuk University, Turkey

ABSTRACT

Renewable energy systems are spread all over the world due to the security problems encountered in accessing fossil fuels, the desire to reduce the environmental damage and to respond to the rapid increase in energy demand. However, the problems are experienced in renewable energy technologies in sustainable supply and reduction of production costs. Obtaining the optimum power distribution planning between photovoltaic, wind, biomass, and other systems depending on the relevant parameters and optimizing the distribution of energy supply-demand planning among the same sources can be applied as an effective solution by using several single optimization methods or new updated hybrid versions of them. In this chapter, common methods were evaluated and an application of crow and particle swarm as a hybrid method was examined in a certain region of Libya for a PV/wind hybrid renewable power system.

DOI: 10.4018/978-1-7998-8561-0.ch008

INTRODUCTION

Among all the various resources pursued on earth, electrical energy is the most crucial. Fossil fuels including coal, crude oil, and natural gas are currently being used to satisfy more than 70% of the planet's overall demand for electrical power (Sinha and Chandel, 2014). As economies and national populations simultaneously grow, demands for electricity increase and the consumption of various fossil fuels accordingly increases. Also. the supplies of these traditional fossil fuels are inherently limited, and the available quantities are steadily continuing to decrease, necessitating urgent attention and long-term solutions to prevent a possible energy crisis. At the same time, these conventional fuels are the direct source of various dangerous emissions, such as greenhouse gases, contributing in turn to global warming (Kuang et al 2016; Sinha and Chandel 2015). These issues are currently being addressed in a variety of ways. One popular strategy is focused on raising public awareness regarding the urgency of decreasing energy consumption in residential and industrial sectors and promoting newer technologies that are more energy efficient. One other strategy entails the promotion of renewable energy systems (RESs) and other related technologies with the goal of increasing their levels of dependability, cost-effectiveness, greenness, and accessibility for the broader population for usage at home. Various countries and territories are exerting considerable effort to improve renewable energy capacities, which are also receiving increased attention from researchers, governments, and many different industries(Menanteau et al., 2010). Alternative energy options such as wind, solar, biofuel, biomass, hydro, and geothermal sources of energy, among others, have been widely used to produce power in recent years. Considering the simultaneous consideration of the reliability, cost, and performance, renewable energy-based hybrid systems (REHSs) are clearly more logical and more feasible than systems utilizing a single source of energy in many applications. REHSs can run off one or more energy sources and can operate independently or in grid-connected mode. Various hybrid system combinations are possible, and these can and should be selected considering the specific need and the resources available for each individual location (Sinha and Chandel, 2014).

Most renewable energy sources, including both solar photovoltaic (SPV) systems and wind turbine generators (WTGs), are both clean and environmentally. According to these investigations, HRESs will provide good efficiency at lower costs in comparison to SPV or WTG systems that stand alone (Dawoud et al 2015; Celik, 2002; Beshr 2013).

Hybrid combinations of SPV with WTG may have disadvantages compared to conventional energy sources if they are not designed carefully and appropriately. For increased performance, it is critically important to address the inconsistent patterns of both wind speed and solar radiation, both of which lead to fluctuations in power. This can be easily resolved, for example, with the use of storage units, perhaps in the form of storage battery banks. When such batteries are present within hybrid systems, the storage of extra power becomes possible and thus the supply of loads in the event of shortages is also possible (Barton and Infield, 2004).

The incorporation of storage batteries into such systems also helps prevent oversizing of the SPV and WTG sources. Even after storage batteries are charged to their full capacity, however, excess power from generation units must still be avoided elsewhere. Reducing the amount of unutilized excess power could lower the cost of energy (COE) (Dufo-lo, 2007). As a result, each RES's optimum capacity must be known to confirm the actual load that can be accommodated.

To operate HRESs, it is necessary to maximize their performances; at the same time, it is necessary to keep both physical and technical constraints in mind. As a result, optimization methods, techniques, and applications have grown in popularity as means of achieving these objectives (Fathima and Palanisamy, 2015).

The main issue with operating grid-connected and islanded microgrids is economic dispatch (ED). This entails calculating the active power output for various generating units within a given system in order to achieve a prespecified objective function and simultaneously satisfy various relevant limitations (Cai et al, 2012). Creative and novel algorithms for the management of such systems are needed to handle dynamic behaviors and environmental restrictions of microgrids and subsequently facilitate their development into more viable and practical alternatives to conventional power systems (Trivedi et al, 2018). A key objective shared by all utility operators is to reduce not just total generating costs but also emissions. The conflicting realities of these two aims must be addressed to reach this goal. As a result, the notion of combined economic emission dispatch (CEED) may be utilized to concurrently reduce both operational costs and emissions from all extant generating units. This conversion of the multi-objective function of minimizing costs and emission levels into a single-objective function using a price penalty factor is one popular technique to tackle CEED problems.

Several optimization strategies designed with the aim of yielding solutions to the CEED problem are proposed in the literature (Parvez 2018; Ahmed 2015). In Reference. (Ri et al, 2007), three alternative methodologies are used to discuss the CEED problem with transmission losses. These are the genetic algorithm, ant colony algorithm, and lambda approach. The gradient method, in contrast, is used in Reference. (Augustine et al, 2012) with the aim of resolving the CEED problem while utilizing a mixture of conventional and renewable energy resources. In Refs. (Trivedi et al, 2018) and (Esmat and Eibakly, 2013), on the other hand, the interior search method and the ant colony algorithm are respectively used with the aim of resolving CEED problems of microgrids.

The harmony search (HS) algorithm, a metaheuristic method, was recently suggested in Reference. (Geem and Kim, 2001). This algorithm was designed considering improvisation processes seen in music and it has demonstrated effective applications for a variety of optimization issues (Gupta and Jha 2018; Sarkhel et al 2018). However, while the HS approach has a high ability to discover higher-performance parts of solution spaces in an acceptable amount of time, it has significant limitations (Ouyang et al, 2017). The first is its sensitivity to control parameter values. Second, due to the selection approach used, it may become caught in local optima. Finally, its improvisation mechanism is deficient in terms of previous knowledge, which reduces the effectiveness of the new harmony vector, also referred to as a solution vector (Ouyang et al, 2017). As a result, these flaws need to be addressed before using the original version of the HS algorithm to handle sophisticated or difficult optimization tasks.

This chapter gives an overview of the use of hybrid optimization methods to size and solve economic dispatch problems in hybrid renewable energy systems. The first step was to explain and describe hybrid renewable energy sources. The CEED (combined economic emission dispatch) model was then evaluated. Then, the four commonly used algorithms, to optimize the operation and modelling of hybrid energy systems, Classical techniques, Metaheuristic techniques, hybrid algorithms and Optimization computer software tools, were described. Finally, the optimization objectives were compared an application was evaluated for a PV/Wind HRES in a certain region of Libya by using Crow and Particle Swarm Optimization methods.

A METHOD TO OPTIMIZE THE SIZE OF RENEWABLE ENERGY SYSTEMS IS NEEDED

Due to their remoteness, many villages around the world may never be connected to the national grid, or towns may be far from the traditional electrical grid due to natural impediments and environmental limits. Remote village electrification using non-conventional energy sources such as solar, micro-hydro, and wind systems becomes important. In terms of cost, efficiency, and reliability, isolated operation of these power units may not be feasible(Ashok, 2007).

Combining these multiple renewable energy sources to construct a hybrid energy system is a potential alternative approach. One ideal sizing technique for the hybrid generation system considering system dependability and economic advantage is required in order to efficiently and economically utilize renewable energy resources(Wu and Chang, 2016). The size optimization approach entails determining the best (or near-best) system for generating a fair amount of renewable energy that matches the load demand distribution at the lowest cost(Nadjemi et al., 2017).

OPTIMIZATION TECHNIQUES FOR SIZING OPERATIONS

The algorithms that are utilized for computing the maximum or minimum values of assorted mathematical functions are known as optimization algorithms. When optimizing a system's architecture, different goals may be taken into consideration. The goals might be increasing system productivity and lowering the cost of production. Optimization approaches and techniques can aid in the resolution of difficult problems. When designing an HRES, we must always keep in mind the efficiency of its components. The key aim is to improve efficiency while lowering costs. These objectives may be successfully fulfilled by optimizing the modeling of the system. The three most popular modeling and optimization approaches for hybrid renewable energy systems are classical algorithms, metaheuristic approaches, and the combination of two or more different optimization techniques.

Optimization algorithms, often called mathematical programming algorithms, are methods for finding the optimal solutions to optimization problems given a set of constraints. While this explanation appears to be simple, it conceals several difficult aspects. For example, a solution could include a range of data kinds, non-linear boundaries could limit the search field, a search space could have numerous viable solutions, the problem's features could change over time, or the optimization process could include competing aims (Engelbrecht, 2007). Optimization algorithms have applications in research, engineering, economics, and business. Optimization is becoming increasingly important in science and technology around the world. Possible areas of application include a transportation problem in which a product must be shipped in specific quantities from many service points to multiple destinations, the goal being minimizing transportation costs; a power circuit state estimation problem; a profit problem involving an investment spread over many areas; or an increase in market players causing a rise in competition, forcing today's society to adapt. In all such situations, researchers seek optimal solutions in the most efficient and effective manner with the most economical process—in short, they seek the best way to do things (Pedregal, 2004).

As stated above, algorithms for computing the maximum or minimum of mathematical functions are known as optimization algorithms. When optimizing a system's architecture, different goals might be taken into consideration. Increasing system productivity while also lowering the cost of production are prime examples of such overlapping goals. Optimization approaches and techniques can aid in the resolution of seemingly conflicting goals. These objectives may be met by optimizing the system's modeling (Siddaiah and Saini, 2016).

As also suggested above, the main approaches to size optimization can be generalized as classic techniques, modern techniques, and software tools. The application of iterative, numerical, probabilistic, logical, or graphical construction approaches is seen in traditional techniques (Sinha and Chandel, 2015). For identification of the best possible solution for the considered problem, these approaches also utilize differential calculus (Siddaiah and Saini, 2016). Artificial and hybrid approaches are applied with modern techniques (Sinha and Chandel 2015; Mahesh and Sandhu 2015). The approaches in this category offer the promise of finding a range of optimal solutions with greater convergence and accuracy while determining global optimum schemes (Upadhyay and Sharma 2014; Mahesh and Sandhu 2015). Computer software tools constitute the third category of primary size optimization techniques. The Hybrid Optimization Model for Electric Renewables (HOMER) (Upadhyay and Sharma 2014 remains the most used software tool for size optimization, although another program called Improved Hybrid Optimization by Genetic Algorithm (iHOGA) is also relatively popular. The optimization techniques described here are graphically mapped in Figure1.

Figure 1. Commonly used optimization techniques

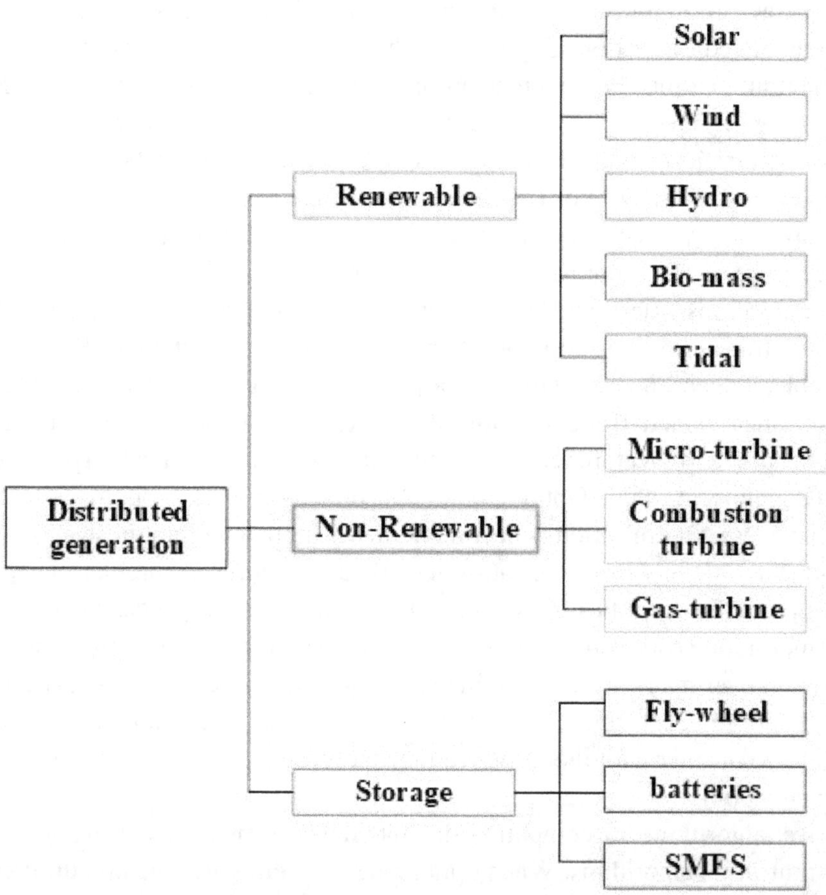

Classical Techniques

Differential calculus is used in traditional optimization algorithms to find the best solutions considering both continuous and differentiable functions. For systems of objective functions that are not actually continuous and/or differentiable, traditional approaches have limitations. For hybrid energy systems, a variety of traditional optimization approaches have been used. Classic algorithms for optimizing HRESs include dynamic programming (DP), the linear programming model (LPM), and nonlinear programming (NLP).

Among these, the LPM may be adopted to investigate situations where objective functions are linear and design variable spaces are defined solely by linear equalities and inequalities. Adoption of the LPM to optimize HRESs has been described in many studies in the literature (Ramakumar 1992; Hennet and Samarakou 1986; Gupta et al 2006; Saini and Sharma 2010; Jyoti and Raju 2011). These studies make use of the LPM's capabilities in undertaking stochastic reliability and economic analysis. The failure of any of the renewable resources, on the other hand, has a negative impact on the overall system's energy delivery capacity (Arabali et al, 2014). In contrast, NLP is usually applied in the investigation of general cases where objective functions, constraints, or both include nonlinear components. Some studies (Ashok, 2007) have used this model. Doing so allows complex problems to be solved using basic operations. However, for numerical methods like NLP, large numbers of iterations are performed, raising the computational difficulty of the problem (Arabali et al, 2014).

Finally, DP is popularly used to examine situations where the optimization approach is focused on breaking the problem down into smaller subproblems. This approach aids in the solution of sequential problems or problems with multiple, connected stages. A key benefit of DP lies in its ability to optimize each level. As a result, it can deal with the complexities of larger structures. The large number of recursive functions encountered in DP, on the other hand, makes the coding and implementation difficult and potentially confusing (Arabali et al, 2014). An example of a study utilizing DP with the goal of HRES optimization can be found in Ref. (Chak and Bengal, 1990).

Metaheuristic Techniques

Since they can produce reliable and effective optimal solutions, metaheuristic search techniques are very popular in efforts aimed at optimizing complex systems such as HRESs. The algorithms falling within this group can be understood as nature-inspired since they focus on natural behaviors. Examples of metaheuristic approaches utilized for HRES optimization include particle swarm optimization (PSO), simulated annealing (SA), the genetic algorithm (GA), and ant colony (AC) algorithms.

Genetic Algorithm (GA)

As of 1975, Holland was among the first to establish heuristic methods. Selection, crossover, mutation, and inheritance are examples of search techniques focused on genetics and natural selection principles (Learning and Publishers, 1988). In comparison to other search techniques designed to operate with one single solution, GA allows a population to evolve into a state that will be able to maximize "fitness" under defined selection rules.

A population of elements would be assimilated to chromosomes, with possible candidates being encrypted for them to evolve into a better state. The solutions to these types of problems are traditionally expressed in binary code. To begin, the initial population is created at random, and the suitability of each candidate is assessed over generations. To create a new population, the selected candidates are changed by mutation. This will be repeated until the algorithm is seen to be satisfied or until reaching the maximum number of iterations.

In the literature, the GA is generally regarded as one of the most widely used optimization strategies for placement and sizing decisions related to distributed generation (DG) (Borges and Falca 2006; Singh and Goswami 2010; Shaaban et al 2013; Singh et al, 2007; Singh and Goswami 2009). It was used in Refs. (Zangeneh et al, 2009) and (Soroudi et al, 2011), for example, with the twofold target of reduced system expansion costs and increased system efficiency. Since these two objectives are incompatible, Pareto-optimal models were preferred to find the dominant solution in a single run.

To address these DG location and sizing problems, new enhanced methods are being proposed, such as that presented in Ref. (Processing, 2007), a study in which the GA was utilized together with the multi-attribute decision-making (MADM) approach to consider various parameters of power systems. Further improved GA methods include the adaptive genetic algorithm (AGA), shown by Ref. (MA et al, 2012) to provide both higher levels of robustness and a higher performance in terms of search ability, as well as the quantum genetic algorithm (Liao, 2012). Table1 sketches a comparison of the major advantages and disadvantages that GAs face.

Table 1. Genetic Algorithm's Benefits and Drawbacks

Benefits	*Drawbacks*
-Have a better chance of finding the global best solution for a large range of functions. -No derivatives are needed. -It is possible to do it for both discrete and continuous variables. -can be used to solve problems that are both complex and poorly described. -Bad strategies have little impact on the result.	-Due to repeated fitness function assessment, it can be time consuming for broad and complex problems. - It is possible that it will be inaccurate.

Particle Swarm Optimization (PSO)

PSO was originally introduced in 1995 as a novel optimization method created by Eberhart and Kennedy. This approach to optimization is based on the social actions of flocking birds and schooling fish, whereby particles move within multidimensional search space and the single intersection of all dimensions forms a particle (Zhu, 2008). The framework can be modified with a collection of arbitrary solutions first and the search for optimization is maintained with the updating of subsequent generations. Particles will evaluate their positions based on their fitness levels at each iteration, and neighboring particles will share the background of their "best" positions to help tailor the final optimized solution (Zhao 2006; Elbaz 2020).

PSO is quite widely used by researchers to solve DG location and sizing problems (El-Zonkoly AM, 2011; Lalitha and Reddy, 2010). For example, in Reference. (Pandi et al, 2013), PSO is applied to determine the best position, form, and size for DG units to provide optimal DG integration while keeping the relevant harmonic limits and security constraints in mind. PSO was also used in References. (Sedighizadeh and Sadighi, 2008) and (Kansal et al, 2013) to boost the voltage profile as well as minimize total harmonic distortion, losses, and costs. When compared to GA-based systems, the results showed that PSO provided higher solution quality while engaging in fewer iterations. Thus, in real-world settings, PSO has faster computational times than the GA and it can be applied to real-world power network scenarios. Table 2 lists the main advantages and disadvantages of PSO-based methods.

Improved PSO (IPSO) (Yustra and Soeprijanto, 2012), binary PSO (BPSO) (Su et al, 2011), social learning PSO (SLPSO) (Arasi and Sasiraja, 2015), PSO with inertia weight (PSO-IW), and PSO with constriction factor (PSOCF) (Ganguly et al, 2012) are some of the novel methods for enhanced PSO application currently being utilized to address DG location and sizing tasks.

Table 2. Particle Swarm Optimization 's Benefits and Drawbacks

Benefits	*Drawbacks*
- It is possible that it will be easy to execute. - There are just a few parameters to tweak. - Having the ability to run parallel computations. - Can be robust - Finding the global optima has a higher likelihood and efficiency. - Can converge quickly. - Avoid overlapping and mutating. -Have short computational time - Can be useful for solving problems where reliable mathematical models are difficult to come by.	- Initial design parameters can be difficult to define. - When dealing with complex problems, it is possible for it to converge prematurely and become stuck in a local minimum.

Ant Colony Optimization (ACO)

Dorigo et al. published the first ACO algorithms in 1996, which were based on the social actions of insects such as ants performed in the quest to find the shortest paths to food (Dorigo et al, 2006). Researchers have found pheromone trails physically left by ants. Some ants use this substance to communicate knowledge about their paths. This method, similarly, to other metaheuristics, begins with the suggestions of random solutions that are assimilated into ants' searches and the trails left behind by their movements. As a result, the shorter the routes, the more trails there are. The subsequent searches will take this knowledge into account. ACO was utilized in both References. (Falaghi et al, 2007) and (Wang et al, 2008) with the aim of solving the problem of locating and sizing DGs from an RES with radial distribution systems together with the simultaneous goal of minimizing total system loss.

Table 3. Ant Colony Optimization's Benefits and Drawbacks

Benefits	*Drawbacks*
§ Can search among a population in parallel. § Can help you find good solutions quickly. § Can adjust to new distances and other adjustments. § Convergence is assured.	-For each iteration, the probability distribution will shift. -Have a challenging theoretical analysis. -Have dependent random decision sequences. -There should be more experimental research than theoretical research.

A reliability index was selected as the objective function in Reference. (Wang et al, 2008) and the ACO algorithm was used for completing various discrete optimization tasks. In comparison to the GA, the results showed that ACO provided more efficient solutions with shorter computational times. Since the solution space to be analyzed is greater, ACO takes longer to converge, but it is still faster than analytical methods. In Table 3, the advantages, and disadvantages of the ACO method are highlighted.

Tabu Search (TS)

TS is another popular metaheuristic technique adopted by researchers for solving optimization problems; it was first proposed by Glover in 1986. This method is based on principles of adaptive memory and receptive exploration, which allow for the cost-effective and efficient searching of the solution space until no improvement is found. TS has had a key role in resolving the issue of finding and sizing DGs. Golshan et al., for example, utilized it based on DG optimum planning with the aim of minimizing both losses and line loadings, but TS does unfortunately have the drawback of requiring the consideration of large numbers of iterations and parameters (Esmail et al 2007; Golshan and Arefifar2006). Table 4 highlights the most important advantages and disadvantages of TS.

Table 4. Tabu Search Genetic's Benefits and Drawbacks

Benefits	*Drawbacks*
-Can be used to solve difficult problems. -Have a clear memory. -It works for both discrete and continuous variables.	-Depending on the Tabu list manipulation technique, this is possible. - Can get stuck in local minima. - Many parameters should be determined. - Have many iterations. -To find a global optimum, you can depend on parameter settings.

Harmony Search (HS)

HS is yet another metaheuristic optimization technique, having been first introduced in 2001. It was primarily designed considering techniques employed by musicians to enhance the harmony of their instruments. Thus, in contrast to the algorithms described above that are focused on observed natural behaviors, HS describes a musical performance process that seeks to improve harmony (Search,2013). In Refs. (Rao et al, 2013) and (Piarehzadeh et al, 2012), HS was used in conjunction with a loss sensitivity factor approach to find the best DG position. The latter of those studies (Piarehzadeh et al, 2012) concluded that using the HS algorithm rather than the PSO algorithm for allocating DGs improved voltage stability. Table 5 highlights the major advantages and disadvantages of this technique.

Table 5. Harmony Search Algorithm's Benefits and Drawbacks

Benefits	*Drawbacks*
§ There are no initial value settings required. § Can work in both discrete and continuous variables. § It is not possible to diverge. § It is possible that you will be able to get away from local optima.	-The ability to search for local information is weak. -It is possible to achieve many iterations. -Have a multimodal problem with a high number of dimensions.

Further Heuristic Methods

In recent years, some authors have introduced new heuristic methods for solving the DGs locating and sizing problem, such as:

- **Artificial Bee Colony (ABC)**, an optimization algorithm based on the honeybee swarm's searching behavior. In Refs. (Abu-mouti and Member, 2011) and (Lalitha et al, 2010), the ABC algorithm was used, and comparing this new method with the older PSO approach revealed that the newer method provided higher quality of solutions and faster convergence.
- **The Cuckoo Search Algorithm (CSA)**, which is based on certain cuckoo species' obligate brood parasitism, which is described by the placement of their eggs in the nests of other host species (Yang et al, 2009). For both DG biomass and solar-thermal units, the CSA was applied in Ref. (Moravej and Akhlaghi, 2013) to improve voltage profile with the simultaneous goal of minimizing power losses.
- **The Shuffled Frog Leaping Algorithm (SFLA)** (Eusuff et al, 2007), as another example of algorithms reflecting natural behaviors, was inspired by frogs' actions while looking for food. The SFLA method has been used with success to resolve problems of DG allocation and sizing. In Ref. (Taghikhani, 2012), for instance, the SFLA was used to optimize device voltage profiles while lowering line loss. This strategy helpfully combines the advantages of both the GA and PSO algorithms.
- **Shuffled Bat Algorithm (SBA)**, which is based on microbat echolocation behaviors. To demonstrate its efficacy, this algorithm was applied to a radial distribution system in Ref. (Yammani et al., 2016), first with 100% base load conditions and then with 120%
- **The Firefly Algorithm (FA)**, which is based on how fireflies communicate among themselves within a courtship system. The flash of a firefly's bioluminescence is used for attracting other fireflies (Yang, 2010). The FA ensured optimum allocation of DG in Ref. (Sulaiman et al, 2012), with goals of minimizing actual and reactive power losses as well as line loading.

Hybrid Algorithms

Combining two or more optimization strategies may make it possible to overcome the limits that are faced when any of the optimization techniques described above are used alone, resulting in more efficient and reliable HRES solutions. Such combinations are referred to as "hybrid" strategies.

Premature convergence is a disadvantage of most evolutionary computational techniques. Significant amounts of time may be consumed for these strategies to emerge from local maxima or minima (Khare and Rangnekar, 2013). Hybrid techniques, however, may be used for solving diverse optimization problems by combining helpful benefits of two or more optimization techniques. Research into these hybrid techniques has received considerable attention in recent years, largely because individual approaches are less efficient and competitive than hybrid strategies.

Many researchers have used hybrid strategies to improve the performances of standalone wind, solar, and solar-wind hybrid systems in recent years. For example, Refs. (Kalogirou, 2004) and (Mellit et al, 2010) reported the use of a neural network and GA to size photovoltaic systems, while Reference. (Khatib et al, 2012) described the outcome of a case study applied in Malaysia to determine the optimal numbers of hybrid device components with key deciding factors being the numbers of wind turbines,

PV panels, and batteries. The authors used a mixture of iterative and GA methods in their research. The iterative method was used to find a range of possible configurations and the GA was used to find the best configuration.

A hybrid technique for sizing a PV-wind-based method was suggested, combining three popular algorithms: chaotic search (CS), SA, and HS. It developed DCHSSA, a discrete chaotic harmony search-based simulated annealing algorithm, by combining those three techniques. The proposed algorithm was reported to be superior to the individual techniques. (Askarzadeh, 2013). Using a mixed multiple-criteria integer programming problem, Ref. (Xu et al, 2005) proposed an improved approach for optimal sizes of the components of a standalone hybrid wind-PV power system. The GA was nominated as a helpful candidate for reduction of overall capital costs when adhering to the LPSP constraint. By examining the relevance of individual algorithms, Another hybrid TS-SA optimization strategy was suggested and the findings showed that the hybrid approach produces better solutions in a shorter amount of time (Katsigiannis et al, 2012). Recently presented studies a hybrid big bang-big crunch algorithm intended for the design of an optimized hybrid solar, wind, and battery system as well (Ahmadi and Abdi, 2016). Ref. (Maleki et al, 2016) proposed a PSO-based Monte Carlo simulation (PSOMCS) for the sizing of off-grid hybrid systems incorporating energy from solar, wind, and battery sources.

Optimization Software Tools

To optimize renewable energy systems, a variety of software methods have been used (Connolly et al, 2010). Hybrid energy systems' design requirements, evaluation criteria, power, and energy management can be better addressed now with the introduction of various software tools:

- HOMER
- HOGA
- HYBRIDS
- MATLAB

The properties of available software tools are summarized in Table 6. When designing an HES, large quantities of variables and goals, such as minimization of total device cost, improved efficiency, reduced emissions, and so on, are taken into account, and the simulation time may increase accordingly. This emphasizes the importance of selecting an appropriate sizing technique.

Among the many software tools available, HOMER program is possibly the most widely used for hybrid energy system sizing. It facilitates quick determination of the best size for energy systems and can also be used for sensitivity analysis to investigate effects of variability or changes to input variables. For example, Hrayshat (Hrayshat, 2009) used HOMER software to perform a comprehensive techno-economic study to develop the best hybrid PV-diesel-battery system for a house in a remote area of Libya (Elbaz and Guneser, 2021).

HOMER Pro is a more advanced version of HOMER that includes features such as an optimizer, a multi-year module, monthly demand limits, advanced battery or load profile options, and the ability to connect to MATLAB software (Hrayshat, 2009).

Table 6. Selected simulation software tools(Dufo-lo, 2009)

Software tools	PV	Genset	Battery	Wind	Mini-hydro	Fuel cell; electrolyzer and H2 tank	Loads	simulation	Control strategies	Multi-objective optimization (economical and technical)
HOMER	√	√	√	√	√	√	√	√	√	√
Hybrid2	√	√	√	√		√	√	√	√	
HOGA	√	√	√	√	√	√	√	√	√	√
MATLAB	√	√	√	√	√	√	√	√	√	√

HYBRID RENEWABLE ENERGY SYSTEMS

Because to population growth, suburbanization, and industrial expansion, energy demand is increasing every day. Because the rate of energy consumption is high, and the supply from renewable energy sources is insufficient to fulfill the load demand, an energy shortfall occurs. Large-scale power generation relies on traditional energy sources that are not only scarce but also poorly distributed across the globe. Due to the increasing depletion of fossil fuels such as coal, oil, and gas, traditional energy sources have a greater influence on the environment, resulting in an increase in CO2 levels, which causes global warming.

The use of renewable energy sources is necessary; however, because renewable energy sources are dependent on environmental conditions such as wind speed and solar irradiance, individual energy sources cannot provide continuous power supply to the load due to their unpredictable and intermittent nature. As a result, renewable energy sources such as wind, solar, hydro, biogas, and fuel cells can be combined to create a hybrid system that is more dependable and environmentally benign.

Despite advances in renewable fields, oil, coal, and natural gas remain the primary energy sources of our planet now (Paliwal et al 2014; Kaldellis et al 2020).

Renewable energy systems are seen as particularly viable options for remote or isolated locations that suffer from limited access to national grids in the presence of problematic technological and economic constraints. They are therefore favored in diverse regions and by many countries. PV, wind, and hydropower or a combination of the three can be used in such a power system. To meet peak hour demand, such a system might include backup devices such as diesel generators or battery banks. Characteristics of renewable energy power systems (REPSs) are described in Table 7.

REPSs that are designed well will be cost-effective and dependable and they will inherently have the power to improve people's lives (Enslin and Africa, 1991). In most relevant situations, solar and wind energies are complementary, but both are unpredictably variable due to the instantaneous variations that occur in the supply of both resources. (Bhandari et al, 2013) reported the development of a new hybrid system that combines wind, solar, and hydro resources, and its potential applications in meeting the needs of a small Nepalese village (Bhandari et al, 2014). (Ahn et al, 2012) looked at various qualities of an off-grid HRES and how they affect system reliability.

Table 7. Characteristics of REPS (Energy, 2017)

Advantages	Disadvantages
-Make use of free resources. For example, the sun and wind may be used as a source of energy. -The cost of operation and maintenance is minimal. -There are no emissions or natural resource waste issues.	-Natural cycles govern the production of renewable energy. -These systems have a higher initial cost than traditional generators of comparable capacity. -Without energy storage, it would be difficult to meet peak loads.

Figure 2. Distributed generation models

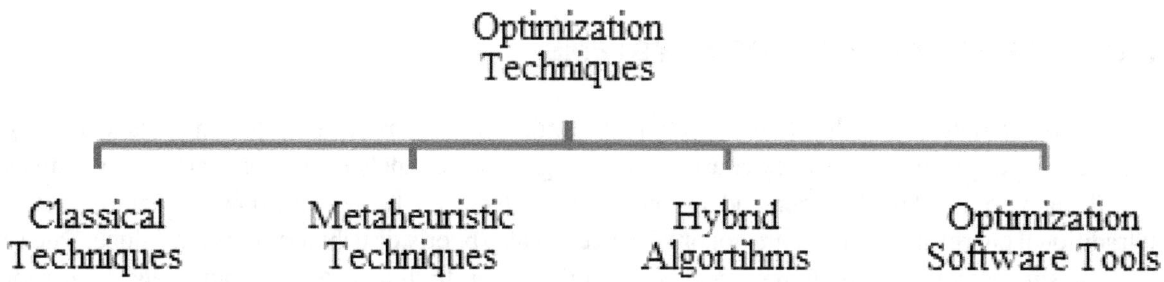

Due to the dependency of these sources of energy on unpredictable environmental conditions, however, such as wind speeds or levels of solar irradiance, one sole intermittent energy source will not be able to sustain the necessary power supplies. To remove this crucial shortcoming, renewable energy sources may be utilized in combination to create hybrid systems that are simultaneously more dependable and environmentally sustainable. "Distributed energy sources" and "distributed generation" are terms used to describe this type of renewable energy (Paliwal et al 2014; Upadhyay and Sharma 2014), as shown in Figure 2.

Since the performances of HRESs are so highly dependent on their individual components, careful modeling of each component offers tools for better understanding the system's efficiency and reliability, as well as assisting in the optimization of HRESs. The current chapter examines an assortment of the available optimum sizing methods together with in-depth requirements for hybrid power systems comprising small PV units or hydro, wind, and storage devices. Information on different optimization strategies and optimal designs has been given for a clear understanding. Wherever possible, mathematical models, statistics, or flow charts have been included.

MODELLING OF RENEWABLE ENERGY RESOURCES

Solar Radiation

Positions on the earth's surface, dates, and times of day all influence the levels of solar radiation that can be feasibly utilized by humanity as a sustainable resource. The actual amount of radiation is determined by these factors.

In practical settings, other variables including altitude above sea level, atmospheric water vapor or contaminants, and cloud cover further reduce those radiation levels and may make them fall below useful limits. While solar radiation is not subject to experiencing turbulence like wind energy is, short-term variation may still be seen in solar power. Most of the time, such variations are linked to cloud movement (Da Rosa and Ordonez, 2021).

Wind

Wind resources are almost entirely driven by the sun's energy, resulting in differential surface heating, but they are also highly location dependent. Average wind speed varies as seasons change across most of the globe, and overall weather conditions and time of day are likely to exert their own effects, as well. The same location may offer several days with relatively higher winds followed by days with lower winds, which will severely interfere in the activity planning of a hybrid system containing wind turbines. "Turbulence" is the term for short-term (seconds to minutes) changes in wind speed and direction (Jevtic et al, 2017).

Hydropower

The volume of non-stagnant water moving through a stream, or a river determines the hydropower resources at a given location, as well as any changes in elevation. The head is normally constant (except for higher water levels during storms), but the volume of water that can be used in any one locale may change dramatically over time.

Rainfall quantities and the exact size of drainage areas upstream of sites, where rain falls, are factors that are effective in determining average discharge. During floods, discharge will be increased, while during droughts, it will decrease. Discharge may also be affected by soil conditions and the landscape. In most cases, however, short-term variations are negligible (Da Rosa and Ordonez, 2021).

Combined Economic Emission dispatch (CEED) Model

The goal of the combined economic and emission dispatch problem is to find the optimal quantity of generated power for the system's generating units while minimizing fuel costs and emissions levels simultaneously while considering various system restrictions.

The emission function is added as a second objective to traditional dispatching issues in the CEED problem. As a result, all such high-emission generation units are given less power, and emissions are lowered. The electric energy industry's contribution to pollution raises concerns about environmental protection and techniques for removing or lowering pollution from power plants, either through design or by operating measures(Taylor et al., 2007). Sulfur dioxide (SO2) and nitrogen oxides (NOx) are the two principal power plant emissions from dispatching (NOx). The amount of SO2 produced is proportional to the amount of fuel consumed. As part of the fuel, sulfur enters the boiler. Some sulfur combines with oxygen from the fuel and combustion air to generate SO2 during the combustion process. The residual sulfur is incorporated into the boiler's bottom ash. The majority of the SO2 is eliminated if stack gas cleaning equipment is available. The leftover SO2 is emitted from the stack. NOx emissions are a little more complicated. NOx is produced when nitrogen from two sources combines with oxygen from the fuel and combustion air. The first source is nitrogen in the air, which creates a thermal NOx emission.

The second source is nitrogen in the fuel, which causes a fuel NOx emission. There is no clear link between the amount of fuel-bound nitrogen and the amount of fuel NOx produced in coal(Lamont, 1995(El-Keib et al., 1994).

In thermal power plants, among the particularly important tasks encountered in planning and operating a power system, the task of reducing fuel costs and pollution emissions is virtually universal. The CEED problem, as introduced above, is solved as an optimization problem in which we will be minimizing the functions of fuel costs and pollutant emissions. The output powers of generators in thermal power plants are adjusted for meeting the device load, which is subject to transmission and operational constraints, to minimize these functions. Many stochastic nature-inspired metaheuristic algorithms to solve the CEED problem have been presented in the literature (Jevtic et al 2017; Chen and Ding 2015) due to the complexity of the objective functions, which take the form of a sum of quadratic, sinusoidal, and exponential functions.

In a thermal power plant, the fuel cost feature, $F_g(P_g)$, of generation unit g can be calculated in two different ways as seen Eq.1 and Eq.2.

1. When the valve point loading effect (VPLE) in the thermal power plant is not considered, a quadratic smooth function is obtained via Eq.(1).

$$F_g\left(P_g\right)=\alpha_g+b_gP_g+c_gP_g^2, g=1,2,\ldots G \tag{1}$$

2. When considering VPLE, consider a more complex, nonsmoothed, and nonconvex function can be calculated as seen on Eq.(2) (Benasla et al, 2014).

$$F_g\left(P_g\right)=\alpha_g+b_gP_g+c_gP_g^2+\left|d_g\sin\left(e_g\left(P_g^{min}-P_g\right)\right)\right| \tag{2}$$

where F_g is expressed in \$/h; P_g is the real power of generation unit g in megawatts; G is the total number of generation units; P_g^{min} is the lower loading limit of generation unit g; cost coefficients a_g, b_g, and c_g; and VPLE coefficients d_g and e_g. And a generation unit's emission function, $E_g(P_g)$, is defined as a sum of quadratic and exponential functions as seen on Eq.(3) (Bhattacharya and Chattopadhyay 2011; Güvenç et al 2012).

$$E_g\left(P_g\right)=\alpha_g+\beta_gP_g+\eta_gP_g^2+\xi_g\exp(\lambda_gP_g) \tag{3}$$

where E_g is expressed in t/h and α_g, β_g, η_g, ξ_g, and λ_g are emission coefficients of the generation unit g. To solve the CEED problem, Eq. (1) or (2) is combined with Eq. (3) using the weighted sum method, i.e. αg, β_g, η_g, ξ_g and λ_g are emission coefficients of the generation unit g, and E_g is expressed in t/h. Eq. (1) or (2) is combined with Eq. (3) using the weighted sum method to solve the CEED (Özyön and Yas, 2014).

$$FE=\omega\sum\nolimits_{g\in G}^{F_g(P_g)}+(1-\omega)\gamma\sum\nolimits_{g\in G}^{E_g}\left(P_g\right) \tag{4}$$

Under device constraints, the combined operation of Eq.(4) is then reduced. In Eq. (4), is the scaling factor, and w is the weight factor, all of which have values between 0 and 1. The limit w = 1 corresponds to fuel cost minimization only, while the limit w = 0 corresponds to pollutant emission minimization only. The objective CEED problem is solved as a single-objective problem using the scaling factor γ. Two constraints are fulfilled in this minimization process as seen below. The transmission system's power equality constraint is as seen on Eq.(5) and the generator unit capacity constraint as seen on Eq.(6), where P_g^{max} and P_g^{min} are the generator unit g's maximum and minimum power values. The transmission system's power loss is expressed using B-loss matrices as seen on Eq.(7), where B_{00}, B_{0g}, and B_{gj} are the coefficients of the *B-loss* matrices (Özyön and Yas, 2014). And one of the generators (e.g., generator G) is chosen to be a dependent generator to fulfill the restriction of Eq.(5). (The slack generator). The value of P_G can be calculated using Eq.(5) as seen on Eq.(8).

$$\sum_{g \in G}^{P_g} - P_D - P_{loss} = 0 \tag{5}$$

$$P_g^{min} \leq P_g \leq P_g^{max} \tag{6}$$

$$P_{loss} = \sum_{g \in G}^{\sum_{j \in G} P_g B_{gj}} P_j + \sum_{g \in G} B_{0g} P_g + B_{00} \tag{7}$$

$$P_G = P_D + P_{loss} - \sum_{g=1}^{G-1} P_g \tag{8}$$

$$\varepsilon = \left| P_{loss}^{(1)} - P_{loss}^{(0)} \right|, \varepsilon \leq \delta \tag{9}$$

By using Eq.(7) and Eq.(8), the values of P_G and P_{loss} can be obtained by setting the initial value of $P_{loss} = P_{loss}^{(0)} = 0$ in Eq.(8) and calculate the initial value $P_G^{(0)}$ from Eq.(8) for the initial value $P_{loss}^{(0)} = 0$. After that the new value $P_{loss}^{(1)}$ can be calculated via Eq.(7). Lastly, it is needed to be checked whether the error value ε is below the specified error tolerance value σ, as seen on Eq.(9).

While obtain the value $P_G^{(1)}$ from Eq.(8) for $P_{Loss} = P_{Loss}^{(1)}$, the power equality constraint of Eq.(5) is met if the condition of Eq.(9) is satisfied. Otherwise, the procedure is repeated. After checking whether the calculated P_G value satisfies the constraint of Eq.(6), the variable P_G^{lim} is defined as seen on Eq.(10). For $P_{Loss} = P_{Loss}^{(1)}$, get the value $P_G^{(1)}$ via Eq.(8). If the condition of Eq.(9) is fulfilled, the power equality restriction of Eq.(5) is satisfied. The procedure is then repeated if necessary. The variable P_G^{lim} is defined seen on Eq.(10)., whether the measured P_G value satisfies the constraint of Eq.(6). P_G is a dependent variable in this equation. Eq.(11), where the quadratic penalty word is applied to the objective function *FE* of Eq.(11) with the penalty factor λp, is the new extended objective function to be minimized.

$$P_G^{lim} = \begin{cases} P_G^{max} \text{if } P_G > P_G^{max} \\ P_G^{min} \text{if } P_G < P_G^{min} \\ P_G \text{if} P_G^{min} \leq P_G \leq P_G^{max} \end{cases} \tag{10}$$

$$FE_p = FE + \lambda_p \left(P_G - P_G^{lim}\right)^2 \tag{11}$$

OPTIMIZATION OBJECTIVES FOR HRES

For the best design and part sizing of HRESs, several factors are considered. Economic and technological criteria may be used to categorize these criteria. To keep HRES costs as low as possible, economic requirements are considered. Technical requirements include reliability, performance, and environmental goals for supplying HRES load demand at optimal reliability levels while maximizing efficiency and reducing emissions of greenhouse gases.

Cost Optimization

Larger volumes of capital costs with smaller volumes of operation and maintenance (O&M) costs are common features of HRESs, necessitating optimization to find the best balance of costs and benefits. Minimizing electricity costs while also considering net present cost (NPC) and all other costs related to hybrid renewable energy systems is part of the cost optimization process.

Net Present Value (NPV)

Since it accounts for the time value of money using discounted cash flows, NPV is a reliable budgeting tool. It involves projecting net cash flows that may occur at some point in the future, discounting these flows with a discount rate, and then subtracting the net start-up investment from the present-day value of these net cash flows using the project risk level, as shown in Eq.(12). *IRR* denotes the internal rate of return, which is estimated to be 10% in this case; *n* denotes the number of years the system is expected to operate (an average of 25 years); and $Cash_{in}$ denotes cash inflow, which can be calculated as seen on Eq.(13) (Gómez et al, 2010).

$$NPV = \left(Cash_{in} * \frac{1-(1+IRR)^{-n}}{IRR} \right) - \left(C_{in} + C_{gp}\right) \tag{12}$$

$$Cash_{in} = kWh_{price} * Load \tag{13}$$

Payback Period

The payback period is the period over which the cash outflow of the initial start-up investment is assumed to have been fully recovered by the inflows that the investment was able to produce. It is a straightforward appraisal method that can be measured as follows as seen on Eq.(14) (Drury et al, 2011).

$$PaybackPeriod = \frac{C_{in}\left(initialinvestment\right)}{Cash_{in}} \tag{14}$$

Cost of Energy (CoE)

The Measurement of the *CoE* can be calculated via the Eq.(15) as seen below.

$$CoE = \frac{\left(\text{Totalcostofgeneratedenergyforoneyear}\right)}{\left(\text{Totalenergysuppliedinoneyear, kWh}\right)} \tag{15}$$

Loss of Power Supply Probability

The LPSP, which can be explained as a load that the system is incapable of fulfilling in the study period divided by the total load, is used to measure the system's reliability in this model. It can be calculated via Eq.(16). The LPSP value is in the range [0, 1], confirming the system's efficiency before the integrated grid-connected solar PV system's total generated power covers the load. A value of 1 indicates that the requisite load is completely unmet, while a value of 0 indicates that the load is always fully met. Permissible LPSP values are usually thought to be 0.05 or 5%. The variable to consider when minimizing the COE with a stable method is NPV. The proposed limitation is seen on Eq.(17), where N_{PVmin} and N_{PVmax} represent the minimum and maximum quantities of PV panels, respectively .

$$LPSP = \frac{\Sigma\left(P_{load} - P_{PV} - P_{GO}\right)}{P_{load}} \tag{16}$$

$$N_{PV\min} \le N_{PV} \le N_{PV\max} \tag{17}$$

CASE STUDY

An Application With Crow and PSO Algorithms for Sizing a Hybrid Renewable Energy System in Libya

The case study presented here involved the design of a hybrid power plant comprising an off-grid wind energy and photovoltaic system with the aim of supplying the demand of a residential building in Libya.

We selected the crow search algorithm to support the goal of minimization of both installation and operational costs by optimally sizing all individuals' parts of the hybrid system. Number of photovoltaic modules, battery capacity, and wind turbine power values were all optimized accordingly. We then conducted a comparison of the performances of the crow algorithm and PSO in consideration of hybrid system design. The crow algorithm demonstrated better results in terms of efficiency for the sizing of lower-cost hybrid power plants that incorporate photovoltaic and wind systems.

Crow Search Algorithm (CSA)

Crows are commonly understood to be intelligent, an assumption that seems to be supported by their brain-to-body mass ratios. They can detect danger, send out warning signals, and recognize faces. After watching crows' behaviors, like living in groups and stealing collectively, the CSA was created. In addition to their other mental capabilities alluded to here, crows also demonstrate so-called awareness probability (AP). Individuals' updating mode can be expressed as Eq.18., where the uniform distribution of r1 and r occurs within the range 0-1, and mem jt symbolizes a given crow's memory location, when there are n crows in dimensional space and X 't denotes the location of crow i. The awareness probability (AP) of the crow j is represented by AP jt. The tth iteration of fl it (provided here especially for crow I is a crow's possible range for flying. Figure 3 graphically illustrates the fundamental basics of the CSA. Although this algorithm has the benefit of providing substantial global search capabilities, its basic form has several restrictions due to the assumption that crow flight is rigid and immobile.

$$X_i^{t+1} = \begin{cases} X_i^t + r_1 * fl_i^t * \left(mem_j^t - x_i^t\right), r \geq AP_j^t \\ arandomposition, other \end{cases} \tag{18}$$

Figure 3. An Individual Crow's Movements

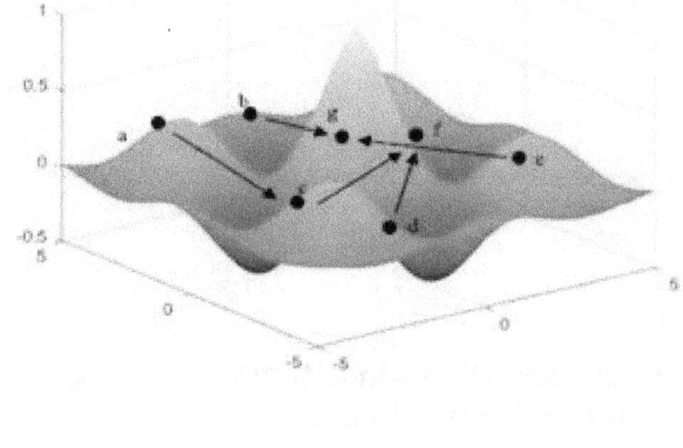

Figure 4. Hybrid PV/Wind HRES design

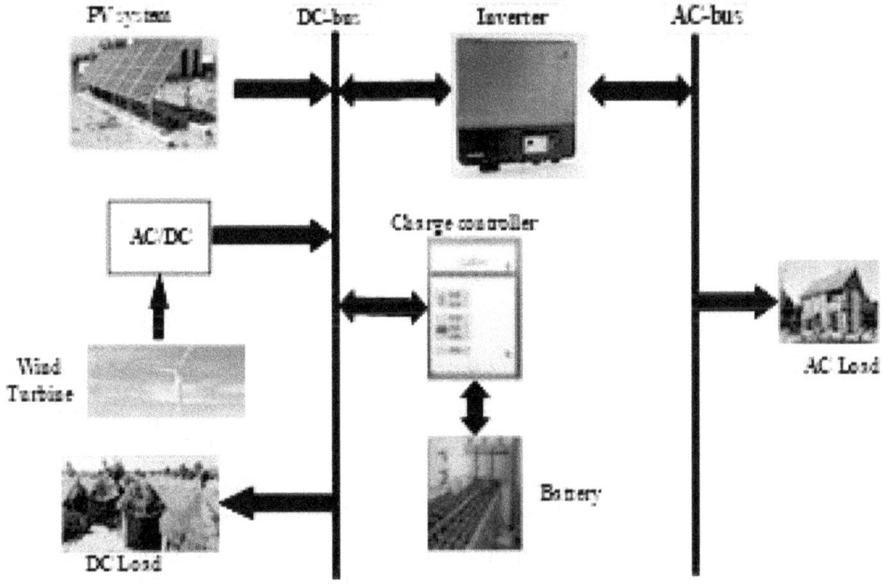

Optimization Problem

A key goal of this investigation is to reduce the device's annual cost (ACS). The components are maintenance costs (C_m), resources (C_a), and substitution costs (C_r), and the mathematical expression for ACS is as seen on Eq.(19)-Eq.(23).,where provided the costs of the relevant wind turbines, PV modules, and battery banks are respectively signified by C_{wt}, C_{pv}, and C_{bat}, while the lifespan of the total considered system is represented by *n*. Battery life is given by LS_{bat} and *CRF* stands for capital recovery factor. The latter may be expressed as seen on Eq.(24) mathematically.

$$ACS = C_a + Cm + C_r \tag{19}$$

$$C_a = CRF*\left(C_{pv_{tot}} + N_{wt_{tot}} + C_{bat_{tot}}\right) \tag{20}$$

$$C_{pv_{tot}} = N_{pv} * C_{pv}S \tag{21}$$

$$C_{wt_{tot}} = N_{wt} * C_{wt} \tag{22}$$

$$C_{bat_{tot}} = \left(\frac{n}{LS_{bat}}\right) N_{bat} * C_{bat} \tag{23}$$

$$CRF = \frac{i(1+i)^n}{(1+i)^n - 1} \tag{24}$$

$$\begin{cases} 0 \le N_{pv} \le N_{pvmax} \\ 1 \le N_{wt} \le N_{wtmax} \\ 0 \le N_{bat} \le N_{batmax} \end{cases} \tag{25}$$

For the case study considered here, a house in Ghiryan, located in a mountainous environment in Libya, has been chosen as a model. Figures 5 and 6 provide illustrations of data on solar irradiation and wind velocity the reference for data form the Centre for Solar Energy Research and Studies in Tripoli, Libya. Table 8 shows the operational expenses of the considered wind turbines, PV panels, and batteries in terms of technical, investment, and maintenance costs. We optimize this system with and without considering the reliability model to determine the CSA's effectiveness for the scale of hybrid renewable systems. We compared those results to the calculated values obtained with the aid of PSO software. Robustness, performance, and convergence were all calculated. The results obtained using the PSO and CSA algorithms are shown in Table 9. The crow algorithm clearly outperformed other optimization strategies. Table 10 demonstrates this in more detail. The cost of PV power generation is lower than that of wind power generation and the relevant limitation may be expressed as seen on Eq.(25)

As seen on Table 11, The able to conclude that the hybrid system offers considerable advantages for both power generation system design and system optimization. Furthermore, the Crow is demonstrated to be a more highly effective algorithm that allows us to obtain cost minimization.

Figure 5. Ghiryan's Annual Average Solar Irradiation Data

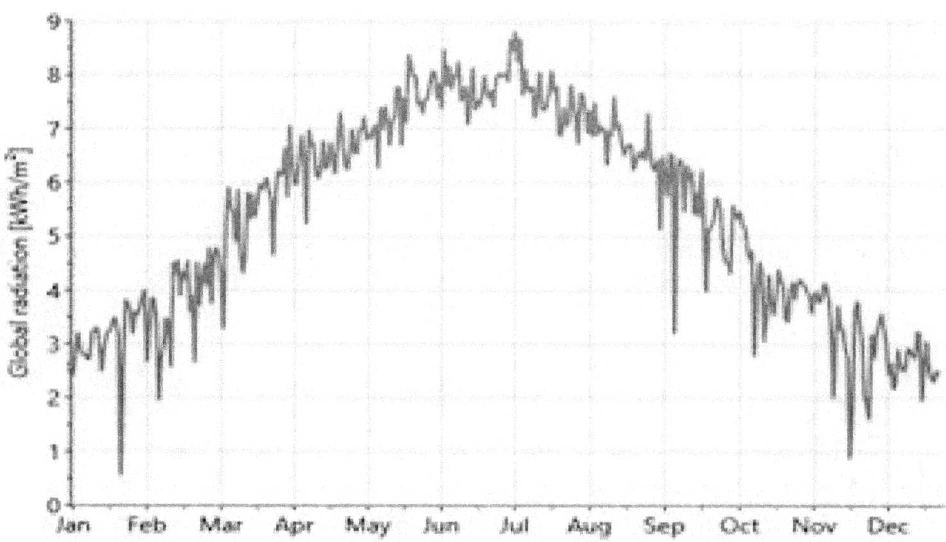

Figure 6. Ghiryan's Average Wind Speeds

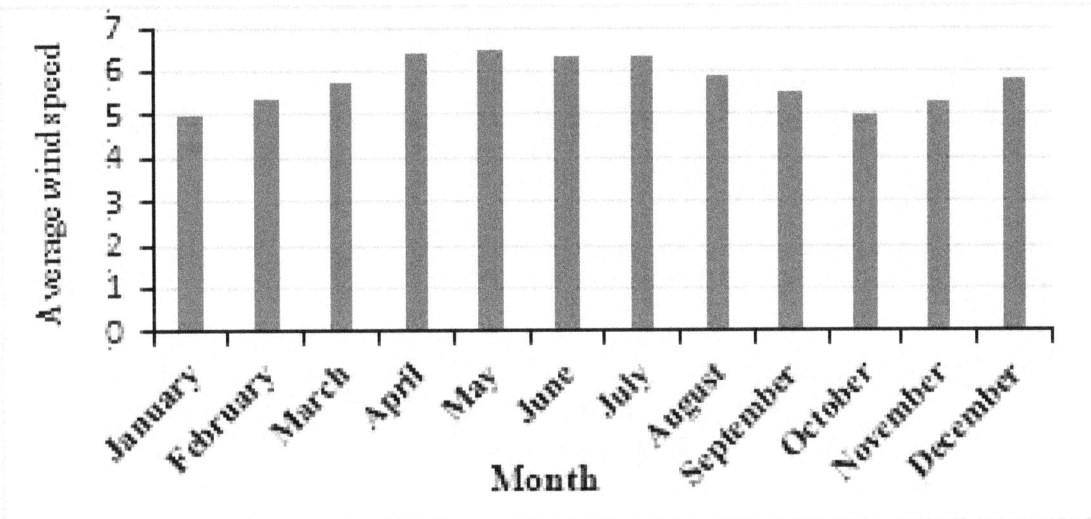

Table 8. Load Estimates for Electrical Equipment in the House

	Cost ($)	Parameters	Maintenance costs ($)
Wind turbine	1690	500 W	16.9
PV panel	260	250 W	2.6
Battery	300	600 Ah	29.6

Table 9. The Performances of CAS and PSO

	PSO	Crow
N_{pv}	22	22
N_w	0	0
N_{bat}	12	10
ACS ($)	14881.65	14458

Table 10. The Performances of PSO and CAS Based on Eq.(25)

	PSO			Crow		
Availability (%)	100	95	90	100	95	90
N_{pv}	22	22	22	22	22	22
N_w	0	0	0	0	0	0
N_{bat}	11	10	10	10	9	9
ACS ($)	14881	14060	13420	14458	13930	13100

Table 11. The compare between of PSO and Crow

	PSO	Crow
N_w	1	1
N_{pv}	22	22
N_{bat}	11	9
ACS ($)	25188.12	24571

CONCLUSION

This chapter has sought to offer a comprehensive review and thorough comparison of the most important size optimization methods being used by researchers for applications in standalone solar and wind-based hybrid energy systems. Among various hybrid energy options, the most preferred combination for use in the context of islands or remote/rural areas is the type of hybrid system based on solar, wind, diesel generator, and battery storage because this combination of resources can provide a continuous power supply with high levels of reliability. However, in such a hybrid system, to reduce costs while ensuring both reliability and social acceptance, it is imperative to identify the optimum sizing of every element.

To better illustrate these dynamics, a case study was presented here using the Crow algorithm for the installation of such a hybrid system in Ghiryan, Libya, in a manner that is simultaneously economically and technically viable. For this case study, we gathered long-term data on solar radiation and wind speed, and we applied PSO and the Crow algorithm separately in search of optimal numbers of wind turbines, batteries, and PV panels. The outcomes of the two techniques were then compared, and it was evident that the nature-inspired Crow, which is based on the behavior of crows in the wild, produced higher accuracy with simpler computations while lowering expenses. The implementation of a dependability model has an impact on prices, appropriate sizes, and loads, according to these studies. Increases in inverter efficiency can result in lower system costs while also enhancing the reliability of the provided load, as shown in the example. In addition, we've shown how the Crow algorithm can help avoid local minima and can be used as a dependable optimization tool.

In this chapter, a diverse array of assessment characteristics including economic, reliability, environmental, and social criteria have also been presented and summarized for use in efforts to resolve the sizing optimization problems of freestanding hybrid solar and wind energy systems. The aim in choosing among these various factors is to find the best combination for application in the context of standalone solar and wind systems. Furthermore, such size optimization challenges are also influenced by metrological data and load profile. The use of anticipated solar, wind, and load profile data in optimization problems will yield improved size optimization results in comparison to a reliance on historical data according to the findings presented in this chapter.

ACKNOWLEDGMENT

We would like to thank the Centre for Solar Energy Research and Studies in Tripoli, Libya, for supports regarding energy consumption and power data.

REFERENCES

Abu-mouti, F. S. (2011). *Sizing in Distribution Systems via Artificial Bee Colony Algorithm.* Academic Press.

Ahmadi, S., & Abdi, S. (2016). Application of the Hybrid Big Bang – Big Crunch algorithm for optimal sizing of a stand-alone hybrid PV / wind / battery system. *Solar Energy, 134*, 366–374. doi:10.1016/j.solener.2016.05.019

Ahmed, N., Bilal, A., Mahmood, A., Ieee, M., Razzaq, S., Zafar, A., ... Sidhu, S. (2015). Combined emission economic dispatch of power system including solar photo voltaic generation. *Energy Conversion and Management, 92*, 82–91. doi:10.1016/j.enconman.2014.12.029

Ahn, S.-H., Lee, K.-T., Bhandari, B., Lee, G.-Y., Lee, C. S.-Y., & Song, C.-K. (2012). Formation strategy of renewable energy sources for high mountain off-grid system considering sustainability. *Journal of the Korean Society for Precision Engineering, 29*(9), 958–963. doi:10.7736/KSPE.2012.29.9.958

Arabali, A., Member, S., & Ghofrani, M. (2014). *Stochastic Performance Assessment and Sizing for a Hybrid Power System of Solar / Wind / Energy Storage.* Academic Press.

Arasi, S. M., & Sasiraja, R. M. (2015). *Optimal location of DG units with exact size for the improvement of voltage stability using SLPSO.* Academic Press.

Arutyunov, V.S., Lisichkin, V. (n.d.). *Energy resources of the 21st century : problems and forecasts . Can renewable energy sources replace fossil fuels?* doi:10.1070/RCR4723

Ashok, S. (2007). *Optimised model for community-based hybrid energy system.* doi:10.1016/j.renene.2006.04.008

Askarzadeh, A. (2013). ScienceDirect A discrete chaotic harmony search-based simulated annealing algorithm for optimum design of PV / wind hybrid system. *Solar Energy, 97*, 93–101. doi:10.1016/j.solener.2013.08.014

Augustine, N., Member, S., Suresh, S., Moghe, P., & Sheikh, K. (n.d.). *Economic Dispatch for a Microgrid Considering Renewable Energy Cost Functions.* Academic Press.

Barton, J. P., & Infield, D. G. (2004). Intermittent. *Renewable Energy, 19*(2), 441–448.

Benasla, L., Belmadani, A., & Rahli, M. (2014). Electrical Power and Energy Systems Spiral Optimization Algorithm for solving Combined Economic and Emission Dispatch. *International Journal of Electrical Power & Energy Systems, 62*, 163–174. doi:10.1016/j.ijepes.2014.04.037

Beshr, E. (2013). *Comparative study of adding PV/wind energy systems to autonomus micro grid.* Academic Press.

Bhandari, B., Lee, K. T., Cho, Y. M., Lee, C. S., Song, C. K., & Ahn, S. H. (2013). Hybridization of Multiple Renewable Power Sources for Remote Village Electrification. *Proc. of the International Symposium on Green Manufacturing and Applications.*

Bhandari, B., Lee, K.-T., Lee, C. S., Song, C.-K., Maskey, R. K., & Ahn, S.-H. (2014). A novel off-grid hybrid power system comprised of solar photovoltaic, wind, and hydro energy sources. *Applied Energy*, *133*, 236–242. doi:10.1016/j.apenergy.2014.07.033

Bhattacharya, A., & Chattopadhyay, P. K. (2011). Solving economic emission load dispatch problems using hybrid differential evolution. *Applied Soft Computing*, *11*(2), 2526–2537. doi:10.1016/j.asoc.2010.09.008

Borges, C. L. T., & Falca, D. M. (2006). *Optimal distributed generation allocation for reliability, losses, and voltage improvement.* doi:10.1016/j.ijepes.2006.02.003

Cai, N., Member, S., Thi, N., & Nga, T. (2012). *Economic Dispatch in Microgrids Using Multi-Agent System*. Academic Press.

Celik, A. N. (2002). *Optimisation and techno-economic analysis of autonomous photovoltaic – wind hybrid energy systems in comparison to single photovoltaic and wind systems.* Academic Press.

Chak, D., & Bengal, W. (1990). *Energy consumption and prospects for renewable energy technologies in an Indian village*. Academic Press.

Chen, G., & Ding, X. (2015). *Optimal economic dispatch with valve loading effect using self-adaptive firefly algorithm*. doi:10.1007/s10489-014-0593-2

Connolly, D., Lund, H., Mathiesen, B. V., & Leahy, M. (2010). A review of computer tools for analysing the integration of renewable energy into various energy systems. *Applied Energy*, *87*(4), 1059–1082. doi:10.1016/j.apenergy.2009.09.026

Da Rosa, A. V., & Ordonez, J. C. (2021). *Fundamentals of renewable energy processes*. Academic Press.

Dawoud, S. M., Xiangning, L., Jinwen, S., Okba, M. I., Khalid, M. S., & Waqar, A. (2015).. . *Feasibility Study of Isolated PV-Wind Hybrid System in Egypt.*, *1093*, 145–151. doi:10.4028/www.scientific.net/AMR.1092-1093.145

Dorigo, M., Birattari, M., & St, T. (2006). *Ant Colony Optimization*. Academic Press.

Drury, E., Denholm, P., Margolis, R., Drury, E., Denholm, P., & Margolis, R. (2011). *The Impact of Different Economic Performance Metrics on the Perceived Value of Solar Photovoltaics*. Academic Press.

Dufo-lo, R. (2007). *Optimization of control strategies for stand-alone renewable energy systems with hydrogen storage.* doi:10.1016/j.renene.2006.04.013

Dufo-lo, R. (2009). *Simulation and optimization of stand-alone hybrid renewable energy systems.* doi:10.1016/j.rser.2009.01.010

El-Keib, A. A., Ma, H., & Hart, J. L. (1994). Environmentally Constrained Economic Dispatch Using the La Grangian Relaxation Method. *IEEE Transactions on Power Systems*, *9*(4), 1723–1729. doi:10.1109/59.331423

Elbaz, A., & Guneser, M. T. (1848). Algorithms to Model and Optimize a Stand-Alone Photovoltaic-Diesel-Battery System. *An Application in Rural Libya.*, *3651*, 523–529.

Elbaz, A., & Güneşer, M. T. (2020). Optimal Sizing of a Renewable Energy Hybrid System in Libya Using Integrated Crow and Particle Swarm Algorithms. Advance online publication. doi:10.25046/aj060130

Energy, G. (2020). *Operation Characteristics of Renewable Energy Sources*. Academic Press.

Engelbrecht, A. P. (2007). *Computational intelligence: an introduction*. John Wiley & Sons. doi:10.1002/9780470512517

Esmail, M., Golshan, H., & Arefifar, S. A. (2007). *Optimal allocation of distributed generation and reactive sources considering tap positions of voltage regulators as control variables*. Academic Press. doi:10.1002/etep

Esmat, A., & Eibakly, A. M. (2013). *A Novel Energy Management System using Ant Colony Optimization for Micro-grids*. Academic Press.

Eusuff, M., Lansey, K., & Pasha, F. (2007). *Shuffled frog-leaping algorithm : a memetic meta- heuristic for discrete optimization Shuffled frog-leaping algorithm : A memetic meta-heuristic for.* doi:10.1080/03052150500384759

Falaghi, H., Member, S., Haghifam, M., & Member, S. (2007). *ACO Based Algorithm for Distributed Generation Sources Allocation and Sizing in Distribution Systems*. Academic Press.

Fathima, A. H., & Palanisamy, K. (2015). Optimization in microgrids with hybrid energy systems – A review. *Renewable & Sustainable Energy Reviews*, *45*, 431–446. doi:10.1016/j.rser.2015.01.059

Ganguly, S., Sahoo, N. C., & Das, D. (n.d.). *Multi-Objective Planning of Electrical Distribution Systems using Particle Swarm Optimization*. Academic Press.

Geem, Z. W., & Kim, J. H. (2001). *A New Heuristic Optimization Algorithm : Harmony Search*. Academic Press.

Golshan, M. E. H., & Arefifar, S. A. (n.d.). *Distributed generation, reactive sources and network-configuration planning for power and energy-loss reduction*. doi:10.1049/ip-gtd

Gómez, M., López, A., & Jurado, F. (2010). Optimal placement and sizing from standpoint of the investor of Photovoltaics Grid-Connected Systems using Binary Particle Swarm Optimization. *Applied Energy*, *87*(6), 1911–1918. doi:10.1016/j.apenergy.2009.12.021

Gupta, A., Saini, R. P., & Sharma, M. P. (2006). *Optimised Application of Hybrid Renewable Energy System in Rural Electrification*. Academic Press.

Gupta, G. P., & Jha, S. (2018). Engineering Applications of Artificial Intelligence Integrated clustering and routing protocol for wireless sensor networks using Cuckoo and Harmony Search based metaheuristic techniques. *Engineering Applications of Artificial Intelligence, 68*(September), 101–109. doi:10.1016/j.engappai.2017.11.003

Güvenç, U., Sönmez, Y., Duman, S., & Yörükeren, N. (2012). Sharif University of Technology Combined economic and emission dispatch solution using gravitational search algorithm. *Scientia Iranica*, *19*(6), 1754–1762. doi:10.1016/j.scient.2012.02.030

Hennet, J. C., & Samarakou, M. T. (1986). Optimization of a combined wind and solar power plant. *International Journal of Energy Research, 10*(2), 181–188. doi:10.1002/er.4440100208

Hrayshat, E. S. (2009). Energy for Sustainable Development Techno-economic analysis of autonomous hybrid photovoltaic-diesel-battery system. *ESD, 13*(3), 143–150. doi:10.1016/j.esd.2009.07.003

Jevtic, M., Jovanovic, N., Radosavljevic, J., & Klimenta, D. (2017). *Moth Swarm Algorithm for Solving Combined Economic and Emission Dispatch Problem*. Academic Press.

Jyoti, R., & Raju, A. B. (2011). *Economic Analysis and Comparison of Proposed HRES for Stand-Alone Applications at Various Places in Karnataka State.*

Kaldellis, J. K., Kapsali, M., Kaldelli, E., & Katsanou, E. (2020). Comparing recent views of public attitude on wind energy, photovoltaic and small hydro applications. *Renewable Energy, 52*, 197–208. doi:10.1016/j.renene.2012.10.045

Kalogirou, S. A. (2004). *Optimization of solar systems using artificial neural-networks and genetic algorithms.* doi:10.1016/S0306-2619(03)00153-3

Kansal, S., Kumar, V., & Tyagi, B. (2013). Electrical Power and Energy Systems Optimal placement of different type of DG sources in distribution networks. *International Journal of Electrical Power & Energy Systems, 53*, 752–760. doi:10.1016/j.ijepes.2013.05.040

Katsigiannis, Y. A., Georgilakis, P. S., Member, S., & Karapidakis, E. S. (2012). *Hybrid Simulated Annealing – Tabu Search Method for Optimal Sizing of Autonomous Power Systems With Renewables.* Academic Press.

Khare, A., & Rangnekar, S. (2013). Review article A review of particle swarm optimization and its applications in Solar Photovoltaic system. *Applied Soft Computing, 13*(5), 2997–3006. doi:10.1016/j.asoc.2012.11.033

Khatib, T., Mohamed, A., & Sopian, K. (2012). Optimization of a PV / wind micro-grid for rural housing electrification using a hybrid iterative / genetic algorithm : Case study of Kuala Terengganu, Malaysia. *Energy and Building, 47*, 321–331. doi:10.1016/j.enbuild.2011.12.006

Kuang, Y., Zhang, Y., Zhou, B., Li, C., Cao, Y., Li, L., & Zeng, L. (2016). A review of renewable energy utilization in islands. *Renewable & Sustainable Energy Reviews, 59*, 504–513. doi:10.1016/j.rser.2016.01.014

Learning, M., & Publishers, K. A. (1988). *Genetic Algorithms and Machine Learning*. Academic Press.

Liao, G. (2012). Electrical Power and Energy Systems Solve environmental economic dispatch of Smart MicroGrid containing distributed generation system – Using chaotic quantum genetic algorithm. *International Journal of Electrical Power & Energy Systems, 43*(1), 779–787. doi:10.1016/j.ijepes.2012.06.040

Ma, Y., Yang, P., Guo, H., & Wu, J. (2012). Power source planning of wind-PV-biogas renewable energy distributed generation system. *Power System Technology, 9*(624), 1.

Mahesh, A., & Sandhu, K. S. (2015). *Hybrid wind / photovoltaic energy system developments : Critical review and findings.* doi:10.1016/j.rser.2015.08.008

Maleki, A., Gholipour, M., & Ameri, M. (2016). Electrical Power and Energy Systems Optimal sizing of a grid independent hybrid renewable energy system incorporating resource uncertainty, and load uncertainty. *International Journal of Electrical Power & Energy Systems*, *83*, 514–524. doi:10.1016/j.ijepes.2016.04.008

Mellit, A., Kalogirou, S. A., & Drif, M. (2010). Application of neural networks and genetic algorithms for sizing of photovoltaic systems. *Renewable Energy*, *35*(12), 2881–2893. doi:10.1016/j.renene.2010.04.017

Menanteau, P., Finon, D., & Lamy, M. (2010). *Prices versus quantities : choosing policies for promoting the development of renewable energy*. Academic Press.

Moravej, Z., & Akhlaghi, A. (2013). Electrical Power and Energy Systems A novel approach based on cuckoo search for DG allocation in distribution network. *International Journal of Electrical Power & Energy Systems*, *44*(1), 672–679. doi:10.1016/j.ijepes.2012.08.009

Nadjemi, O., Nacer, T., Hamidat, A., & Salhi, H. (2017). Optimal hybrid PV / wind energy system sizing : Application of cuckoo search algorithm for Algerian dairy farms. *Renewable & Sustainable Energy Reviews*, *70*, 1352–1365. doi:10.1016/j.rser.2016.12.038

Optimal placement of multi-distributed generation units including different load models using particle swarm optimization. (2011). *Swarm and Evolutionary Computation*, *1*(1), 50–59. doi:10.1016/j.swevo.2011.02.003

Ouyang, H., Gao, L., Li, S., Kong, X., Wang, Q., & Zou, D. (2017). Improved Harmony Search Algorithm : LHS. *Applied Soft Computing*, *53*, 133–167. doi:10.1016/j.asoc.2016.12.042

Özyön, S., & Yas, C. (2014). *Electrical Power and Energy Systems Artificial bee colony algorithm with dynamic population size to combined economic and emission dispatch problem.* doi:10.1016/j.ijepes.2013.06.020

Paliwal, P., Patidar, N. P., & Nema, R. K. (2014). Planning of grid integrated distributed generators : A review of technology, objectives and techniques. *Renewable & Sustainable Energy Reviews*, *40*, 557–570. doi:10.1016/j.rser.2014.07.200

Pandi, V. R., Zeineldin, H. H., & Xiao, W. (2013). *Distributed Generation Resources Considering Harmonic and Protection Coordination Limits*. Academic Press.

Parvez, F., Vasant, P., Kallimani, V., & Watada, J. (2018). A holistic review on optimization strategies for combined economic emission dispatch problem. *Renewable and Sustainable Energy Reviews*, *81*(July), 3006–3020. doi:10.1016/j.rser.2017.06.111

Pedregal, P. (2004). Γ-convergence through Young measures. *SIAM Journal on Mathematical Analysis*, *36*(2), 423–440. doi:10.1137/S0036141003425696

Piarehzadeh, H., Khanjanzadeh, A., & Pejmanfer, R. (2012). Comparison of harmony search algorithm and particle swarm optimization for distributed generation allocation to improve steady state voltage stability of distribution networks. *Research Journal of Applied Sciences, Engineering and Technology*, *4*(15), 2310–2315.

Rao, R. S., Ravindra, K., Satish, K., & Narasimham, S. V. L. (2013). *Power Loss Minimization in Distribution System Using Network Recon figuration in the Presence of Distributed Generation.* Academic Press.

Saini, R. P., & Sharma, M. P. (2010). Integrated renewable energy systems for off grid rural electrification of remote area. *Renewable Energy*, *35*(6), 1342–1349. doi:10.1016/j.renene.2009.10.005

Sarkhel, R., Das, N., Saha, A. K., & Nasipuri, M. (2018). Engineering Applications of Artificial Intelligence An improved Harmony Search Algorithm embedded with a novel piecewise opposition based learning algorithm. *Engineering Applications of Artificial Intelligence*, *67*(March), 317–330. doi:10.1016/j.engappai.2017.09.020

Sedighizadeh, M., & Sadighi, M. (n.d.). *A Particle Swarm Optimization for Sitting and Sizing of Distributed Generation in Distribution Network to Improve Voltage Profile and Reduce THD and Losses.* Academic Press.

Shaaban, M. F., Member, S., Atwa, Y. M., Member, S., El-saadany, E. F., & Member, S. (2013). *DG Allocation for Bene fi t Maximization in Distribution Networks.* Academic Press.

Siddaiah, R., & Saini, R. P. (2016). A review on planning, con fi gurations, modeling and optimization techniques of hybrid renewable energy systems for off grid applications. *Renewable & Sustainable Energy Reviews*, *58*, 376–396. doi:10.1016/j.rser.2015.12.281

Singh, D., Singh, D., & Verma, K. S. (2007). *GA based Optimal Sizing & Placement of Distributed Generation for Loss Minimization.* Academic Press.

Singh, R. K., & Goswami, S. K. (2009). Optimum Siting and Sizing of Distributed Generations in Radial and Networked Systems. *Electric Power Components and Systems Optimum Siting and Sizing of Distributed Generations in Radial and Networked Systems Optimum Siting and Sizing of Distributed Generations in Radial and Networked Systems*, *5008*(2), 127–145. Advance online publication. doi:10.1080/15325000802388633

Singh, R. K., & Goswami, S. K. (2010). Electrical Power and Energy Systems Optimum allocation of distributed generations based on nodal pricing for profit, loss reduction, and voltage improvement including voltage rise issue. *International Journal of Electrical Power & Energy Systems*, *32*(6), 637–644. doi:10.1016/j.ijepes.2009.11.021

Sinha, S., & Chandel, S. S. (2014). Review of software tools for hybrid renewable energy systems. *Renewable & Sustainable Energy Reviews*, *32*, 192–205. doi:10.1016/j.rser.2014.01.035

Sinha, S., & Chandel, S. S. (2015). Review of recent trends in optimization techniques for solar photovoltaic – wind based hybrid energy systems. *Renewable & Sustainable Energy Reviews*, *50*, 755–769. doi:10.1016/j.rser.2015.05.040

Soroudi, A., Ehsan, M., & Zareipour, H. (2011). A practical eco-environmental distribution network planning model including fuel cells and non-renewable distributed energy resources. *Renewable Energy*, *36*(1), 179–188. doi:10.1016/j.renene.2010.06.019

Su, S., Lu, C., Chang, R., & Gutiérrez-alcaraz, G. (2011). *Distributed Generation Interconnection Planning : A Wind Power Case Study.* Academic Press.

Sulaiman, M. H., Mustafa, M. W., Azmi, A., Aliman, O., & Rahim, S. R. A. (2012). *Optimal Allocation and Sizing of Distributed Generation in Distribution System via Firefly Algorithm*. Academic Press.

Taghikhani, M. A. (2012). *DG Allocation and Sizing in Distribution Network Using Modified Shuffled Frog Leaping Algorithm*. Academic Press.

Taylor, P., Balamurugan, R., & Subramanian, S. (2007). *Electric Power Components and Systems A Simplified Recursive Approach to Combined Economic Emission Dispatch*. doi:10.1080/15325000701473742

Trivedi, I. N., Jangir, P., Bhoye, M., & Jangir, N. (2018). An economic load dispatch and multiple environmental dispatch problem solution with microgrids using interior search algorithm. *Neural Computing & Applications*, *30*(7), 2173–2189. doi:10.100700521-016-2795-5

Upadhyay, S., & Sharma, M. P. (2014). A review on con fi gurations, control and sizing methodologies of hybrid energy systems. *Renewable & Sustainable Energy Reviews*, *38*, 47–63. doi:10.1016/j.rser.2014.05.057

Wang, L., Member, S., & Singh, C. (2008). *Reliability-Constrained Optimum Placement of Reclosers and Distributed Generators in Distribution Networks Using an Ant Colony System Algorithm*. Academic Press.

Wu, Y., & Chang, S. (2016). *Review of the Optimal Design on a Hybrid Renewable Energy System 2 Case Studies on Small-Scale Hybrid Generation Systems*. Academic Press.

Yammani, C., Maheswarapu, S., & Matam, S. K. (2016). *Optimal placement and sizing of distributed generations using shuffl ed bat algorithm with future load enhancement*. Academic Press. doi:10.1002/etep

Yang, X. (2010). *Firefly algorithm, stochastic test functions and design optimisation*. Academic Press.

Yang, X., Deb, S., & Behaviour, A. C. B. (2009). *Cuckoo Search via Levy Flights*. Academic Press.

Yustra, M. A., & Soeprijanto, A. (2012). Optimal distributed generation (DG) allocation for losses reduction using improved particle swarm optimization (IPSO) method. *Journal of Basic and Applied Scientific Research*, *2*(7), 7016–7023.

Zangeneh, A., Jadid, S., & Rahimi-kian, A. (2009). Promotion strategy of clean technologies in distributed generation expansion planning. *Renewable Energy*, *34*(12), 2765–2773. doi:10.1016/j.renene.2009.06.018

Zhao, B. (2006). *A Survey on Application of Swarm Intelligence Computation to Electric Power System*. Academic Press.

Zhu, Z. (2008). *Computer vision research progress*. Nova Publishers. doi:10.5772/72

Chapter 9
Multi-Objective Optimal Performance of a Hybrid CPSD-SE/HWT System for Microgrid Power Generation

Bashar Shboul
The University of Sheffield, UK & Al al-Bayt University, Jordan

Godfrey T. Udeh
The University of Sheffield, UK & University of Port Harcourt, Nigeria

Ismail Al-Arfi
The University of Sheffield, UK

Lin Ma
The University of Sheffield, UK

Stavros Michailos
The University of Sheffield, UK

Kevin Hughes
The University of Sheffield, UK

Derek Ingham
The University of Sheffield, UK

Mohamed Pourkashanian
The University of Sheffield, UK

ABSTRACT

A new integrated hybrid solar thermal and wind-based microgrid power system is proposed. It consists of a concentrated parabolic solar dish Stirling engine, a wind turbine, and a battery bank. The electrical power curtailment is diminished, and the levelised cost of energy is significantly reduced. To achieve these goals, the present study conducts a dynamic performance analysis over one year of operation. Further, a multi-objective optimisation model based on a genetic algorithm is implemented to optimise the techno-economic performance. The MATLAB/Simulink® software was used to model the system, study the performance under various operating conditions, and optimise the proposed hybrid system. Finally, the model has been implemented for a specific case study in Mafraq, Jordan. The system satisfies a net power output of 1500 kWe. The developed model has been validated using published results. In conclusion, the obtained results reveal that the optimised model of the microgrid can substantially improve the overall efficiency and reduce the levelised cost of electricity.

DOI: 10.4018/978-1-7998-8561-0.ch009

INTRODUCTION

Nowadays, small-scale decentralised distributed generation systems are on track to become the foundation of the worldwide energy network and it is a promising alternative solution to the typical large-scale centralised power plants. Consequently, microgrid systems comprising of a hybrid renewable energy system (HRES) can play a significant role in satisfying the energy demands of remote regions and resolving the energy price inflation and environmental problems posed by the use of fossil fuels in energy production (Motevasel et al., 2013). Therefore, microgrids with HRES would help to eliminate over a hundred million tonnes of CO_2 emissions from the atmosphere each year by providing a reliable, sustainable, and cost-competitive renewable energy supply, thus meeting the global energy needs without compromising the planet's well-being. For instance, integration of HRES with energy storage systems, such as batteries and traditional power systems, such as a boiler or diesel engine into microgrid is considered the popular way of increasing the reliability to meet the energy demand (Belfkira et al., 2011; Kumar et al., 2013), as depicted in Figure 1 (Kabalci, 2021). In consequence, the HRES microgrids are more reliable and economical when compared to the single renewable energy system (Dufo-López et al., 2011).

Figure 1. Microgrid with its multiple demand and source connections (Kabalci, 2021).

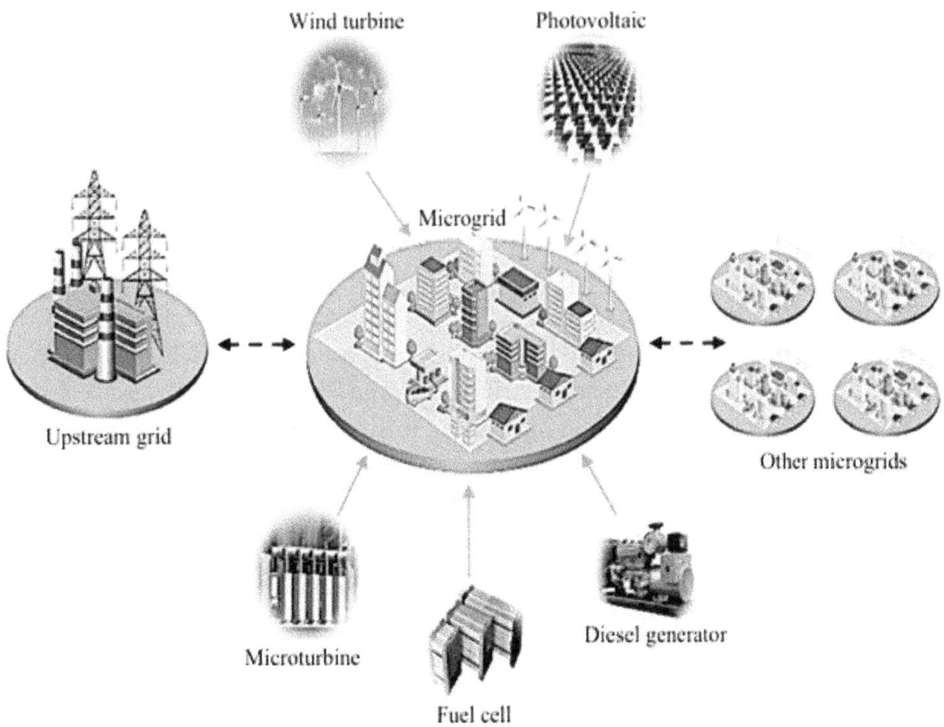

Solar and wind energy resources are widely used in microgrids. Nevertheless, these resources need proper management in order to facilitate their power operations to mitigate the implications of the intermittent output. To fully utilise this process, the sizing optimisation methodologies should be applied by a suitable selection of the governing parameters to obtain an optimal hybrid system design. Therefore, the optimised system will be economical, efficient and reliable.

Generally, it has been found that two generic optimisation techniques have been adopted in the design of microgrid systems: heuristic and classical techniques (Singh et al., 2016). The former technique is employed once long-term weather data for a given location is known but this is not always available, whilst the latter is typically employed in the design of microgrids if sufficient information of long-term weather data for the design location is available (Badwawi et al., 2015). Traditional or classical optimisation methods that are deployed in the microgrids design are an iterative approach, probabilistic approach, graphical construction methods, linear programming, trade-off method, and the least square method (Khan et al., 2018), (Rojas-Zerpa & Yusta, 2015). Various heuristic approaches have been employed in the techno-economic design optimisation of microgrid systems, such as genetic algorithm (GA), particle swarm optimisation (PSO), simulated annealing (SA), ant colony algorithms (ACA), bacterial foraging algorithm (BFO), artificial bee colony algorithm (ABC), biogeography-based optimisation (BBO), and artificial neural networks (ANN) (Bernal-Agustín et al., 2006; Cristóbal-Monreal & Dufo-López, 2016; Diaf et al., 2008; Khan et al., 2018; Olatomiwa et al., 2016; Singh et al., 2016). Erdinc and Uzunoglu (Erdinc & Uzunoglu, 2012) elaborate various heuristic optimisation techniques reported in the literature that contribute significantly to size hybrid microgrid systems, such as GA, PSO, SA.

Numerous studies have been extensively investigated in the literature to design microgrid systems for remote regions over the world. Traditionally, developing mathematical models of each subsystem and subsequently applying optimisation methods to size the entire system is adopted to design microgrid systems. In this regard, Halabi and Mekhilef (Halabi & Mekhilef, 2018) summarised the commonly used techniques in optimising HRES. Petrescu et al. (Petrescu et al., 2010) have investigated an optimisation method of a solar Stirling engine (SE) power plant to supply the required electrical energy demand for residential buildings using two sources, which are the parabolic dish mirror and the hydrogen/oxygen fuel cell. In particular, an extensive review of optimisation methods employed in the design of hybrid solar-wind systems has been performed but, in particular, the Photovoltaic (PV)/wind system has been overlooked in (Singh et al., 2016), (Badwawi et al., 2015). To obtain more reliability, a tri-hybridisation was implemented by Su Guo et al. (Guo et al., 2020). They proposed a new hybrid PV/Wind/thermal energy storage (TES) power system with an electric heater. The PSO algorithm was utilised to minimise the levelised cost of electricity (LCOE) and maximise the utilisation rate of transmission channels. In addition, Solar energy-driven multigeneration systems coupled with other renewable energies have been investigated comprehensively in Ref. (Mohammadi et al., 2020). They recommended, in the future, an investigation for such as the optimisation of solar energy-driven multigeneration systems.

For the hybrid systems that consist of concentrated solar power (CSP) and wind, there have been several attempts to optimise the hybrid CSP/wind system with the aim of minimising the power supply curtailment as the wind and solar energy usually do not peak simultaneously. To illustrate this isue, Yang et al. (Y. Yang et al., 2018) proposed a new hybrid system that includes CSP/wind/electric heater being employed with TES. This hybrid system is designed to optimise the profit under technical limitations as a mixed-integer linear programming problem. The proposed method has substantially reduced the wind curtailment by more than 90% over 151 days. Zeyu Ding et al. (Ding et al., 2019) have developed an optimisation technique based on a PSO algorithm to find the optimal design of the hybrid CSP/wind system coupled to TES. Such CSP/wind hybridisations are also seen in a recent study in the literature, Keyif et al. (Keyif et al., 2020). Keyif (Keyif et al., 2020) performed a non-linear optimisation model measuring the critical component investment costs and operational flexibility in the plant configuration.

The GA method is widely used in microgrid scheduling optimisation to investigate the optimum operating parameters. It is essential to combine the entire simulation model with a suitable optimisation method. Yang Hongxing et al. (H. Yang et al., 2009), (H. Yang et al., 2008) introduced a multi-objective GA method for identifying the stand-alone hybrid PV/wind system optimum configuration with the lowest cost based on the power supply probability (LPSP) and the annualised cost of the system (ACS). This developed model was implemented to provide electricity for a telecommunication relay station along the southeast coast of China. Similarly, Bilal et al. (Ould Bilal et al., 2010) have adopted two principles, namely the minimisation of ACS and LPSP with the use of GA on the northern coast of Potou, Senegal. Koutroulis et al. (Koutroulis et al., 2006) presented a minimum cost objective optimisation-based GA methodology for the optimal sizing of autonomous PV/wind systems to supply power for a residential household. For desalination purposes, Koutroulis and Kolokotsa (Koutroulis & Kolokotsa, 2010) have applied a GA methodology based on the total cost function minimisation for the optimal sizing of the PV/wind generator. Daming Xu et al. (Xu et al., 2005) investigated the GA approach of the sizing stand-alone hybrid PV/wind power systems. The objective of sizing these systems is to reduce the total capital cost, subject to the constraint of the LPSP. Bakir and Kulaksiz (Bakir & Kulaksiz, 2020) optimised the gain parameters of four PI controllers for the hybrid microgrid PV/wind system, which was modelled in MATLAB/Simulink® to examine the voltage profiles at the output. Two optimal sizing algorithms are used in the analyses are the Bacteria Foraging Algorithm (BFA) and GA.

According to Tafreshi et al. (Tafreshi et al., 2010), a GA method was also developed in the MATLAB® toolbox to find the optimum configuration for the hybrid PV/wind/biogas system. Furthermore, Kalantar and Mousavi (Kalantar & Mousavi G., 2010) have performed a GA method based on an economic analysis, i.e. ACS for decentralised the hybrid PV/ wind/microturbine/lead-acid battery storage system. Lagorse et al. (Lagorse et al., 2009) carried out the optimal sizing method for the multisource tri-hybrid PV/wind/fuel cell using GA based on the LCOE and meteorological features of the installed region and the consumption behaviour.

Other studies have been conducted by using GA to examine the reliability and dispatchability of a hybrid PV plant with a CSP plant. Starke et al. (Starke et al., 2018) have implemented a multi-objective optimisation approach based on the GA for evaluating the optimal design for hybrid CSP, including a central receiver system and parabolic trough collectors, and PV plants in Chile. The three objective functions were considered in this analysis: LCOE, overall investment and capacity factor. Moreover, four variables were adopted in terms of the design variables, namely: the solar field size, thermal storage capacity, PV power ratio and PV tilt angle. In addition, Liu et al. (Liu et al., 2019) utilised GA-PSO to obtain a minimised LCOE of the hybrid CSP-PV employing TES.

Based on the literature review, it is clear that there exists a lack in the literature of studies dealing with the optimisation analysis of the CPSD-SE/HWT integrated solution. To fill this knowledge gap, in this work, a concentrated parabolic solar dish Stirling engine (CPSD-SE) and a horizontal axis wind turbine (HWT) are integrated to generate power for a low to medium scale microgrid application. The predicted power that is generated by the system is in the range of 100 kWe and 1500 kWe, and the system performance throughout one year has been investigated dynamically via rigorous modelling. In addition, a techno-economic sensitivity analysis has been carried out to study the performance of the integrated hybrid system under the meteorological data for the city of Mafraq, Jordan using MATLAB/Simulink®. The main aim of the study is to carry out a post-design analysis of the new hybrid CPSD-SE/HWT system and calculate the generated power and efficiency. Furthermore, a multi-objective optimisation-based GA approach has been applied in which the $LCOE_{tot}$ and the energy efficiency of the system are simultaneously optimised. Optimal configuration and operating conditions in dispatch strategies are discussed in this work. Therefore, the main contribution of this work is to optimise the new hybrid CPSD-SE/HWT for power generation.

METHODOLOGY AND PROCEDURE

In this work, the performance of the system based on the operating conditions is considered in Mafraq, Jordan as a case study. The data for the study region in Jordan is provided by the Solargis™ satellite-driven data. In the performance model, the efficiency and the generated power of the existing system are measured depending on the predesign analysis of a published study (Shboul et al., 2021a). Further, a battery bank has been used as a recovery source of power to overcome the fluctuations in the solar and wind energy generation. The batteries are employed under the control of environmental operating conditions. CO_2 is used as a primary working gas in the SE and the predicted generated power is in the range of 100 kWe and 1500 kWe.

In general, the performance modelling and optimisation of the CPSD-SE/HWT using MATLAB/Simulink® is used by implementing the following steps:

- Specify the input and outputs variables of each sub-system as well as the assumptions that have been considered in this work. Table 1 shows the input and outputs variables of the model as well as the assumptions that have been considered in this work.
- The overall mathematical model of each unit has been carried out by implementing the fundamental thermodynamic energy balance equations that govern the operation of each investigated technology. The mathematical model of the integrated system is presented in Tables 11, 12, and 13 in the Appendix.
- The models have been applied and solved in the MATLAB/Simulink® software based on the operating conditions, namely; the global solar radiation (GSR) and the average wind speed limit (AWS).
- The simulation results obtained by the proposed dynamic model of each sub-system is validated by comparing with the findings of previous related studies in the literature.
- To test the developed model and measure the dynamic performance of the system, the model has been implemented in Mafraq, Jordan, and the meteorological data is obtained from the Solargis™ toolbox.
- The multi-objective optimisation using GA methodology will be performed using Multi-objective GA Solver, which is built on the optimisation tool. The stepwise procedures for implementation of the GA optimisation in the MATLAB/Simulink® is shown in Figure 2.

Table 1. Data information related to the proposed system units

Input parameters			Output parameters		Assumptions		
Description	**Value**	**Unit**	**Description**	**Unit**	**Description**	**Value**	**Unit**
CPSD-SE							
Dish diameter	11	m	Top temperature	°C	Ambient temperature	25	°C
Receiver diameter	0.3	m	Engine efficiency	%	Atmospheric pressure	1.0132	bar
Number of dishes	60	-	Total efficiency	%	Receiver efficiency	80	%
working gas	CO_2		Engine and total powers	kWe	Solar flux limit	500-1000	W/m²
Rim angle	37	degree	Compression ratio	-	Operating hours	10	h
Engine piston diameter	5.5	cm	Pressure ratio	-			
Piston stroke	5	cm	Mean effective pressure,	bar			
Engine speed	1800	rpm	Top temperature	°C			
Number of cylinders	4	-	Engine efficiency	%			
Generator efficiency	95	%					
Mirrors efficiency	97	%					
HWT							
Air speed ratio	0.35	-	Wind speed at blades	m/s	Ambient temperature	25	°C
Rotor diameter	47	m	Power coefficient		Atmospheric pressure	1.0132	bar
Number of modules	15	-	Wind power	kW	Wind speed limit	1.5-25	m/s
Generator efficiency	97	%	Mechanical power	kW			
Rotational loss	10.1-12	%	Generator power	kWe			
Power factor lag	0.9	-	Total farm power	kWe			
Battery bank							
Depth of discharge	0.8	-	Total power	kWe	Operating hours	24	h
Battery voltage	80	V	Battery storage	kWh	Load voltage	200	V
Battery current	10	A			Number of cloudy days	2	day
Battery Efficiency	75	%					
Number of batteries	1900	-					

Figure 2. The GA implementation steps in MATLAB/Simulink®.

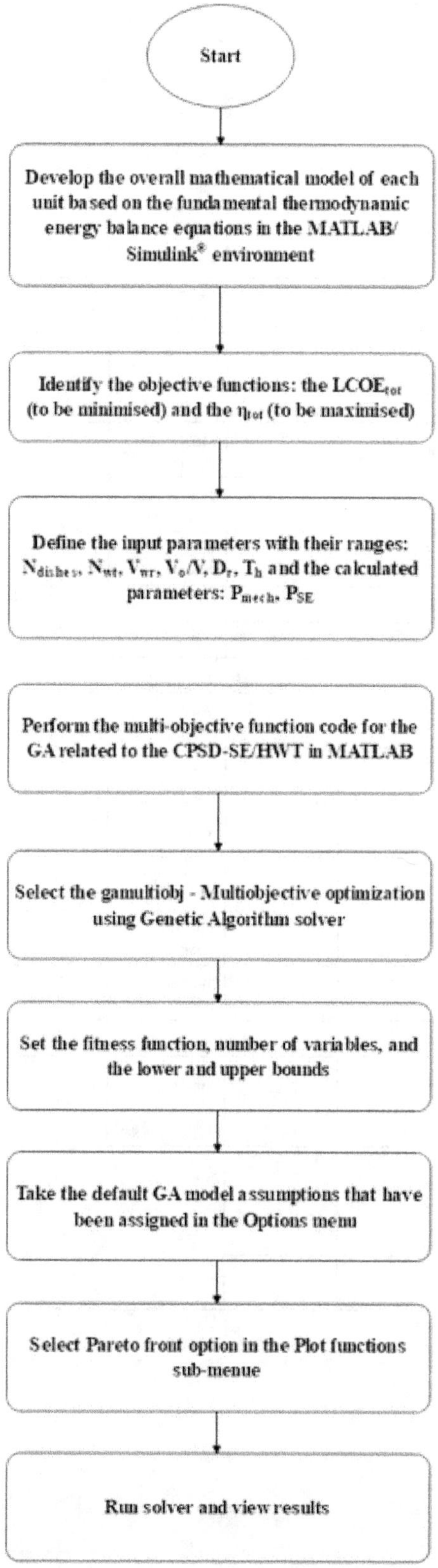

Site Location

The CPSD-SE system has economic feasibilities only for locations with direct normal irradiation (DNI) values higher than 5.5 kWh/m²/day or 2000 kWh/m²/year (Hirbodi et al., 2020). Jordan with an estimated average DNI of 2700 kWh/m² and approximately 300 clear sunny days per annum with 3311 hours and this is one of the most appropriate regions for the installation of CPSD-SE plants in the MENA region (The Middle East and North Africa). In addition, Jordan has a high annual average wind speed and this is higher than 7 m/s in some locations (with highs of 10 m/s) (H. H. Ali et al., 2020; Shboul et al., 2021). Consequently, the hybrid CPSD-SE/HWT is a suitable option for electricity production in Jordan. As shown in Figure 3, the northern region of Jordan, such as Mafraq with a daily average of DNI between 6.54 and 7.29 kWh/m² and an annual average wind speed of 4.72 m/s, has abundant wind speed and solar irradiations (*Global Solar Atlas*, 2020; *Global Wind Atlas*, 2020). Further, based on the study of Shboul et al. (Shboul et al., 2021), the city of Mafraq is one of the best regions in Jordan for exploiting the wind and solar potential as well as deploying hybrid CSP and wind plants. Thus, the city of Mafraq with latitude and longitude of 32.2° N and 36.84° E, respectively, has been selected as the case study location to investigate the stand-alone hybrid CPSD-SE/HWT plant in this research work.

Meteorological Data

In general, the climate conditions have a considerable influence on the performance and operation of microgrid power plants. The essential meteorological parameters to simulate the CPSD-SE/HWT plants in the adopted location by the MATLAB/Simulink® software are solar irradiation, wind speed, air temperature, relative humidity, and atmospheric pressure. In this study, these data were obtained from a newly developed ANN forecasting model for solar radiation and wind speed prediction based on the Solargis™ data (Shboul et al., 2021). This highly accurate model (error less than 3%) covers a wide time span of 20 years, and they have been validated to check the reliability and avoid errors in the results. The validated data have been used as input parameters in the MATLAB/Simulink® software.

Figure 3. Wind speed and DNI distribution in Mafraq (Global Solar Atlas, 2020; Global Wind Atlas, 2020).

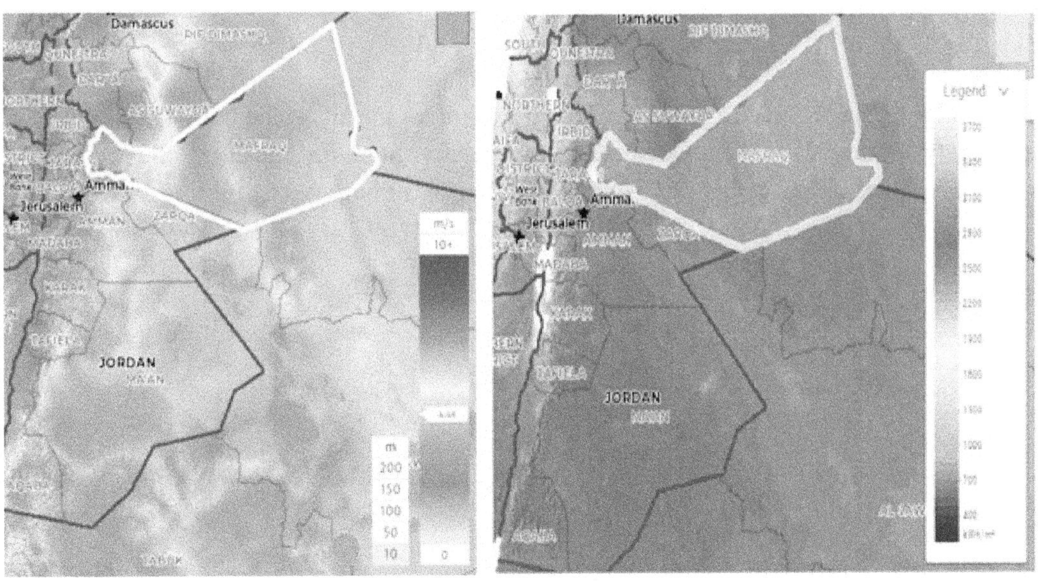

The monthly average of the meteorological data of Mafraq, including solar radiation, wind speed, air temperature, relative humidity, and atmospheric pressure, is indicated in Figures 4 and 5. As shown in Figure 4-a, the estimated solar data are the global horizontal irradiation (GHI), the DNI, and the diffuse horizontal irradiance (DIF). According to Figure 4-a, b, Mafraq city has the highest value of solar radiation (DNI and GHI) and wind speed and these occur in June and July, respectively. Also, it shows the lowest values of solar radiation (DNI and GHI) and the wind speed occur in January and December, respectively. In addition, the DIF meets its maximum and minimum values in June and December, respectively.

Figure 4. Monthly average data of (a) solar radiation and (b) wind speed.

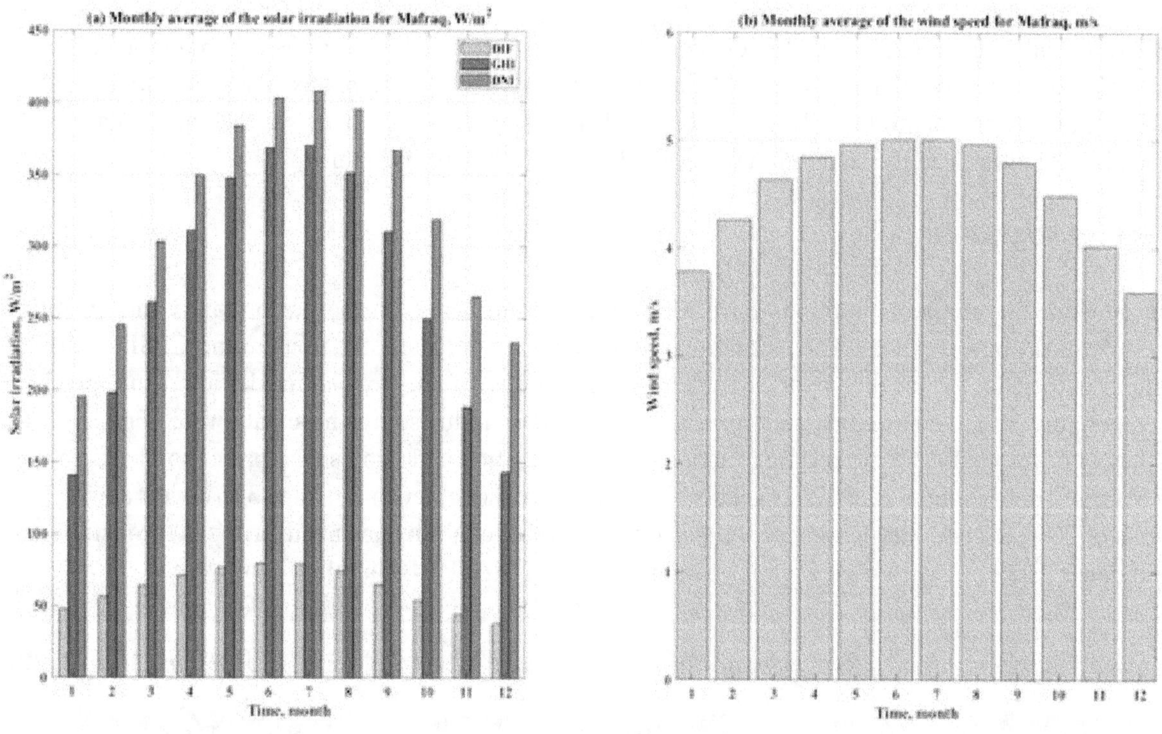

Based on Figure 5-a, the maximum and minimum values of the air temperature are in July and January, respectively. Figures. 5-b and 5-c illustrate that the relative humidity and atmospheric pressure reach their maximum values in December and January and their minimum values in June and July.

Figure 5. Monthly average of the meteorological data for Mafraq, Jordan.

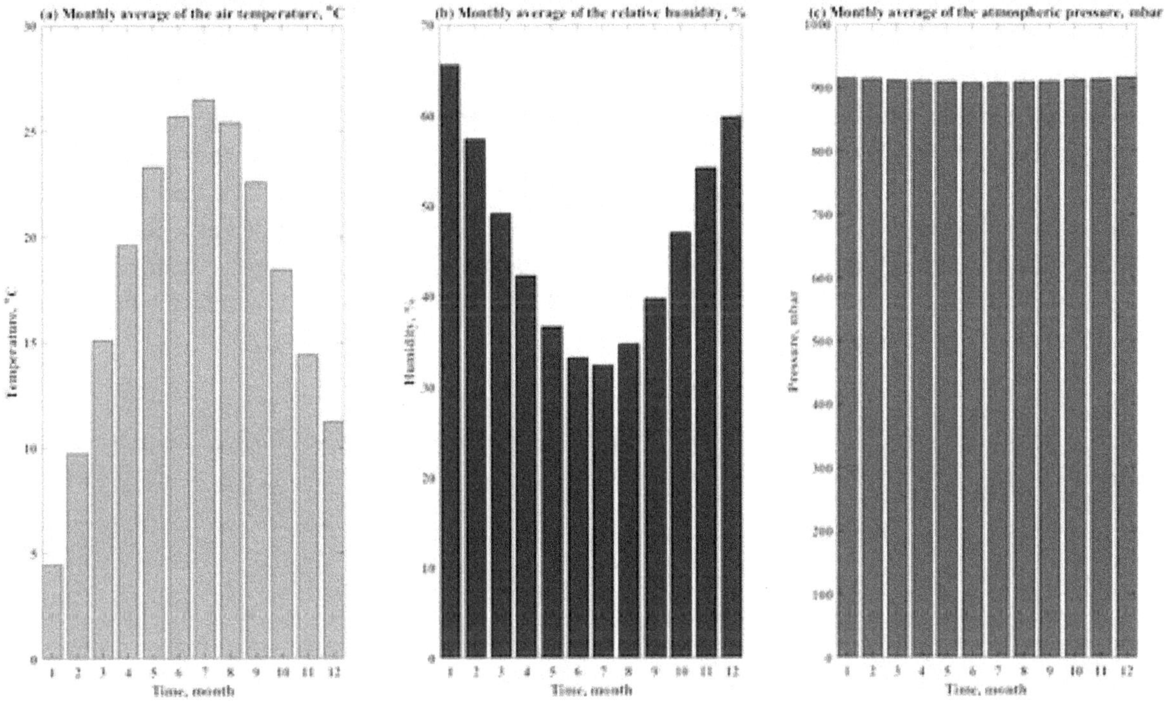

Figure 6. Data signal entry for one year in the ANN model – adapted from (Shboul et al., 2021).

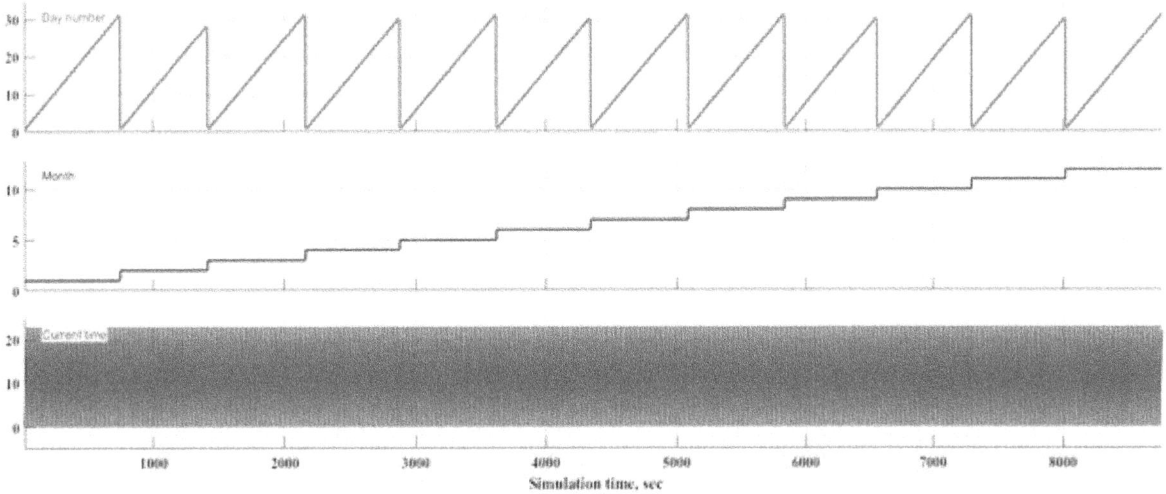

In order to determine the hourly meteorological variables, a signal builder has been developed as the input parameter to the ANN model in the MATLAB/Simulink®. For this purpose, it is primarily used to analyse the CPSD-SE/HWT plant output variations each hour throughout one year, as shown in Figure 6.

Figure 7-a shows the hourly solar radiation (GHI, DNI, and DIF) and the wind speed recorded for the Mafraq City over one year. The annual solar irradiance for GHI, DNI, and DIF at the site varies in the range of 200 to 1015 W/m^2, 300 to 875 W/m^2, and 152 to 189 W/m^2, respectively. It should be noted that the annual GHI is close to the global maximum of GHI. Therefore, this can be traced back to the coordinates of the location in the Sunbelt region. The hourly wind speed recorded at a height of 10 m for the selected location is shown in Figure 7-b. According to Figure 7-b, the average wind speed at the site is approximately between 2.3 m/s and 5.8 m/s and this wind energy potential is sufficient to generate a reasonable amount of electricity.

Figure 7. (a) Hourly GHI, DNI, and DIF and (b) average of hourly wind speed during a year for Mafraq, Jordan.

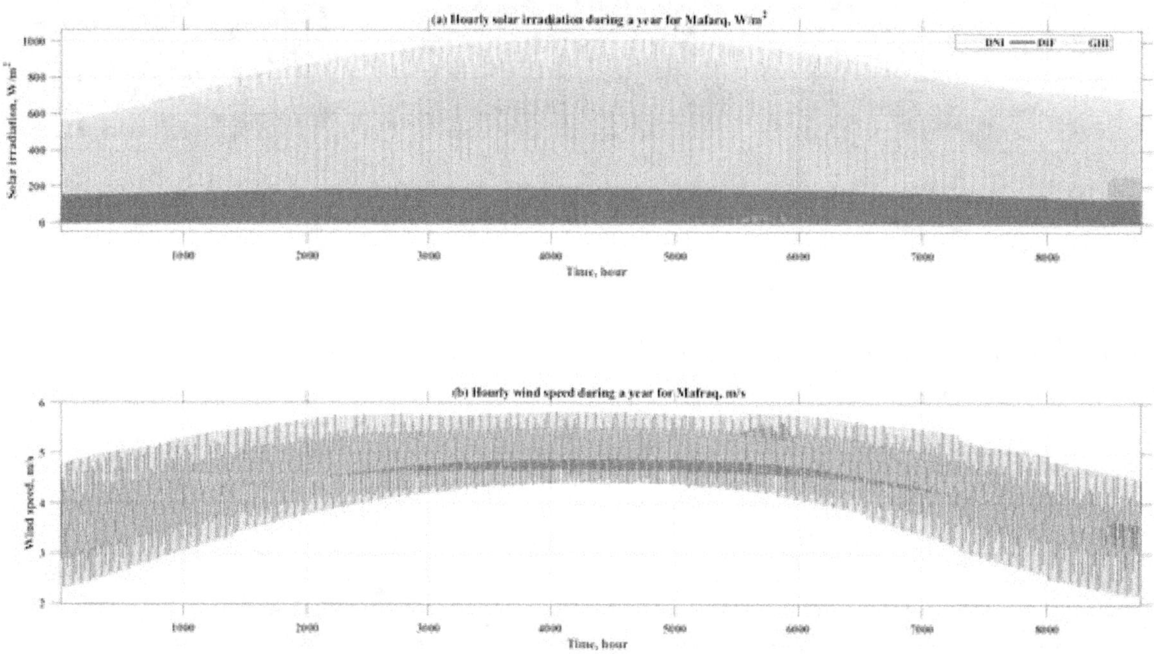

The ambient temperature, relative humidity, and air pressure are critical climatic parameters and these have a direct effect on the characteristics of the CPSD-SE system and power load. Figures 8-a, b, and c demonstrate the annual profile of hourly meteorological parameters for Mafraq City. Figures 8-a to 8-c show that the mean temperature, humidity, pressure of Mafraq City varies from less than -3.2oC, 17.73%, and 907.3 mbar by more than 33.3°C, 83.4%, and 916.5 mbar, respectively.

Electric Load Data

The hourly electrical load of the selected case study represents the load power variation for typical residential buildings in remote regions. The load demand was obtained at an hourly interval throughout the year. Also, the load has been estimated using the random fluctuation mode based on the computational code to approximate real-time operation. The annual and monthly fluctuations of the hourly load consumption is depicted in Figures 9 and 10, respectively. To illustrate this model, an array of random floating-point numbers that are drawn from a uniform distribution has been created. By default, "*rand*" returns normalized values (between 0 and 1) that are drawn from a uniform distribution. To change the range of the distribution to a new range, (a, b), multiply each value by the width of the new range, (b – a) and then shift every value by a. To utilise this proceedure, it should initialise the random number generator to make the results repeatable with equal probability. In this case, the random number is denoted *n*, which is the number of hours either in the month or during the year. To illustrate, the "*n*" values in the month are 672, 720, and 744, which comes from the multiplication of 24 hours by the number of days in that month and thus indicating the total number of hours in a month. Subsequently, the monthly average of the hourly electrical load would be calculated by multiplying the random hours' number with the power range from 100 to 1500 kWe. Similarly, the hourly electrical load during a year could be calculated by multiplying the adopted power range with 8760 random hours. Table 16 in the Appendix shows the numerical code that was used to estimate the electrical load.

Figure 9. Hourly electrical load for the addressed location during a year.

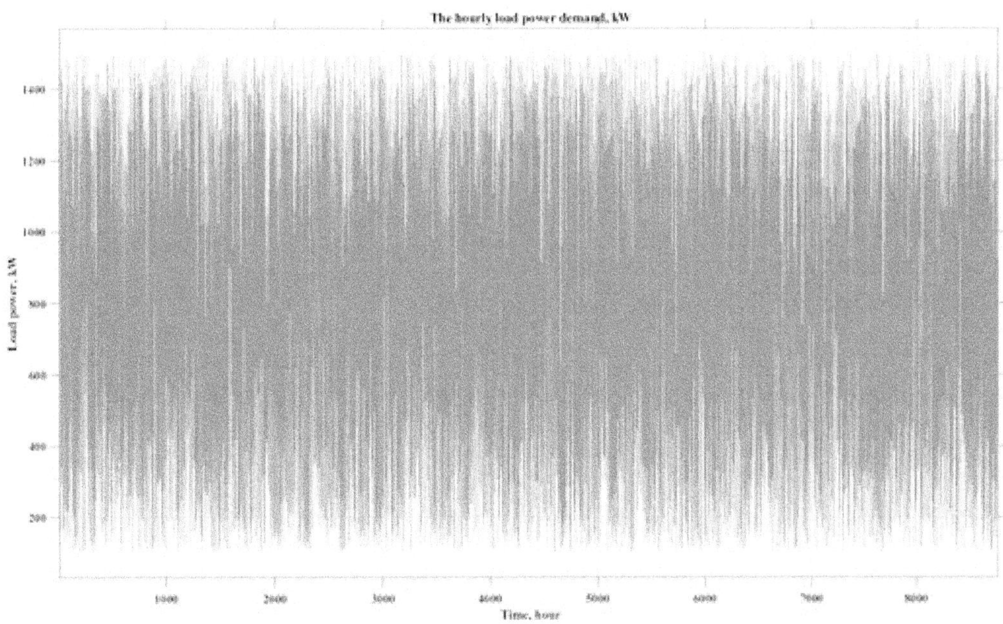

Figure 10. Monthly average of the hourly electrical load for the addressed location.

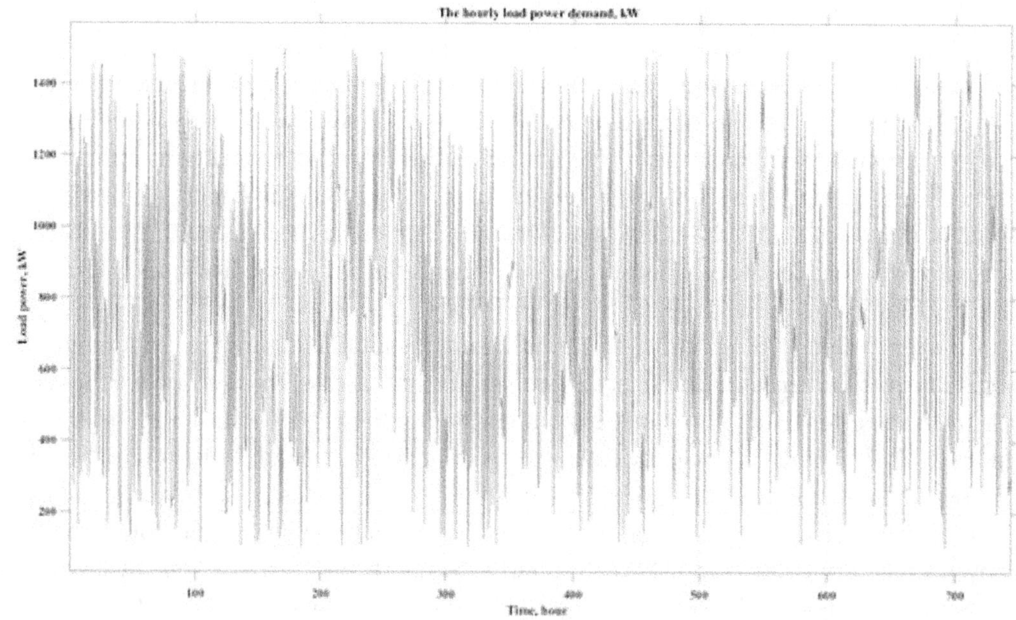

The Proposed CPSD-SE/HWT System

The designed system is suitable for power generation on a residential scale of arid and semi-arid regions. Figure 11 shows a schematic diagram of the proposed hybrid CPSD-SE/HWT stand-alone microgrid power system for the residential buildings described herein. The system main units are as follows:

- ANN solar/wind forecasting model.
- CPSD-SE as a prime power source.
- HWT as a recovery unit.
- Battery bank as an energy storage system.
- Control unit for power disruption & application load.

Figure 11. Schematic diagram of the proposed new hybrid CPSD-SE/HWT system – adapted from (Shboul et al., 2021).

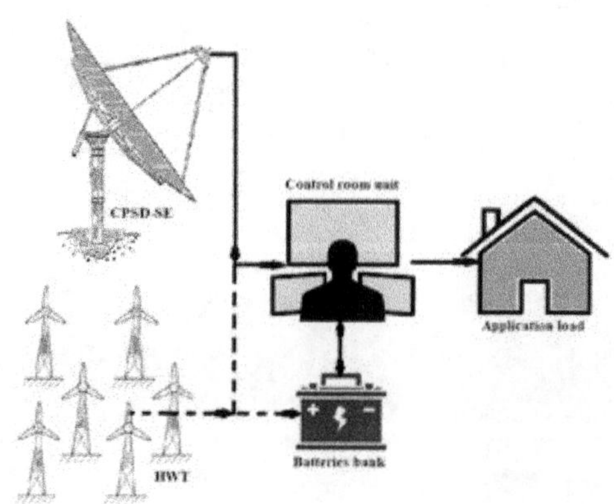

The CPSD-SE will serve as the main source of power generation, while the batteries bank and HWT will provide backup power in case that the primary source of electricity becomes unavailable. This maintains power production continuity. The CPSD-SE system consists of a collection of parabolic dish collectors and a power conversion unit comprised of a Stirling engine (SE), a thermal receiver and an alternator. The primary characteristic that differentiates the CPSD-SE from other solar technologies, such as concentrated solar power (CSP) and photovoltaic (PV), is its capacity to directly convert solar radiation into electrical and thermal energy. In a typical CPSD-SE, three main forms of energy conversion are produced: solar, thermal, and electrical energy. Figure 12 shows the thermodynamic balance diagram of the CPSD-SE diagram (Zayed et al., 2020). In principle, the CPSD-SE mirrors reflect incident sunlight to a focal point on a thermal receiver (i.e., the SE's hot chamber), converting the concentrated radiation into thermal energy to generate heat at a high temperature. The absorbed heat is then transferred to the SE's working fluid via the heater driving an electric generator. In three different mechanisms, the receiver loses heat by conduction through its walls: convection to the surrounding air, radiation from the aperture opening to the atmosphere. In the absence of sunshine, the power demand is provided by the HWT. Moreover, the batteries are being discharged when the CPSD-SE and HWT cannot generate the electricity. To achieve optimum performance and optimum sizing, GA is used for that purpose.

Figure 12. Thermodynamic balance diagram of the CPSD-SE system (Zayed et al., 2020).

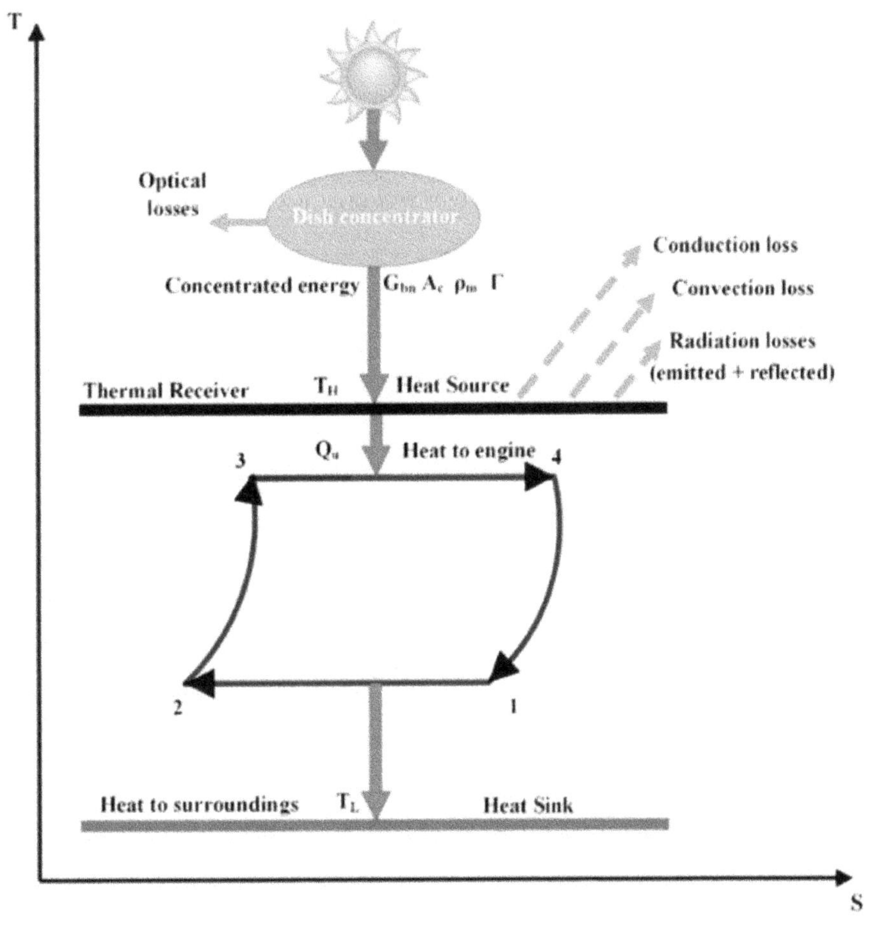

The ideal Stirling cycle is fundamentally composed of four processes; two isothermal and two isochoric processes as can be seen in Figure 13 (Yunus A. Çengel, 2015). These four processes are as follows:

1. 1-2 is an isothermal expansion process. The working fluid is expanded at constant temperature T_H, and after that heat, q_{in} is absorbed from the external heat source.
2. 2-3 is an isochoric heat removal process. The working fluid releases heat is transferred through to the hot side of the engine. The temperature of the working fluid is gradually decreased and this causes a pressure drop.
3. 3-4 is an isothermal compression process. The working fluid is compressed at constant temperature T_L, and then heat q_{out} is released to the heat sink.
4. 4-1 is an isochoric heat addition process. The working fluid absorbs heat and this is transferred from the regenerator to the cold side of the engine. The temperature of the working fluid is gradually increased and this causes an increase in the pressure.

It should be noted that the amount of sunlight incident on the reflectors will be utilised exclusively for electricity generation, with no consideration given to thermal output. To summarise, the developed thermodynamic model based on energy balance was obtained by adopting the following assumptions:

- The heat lost by the receiver via conduction is disregarded.
- The optical heat loss is disregarded.
- The dissipated heat from the SE is disregarded.
- The receiver and generator efficiencies are given.

Figure 13. P-v and T-s diagram of the Stirling cycle (Yunus A. Çengel, 2015).

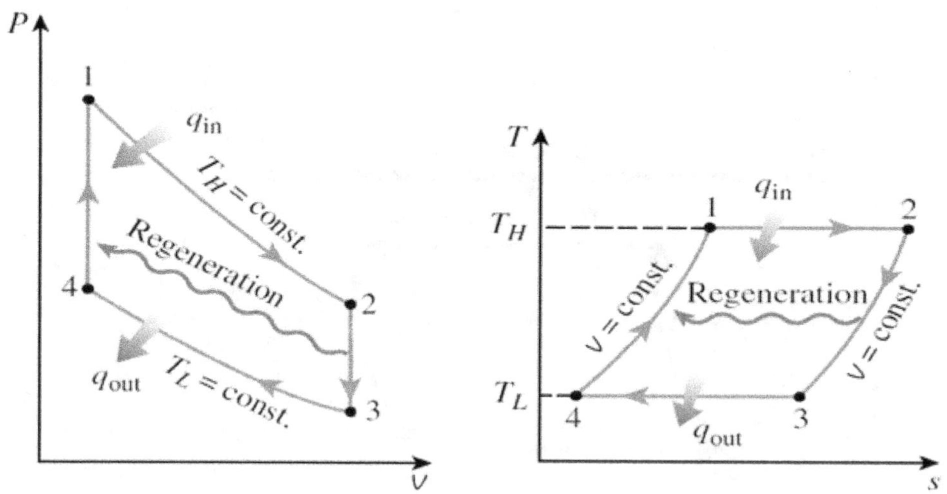

Simulation Tool Selection

In this research work, the MATLAB/Simulink® software is applied to perform the performance and optimisation analysis of the proposed hybrid CPSD-SE/HWT. The main reason for the selection of the MATLAB/Simulink®, for hybrid CPSD-SE/HWT modelling and optimisation purposes are: (i) it includes Simulink blocks (drag & drop ability) with appropriate connections based on the proposed design; (ii) each block contains a variety of MATLAB® command functions, and the system is solved iteratively; and (iii) the optimisation Toolbox™ is made up of multiple solvers for the optimisation techniques to minimise or maximise the objectives while satisfying the constraints. The friendly graphical user interface is powerful for setting up and running optimisation problems, including parameter estimation, component selection, and parameter tuning. This tool has numerous significant applications in the design optimisation, such as an energy management system and production planning. Accordingly, these features facilitate the user to be able to execute further calculations, such as sizing and economic calculations.

Economic Performance Analysis

The economic considerations are vital for assessing the competitiveness of the electricity production from power generation systems. The economic performance model is developed in the MATLAB/Simulink® using a set of correlations based on four economic indicators, namely the total levelised costs of energy, $LCOE_{tot}$, *\$/kWh*, total hourly cost, THC_{tot}, \$/h, annual electricity savings, *AES*, *\$*, and the payback period, *year*. In this regard, the economic parameters of the CPSD-SE/HWT system, these were calculated using Eq. (1) to (4). The detailed cost analysis model that has been considered in this study and this is described in Table 14 in the Appendix (Shboul et al., 2021a). In this work, the batteries are being charged when the electricity production is in surplus from the CPSD-SE and HWT, while on the another hand, the batteries bank would deliver electricity to the application load when there is a shortage in the availability of wind and sunlight. Moreover, the key inputs comprise the following economic indicators: the plant lifetime, interest rate, power cost, variable operating cost and fixed charge rate. Consequently, the subsequent assumptions have been undertaken: plant lifetime is 25 years, the interest rate is 5%, variable operating cost for the solar dish and turbine are 0.06 \$/kWh and 0, respectively, the fixed charge rate is 0.098, battery lifetime is 5 years, turbine power cost is 1628 \$/kWe, battery cost is 100 \$/unit, dish cost is 300 \$/m^2, engine cost is 370 to 400 \$/kWe, receiver cost is 185 \$/kWe, and site cost is 2.2 \$/m^2. Table 2 displays the economic assessment variables and assumptions. The $LCOE_{tot}$ is calculated using the following expression (Shboul et al., 2021a):

$$LCOE_{tot} = \frac{THC_{tot}}{Loadfactor \times P_{tot}} + VOC_{tot} \tag{1}$$

where THC_{tot} is the total hourly costs, \$/h, VOC_{tot} is the total variable operating costs, \$/kWh, P_{tot} is the total plant power, kWe and the load factor is 0.9.

The THC_{tot} is calculated as follows (Shboul et al., 2021a):

$$THC_{tot} = \frac{ATC_{tot}}{8760} \tag{2}$$

where ATC_{tot} represents the total annual cost of the power plant, \$/y, including the annual cost of the batteries, HWT and CPSD-SE.

The AES can be determined according to e following expression (Saffari et al., 2018):

$Electricity\ cost = LCOE_{tot} \times Annual\ electricity\ generated$ (3).

where the annual electricity generated is in kWh.

The payback period is used to estimate the financial competitiveness, which can be calculated by the following equation (Wang et al., 2015):

$$Paybackperiod = \frac{TCC_{tot}}{Electricitycost} \quad (4)$$

where TCC_{tot} is the total capital cost of the system units, \$.

Table 2. Economic parameters for all units that have been considered in the cost analysis model

Parameter	Value	Reference
Input economic parameters		
Interest rate, %	5	(Nafey et al., 2010)
Battery lifetime, year	5	(Nafey et al., 2010)
Plant lifetime, year	25	(Nafey et al., 2010)
Fixed charge rate	0.098	(SAM, 2021)
Electricity sale price, \$/kWh	0.183	(Helioscsp, 2014)
Annual electricity generation, kWh	22.28	Present model
Direct costs		
Battery cost, \$	100	(Shboul et al., 2021a)
Normalised capital cost of the wind turbines, \$/kWe	1628	(Shboul et al., 2021a)
Stirling engine cost, \$/kWe	370	(Shboul et al., 2021a)
Receiver cost, \$/kWe	185	(Shboul et al., 2021a)
Dish concentrator cost, \$/m^2	300	(Shboul et al., 2021a)
Indirect costs		
Site cost, \$/m^2	2.2	(Shboul et al., 2021a)
Wind turbine cost-share, %	65	(Shboul et al., 2021a)
Construction cost share (civil works), %	16	(Shboul et al., 2021a)
Other capital cost-share, %	5	(Shboul et al., 2021a)
Construction, procurement, and engineering cost-share, %	16	(Shboul et al., 2021a)
Contingency cost-share, %	10	(Shboul et al., 2021a)
Other capital cost-share, %	3	(Shboul, et al., 2021a)
Variable operating costs		
Variable operating cost of the batteries, \$/kWh	0.07	(Shboul et al., 2021a)
Variable operating cost of the wind turbines, \$/kWh	0	(Shboul et al., 2021a)
Variable operating cost of the CPSD-SE, \$/kWh	0.06	(Shboul et al., 2021a)

The Multi-Objective GA Methodology

The GA is an optimisation technique known as the evolutionary algorithm used to solve complex, large-scale optimisation problems in various fields based on the mechanism of the natural selection process that mimics biological evolution. GAs search for the optimum solution from one of the candidate solutions that is an array of decision-variable values. These random solutions that are tested against the objective function are called a population. Each individual in the population is called a chromosome. Several populations evolve through successive iterations, namely the selection, crossover and mutation in a GA run and all of these populations are referred to as a generation. In general, with each newer generation, improved solutions (i.e., decision-variable values), which are nearer to the optimal solution than the preceding generation are formed. In the GA context, the set of alternative solutions (array of decision-variable values) is referred to as a chromosome, and each decision-variable value represent in the chromosome is designated by genes (Rani et al., 2013). The size of the population is the number of chromosomes present in a population. The GA mechanism is briefly outlined in Figure 14.

Figure 14. Flow diagram of the overall GA process (Dincer et al., 2017).

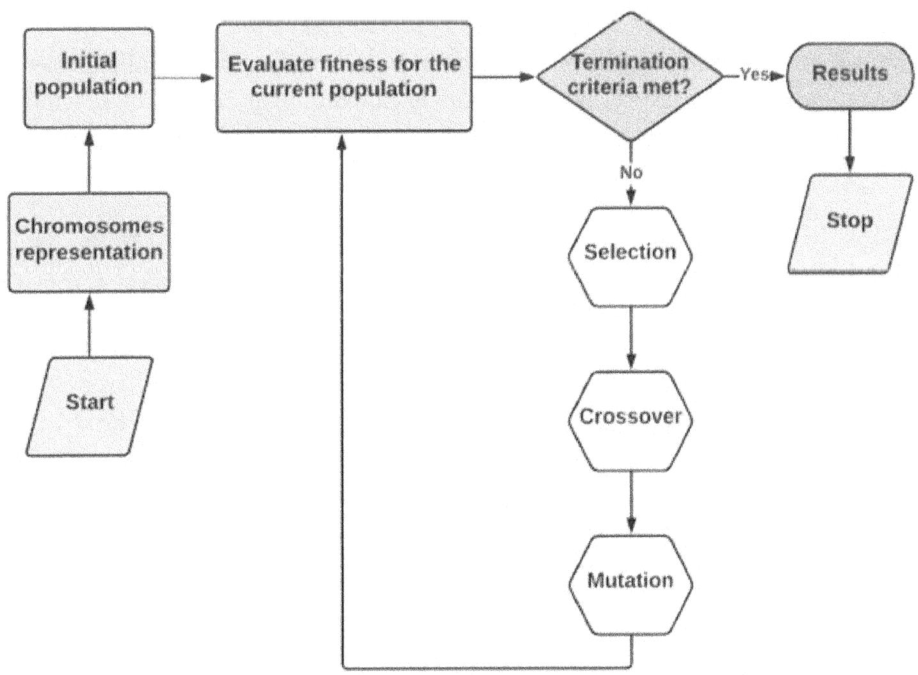

The GA offers a number of advantages over traditional optimisation techniques, which can be listed as follows:

- The GA can be used with continuous as well as discrete variables.
- The GA is capable of dealing with a high number of variables.
- The GA could deal with numerical, experimental, and analytical objective functions.

- The GA technique does not require derivative information.
- The GA would save the overall computational time.
- The GA used to solve stochastic optimisation problems that could be stuck to the optimum.

In this work, it is particularly important to assign the main multi-objective GA model criteria regarding the main process as shown in Figure 14. Generally, a multi-objective optimisation problem can either minimise or maximise the objective function. Unlike single-objective optimisation, multiple objectives are being implemented that require different constraints, all of the possible solutions must be accomplished at once, including the optimum one. A multi-objective optimisation problem can be formulated as follows (Dincer et al., 2017):

Minimise/ maximise: $\{f_n(x)\ \ n=1,2,\ldots N$

$$\text{Subject to: } \begin{cases} g_i(x) > 0 j = 1,2,\ldots J \\ h_k(x) = 0 k = 1,2,\ldots K \\ x_i^{(L)} \leq x_i \leq x_i^{(U)} i = 1,2,\ldots n \end{cases}$$

In this case, we need to the solve vector of the n decision variables or design parameters to find x, with one for each variable. The last set of constraints is called the variable bounds, which restrict the searching bound. Any solution exists inside a lower bound ($x_i^{(L)}$) and upper bound $\left(x_i^{(U)}\right)$ of the decision variables.

Objective Functions

The main purpose of hybridising CPSD-SE/HWT is to minimise the overall plant costs and achieve a higher average overall annual plant efficiency (η_{tot}), thus increasing the competitiveness of solar/wind electricity. In this context, the multi-objective GA optimisation procedure for the CPSD-SE/HWT power plant proposed herein considers two objective functions: the $LCOE_{tot}$ (to be minimised) and the η_{tot} (to be maximised). The multi-objective function code for the GA related to the CPSD-SE/HWT is performed as presented in Table 15 in the Appendix.

Optimisation Framework

For the hybrid CPSD-SE/HWT, the model will consider 6 main inputs and 2 calculated parameters (intermediate inputs) for the target optimisation of two outputs. As mentioned earlier in section 2.5.1, the main target of this model is to minimise the levelised cost of electricity, $LCOE_{tot}$, \$/kWh and maximise the average overall annual efficiency, η_{tot}, %, as illustrated in Eqs. (1) and (5) (Shboul et al., 2021a). The detailed multi-objective GA model that has been considered in this study is presented in Table 15 in the Appendix. The main inputs of the optimisation function are the number of dish units, N_{dishes}, number of wind turbines, N_{wt}, average wind speed, V_{wr}, m/s, air speed ratio, V_o/V, rotor diameter, D_r, *m*, and top cycle temperature, T_h, *°C*. The calculated parameters are the mechanical power, P_{mech}, *kWe*, Stirling engine power, P_{SE}, *kWe*. The model assumed that the number of batteries is 1900, collector area, A_c,

m^2, the normalised capital cost of the wind turbine, *\$/kW,* solar radiation, *DNI, W/m²* and lower cycle temperature, T_l, *°C* are equal to 95 m², 1628 \$/kW, 875W/m² and 25 °C, respectively. Also, the generator efficiency, η_{gen}, receiver efficiency, η_r, and mirror efficiency, η_c, are equal to 95%, 97%, and 80%, respectively. Moreover, the model assumed that air pressure, P_{air}, bar, is 1.0132, and site elevation, H_s, m is 10. Table 3 summarises the parameters related to the optimisation model.

$$\eta_{tot} = \frac{P_{tot} \times 1000}{\left(I_s \times A_{c,tot}\right) + \left(P_w \times N_{wt}\right)} \tag{5}$$

where P_{tot} is the total plant power, *kWe* ($P_{mech}+P_{SE}$), I_s is the solar irradiation, *W/m²* (equals to the DNI), $A_{c,tot}$ is the total plant power of the CPSD-SE, *kW*, P_w is the wind power, *W* and N_{wt} is the number of wind turbines.

Table 3. The developed GA multi-objective functions related to the entire model optimisation

Parameter	Symbol	Unit	Range
Input parameters and ranges			
Number of dishes	N_{dishes}	#	20-80
Number of wind turbines	N_{wt}	#	1-20
Average wind speed	V_{wr}	m/s	2-6
Air speed ratio	V/V_o	-	0.2-0.6
Rotor diameter	D_r	m	60-140
Top cycle temperature	T_h	°C	200-800
Calculated parameters			
Wind turbine mechanical power	P_{mech}	kWe	1.77-1170.33
Stirling engine power	P_{SE}	kWe	11.34-22.13
Outputs			
Total levelised cost of energy	$LCOE_{tot}$	\$/kWh	
Average overall annual efficiency	η_{tot}	%	

SIMULATION RESULTS AND DISCUSSION

In this part of the chapter, first, the validation of each system unit is conducted individually via the comparison with actual power plants and both theoretical and actual published works. Also, the dynamic performance analysis is performed to obtain the energy yield and efficiency of the CPSD-SE/HWT via the MATLAB/Simulink® environment. Steady-state conditions are primarily applied to all runs and the design specifications and operating conditions, as presented in Table 1, are entered into the model. The MATLAB/Simulink® can be considered an ideal platform for an economic potential assessment of the plant. Then, a sensitivity analysis of the key parameters, that affect the performance of the system components, is carried out. Finally, according to the GA method, the optimum energy system is identified based on multi-objective functions.

Mathematical Model Validation

The simulation results, estimated in this analysis for the developed hybrid CPSD-SE/HWT model in MATLAB/Simulink®, are compared and validated with both the simulated published data and actual power plants. Tables 4 and 7 provide a comparison of the findings evaluated from this study, other commercial models and published results, which indicates reasonable variations. Clearly, the results obtained from this research demonstrate that the MATLAB/Simulink® software is very reliable; consequently, it may be used to generate realistic findings in additional analyses in the discussion section by simulating the CPSD-SE and HWT and optimising the developed model.

As with other models, the model validation is carried out to confirm the model adequacy by comparing the results of data sets provided either independently or experimental, which align with the simulated scenario, to those estimated by the present model. Therefore, the operational model validation method is selected to determine whether the model findings agree with the observed data. Consequently, a deterministic model has been developed in this study using the dynamic programming methodology. Furthermore, two distinct aspects are taken into consideration during the model validation involving the input parameter and operating condition values and also the assumptions. To illustrate, the design specifications of the HWT and CPSD-SE then these are presented in Tables 4 and 6, respectively. In addition, two approaches are applied following the various attributes of the model, including theoretical results and real system measurements. Nevertheless, in practice, full validation of the entire model would be challenging, particularly whether the system is being modelled does not yet exist. In this work, the validation will focus on the output of each unit separately.

HWT Model Validation

The specifications of the selected HWT model (The Wind Power, 2020) are listed in Table 4 and is briefly described below. The simulation results obtained by the proposed dynamic model of the HWT system is validated by using results from other existing models, and the results from this model are compared with other types of commercial HWT models. These models include the GAMESA model in the Ma'an Part I wind farm in Ma'an, Jordan and the ENCORN model in the Feldheim wind farm in Germany (The Wind Power, 2020). The data is obtained by inputting the same geometrical design parameters and operating conditions. As shown in Table 5, the comparison reveals an excellent agreement between the developed model and the selected wind farms. It is shown that the output power per HWT unit of the developed model has errors of approximately 0.55% and 1.99%, compared to the Ma'an Part 1 and Feldheim wind farms, respectively.

Table 4. Specifications of the selected models

Description	Wind farm	Ma'an Part 1	Feldheim
General data	Country/zone	Jordan/Ma'an	Germany
	Manufacturer	GAMESA	ENCORN
	Model	G97/2000	E115 3.000
	Number of Turbines	33	3
	Total nominal power	66000 kW	9000 kW
Operating data	Rated power	2000 kW	3000 kW
	Cut-in wind speed	3 m/s	2 m/s
	Rated wind speed	14 m/s	11.5 m/s
	Cut-out wind speed	25 m/s	25 m/s
Rotor	Rotor diameter	97 m	115.7 m
	Swept area	7390 m^2	10515.5 m^2
	Rotational speed	9.6-17.8 rpm	12.8 rpm
Generator	Generator efficiency	0.97	0.97
	load factor	0.95	0.95
Tower	Hub heights	78-120 m	92,135, and 149 m
Power curve	Wind speed	0-25 m/s	0-25 m/s
	Power coefficient	0.18	0.33

Table 5. Data validation results of the HWT model

Description	The developed model	Ma'an (Gamesa G97/2000)	Error (%)	The developed model	Feldheim (Encorn E115 3.000)	Error (%)
Module power	2011	2000	0.55	3000	3061	1.993
Hub height	121.3	120	1.083	149	144.6	3.043
Rotor swept area	7389.81	7390	0.00257	10515.5	10513.72	0.0169

CPSD-SE Model Validation

In this study, the proposed mathematical CPSD-SE model was developed using the MATLAB/Simulink® toolbox to carry out the performance analysis. Table 6 lists the geometrical and operating parameters of the CPSD-SE models used for validation. To validate, the simulation results of the present model then these results are compared with the findings of previous related studies from Refs. (v. Siva Reddy, 2012; Zayed et al., 2020) under the same operating conditions. The comparison shows that the same major parameters of the two models have an excellent agreement, as indicated in Table 7. For instance, the Stirling efficiency values have a percentage of deviation of approximately 4.98% and 16%, when compared with Refs. (v. Siva Reddy, 2012; Zayed et al., 2020), respectively. These deviations in the simulation results of the two different systems could be caused by adopting different parameters in the different modelling techniques. Overall, it can be concluded from the high validity of the obtained results that this developed model is a reliable tool for simulating the performance of several CPSD-SE commercial prototypes.

Table 6. Design specifications of the 25 kWe CPSD-SE system and 50MWe Jodhpur power plant

Specifications	Unit	Jodhpur power plant	Mohamed E. Zayed
Operating conditions			
Ambient Temperature	°C	25	30
Solar radiation	W/m²	1000	900
Stirling engine			
Stirling engine power	kWe	25	25
Working fluid	-	H_2	H_e
Receiver gas temperature	°C	810	650-900
Engine speed	rpm	1800	1800
No. of cylinders	-	4	4
Bore and stroke	mm	44 × 57	55 × 40
Dish concentrator			
Rim angle	degree	39^0	45^0
Intercept factor	-	0.92	0.90
Mirror reflectivity	-	0.92	0.92
Un-shading factor	-	0.98	0.97
Cavity Receiver			
Cavity absorptivity	-	0.94	0.96
Receiver efficiency	%	93.89	82.336
Optical efficiency	%	77.88	66.13
Alternator			
Generator efficiency	%	92-94	92.5

Table 7.

Description	The developed model	Actual published data at Jodhpur power plant	Error (%)	The developed model	Simulated published data by Mohamed E. Zayed	Error (%)
Total plant power, kW	50000	50000	0	25	25	0
Stirling efficiency, %	36.24	38.14	4.982	37.08	31.96	16.020
Aperture diameter, m	10.95	10.57	3.595	12.49	12.5	0.080
Projected area, m²	94.23	91.01	3.538	122.5	122.75	0.204
Concentrator efficiency, %	82.95	82.94	0.012	80.32	80.316	0.00498
Focal length, m	7.485	7.45	0.470	7.511	7.54	0.385
Peak net efδciency, %	26.53	29.68	10.613	22.68	19.55	16.010
Rated output Power, kW	25	24.5	2.041	25	25	0

Performance Analysis

As aforementioned, to test the developed model and measure the performance of the system, the model has been implemented in a specific location, namely Mafraq, Jordan, and the meteorological data is obtained from the Solargis™ toolbox (Solargis, 2019). Figures 15-a, b show the annual solar radiation and wind speed data variation based on the hourly resolution in Mafraq. The wind speed varies between 2.3 m/s and 5.8 m/s, while solar radiation values range between 300 W/m² and 875 W/m².

Figure 15. Solar and wind profile of Mafraq throughout one year.

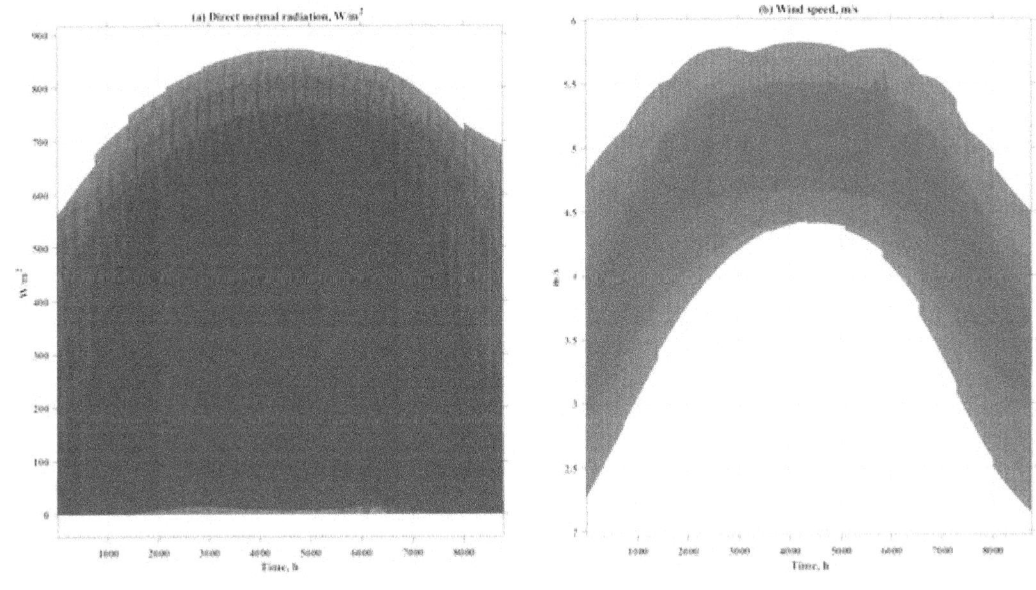

The MATLAB/Simulink® signal builder has been developed to represent the dynamic input battery model in order to specify the battery capacity as a function of the time throughout the year. The purpose of these generated Figures is to understand how 1-hour values spread over the year that assigned the battery charging and recharging cycle as depicted in Figure 16.

The control unit is in charge of distributing the load among the system units and the wind speed and solar radiation are the key variables that shape the load distribution. Figure 17 is the pseudo-algorithm that demonstrates the operation of the load distribution. If radiation from the sun exceeds the solar irradiance limitation (500 W/m²), the signal prompts the CPSD-SE to operate without the assistance of wind turbines and/or the batteries bank. In case of the solar radiation is below the assigned solar limit and the average wind speed goes up the wind limitation (assumed to be 1.5 m/s), the HWT will enter into service instead of the CPSD-SE and/or the battery bank. In addition, the battery bank will serve the electrical demand when the charge rate is lower than 1, otherwise, the battery is charging along with HWT and CPSD-SE are operating.

Figure 16. Charge/discharge rate signal of the batteries along (a) one day, (b) one month.

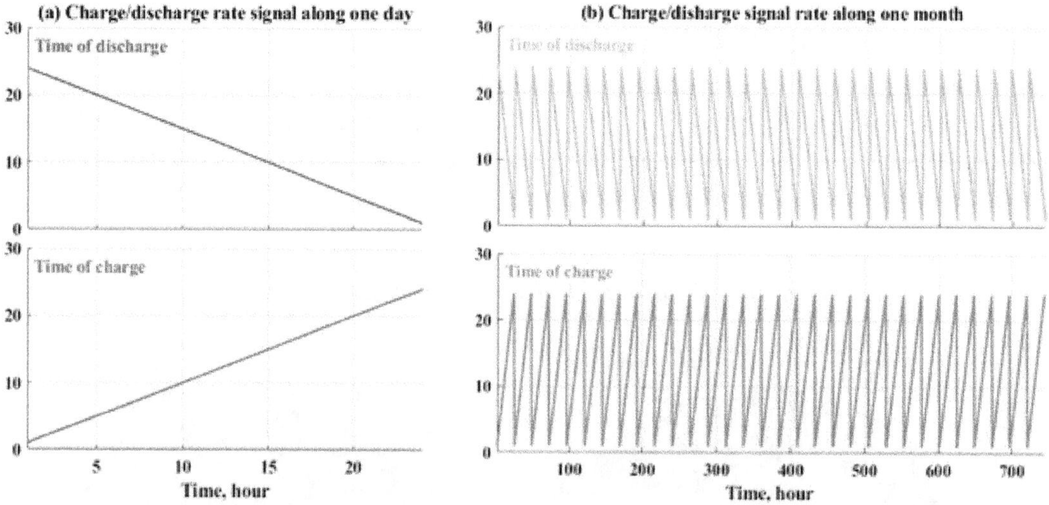

Figure 17. Flow chart of the control unit.

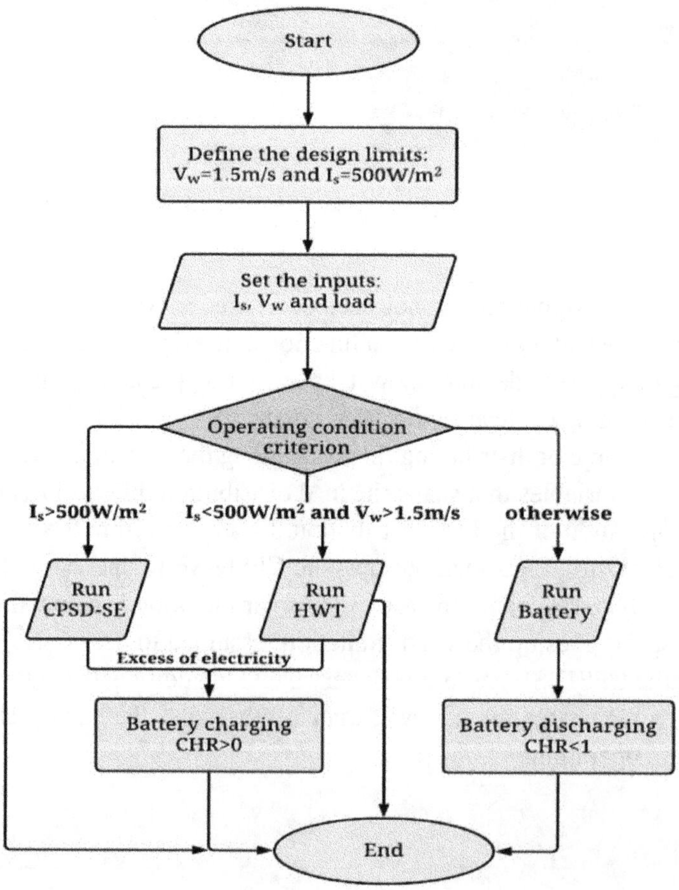

Figure 18 shows the hourly behaviour across one year (2018-2019) regarding the wind farm performance. The power developed was relatively high, depending on the wind speed variation. Therefore, the maximum total power generated by the wind farm marginally exceeds 890 kWe during the spring & summer times. The power coefficient (C_p) is in the range of 0.2 and 0.3, which is considered relatively high. The axial force is in the range of 19.3 and 32.4 kN/module and the net power from the module generator does not exceed 65 kWe.

Figure 18. Data results related to the wind speed effect throughout one year (Mafraq, Jordan case study).

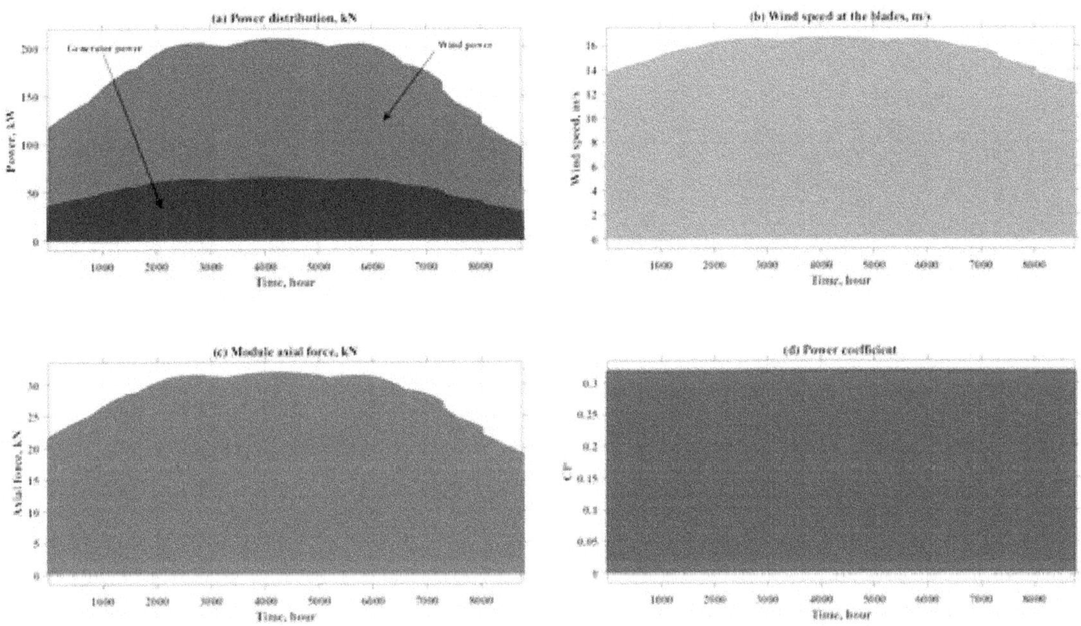

Figure 19 shows the results of the CPSD-SE unit. The power range is estimated to be about 1200 kWe. As an anticipated reflection of the solar radiation effect, Figure 19-b shows the top engine temperature throughout the year. The maximum allowable temperature is recorded during the summertime and it is between 590°C and 605°C. During the winter, the temperature drops, and it is between 460°C and 515°C. The CPSD-SE efficiency is in the range of 30% to 33% which is considered high when compared to the PV or solar gas turbine cycle. The engine compression ratio (CR_{SE}) and the pressure ratio (PR_{SE}) are attractive and in the range of 7.5-9.2 and 19-27.2, respectively, and this is because of the use of CO_2 instead of air that is commonly utilised (Sharaf Eldean et al., 2017). In addition, the fluctuations in the solar radiation have significantly affected the mean effective pressure, which is relatively high and equal to about 10.2, as indicated in Figure 19-f.

Figure 19. Data results related to the solar radiation effect throughout one year (Mafraq, Jordan case study).

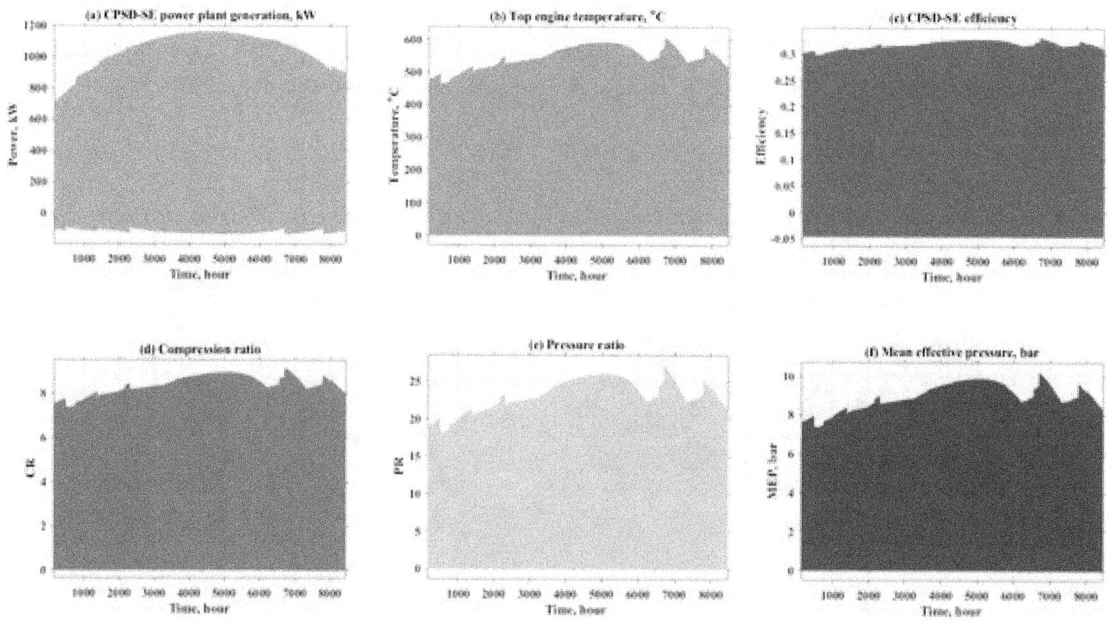

To examine the monthly performance, the predicted net average power production and the overall efficiency of the integrated power plant are shown in Figure 20. The results demonstrate that the high net power and overall efficiency values of the hybrid CPSD-SE/HWT system are obtained in the summertime between May and September.

Figure 20. Monthly net average output power and overall efficiency of the hybrid CPSD-SE/HWT system.

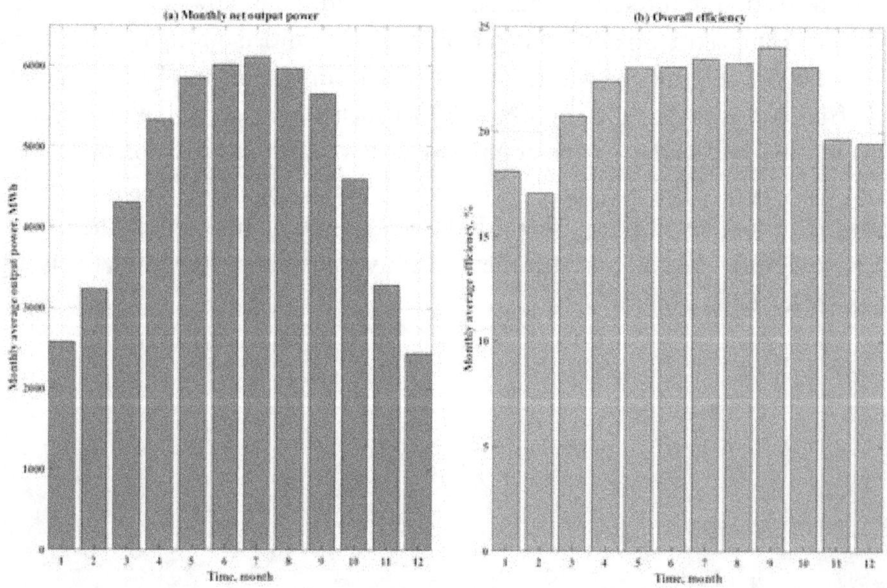

It can be seen that the monthly peak predicted the output power and the overall efficiency for the proposed power plant to be 6113.35 MWh and 24.03% in July and September, respectively. While the lowest power and efficiency values are recorded in the wintertime. The lowest output power is found to be about 2441.21 MWh in December, and the lowest overall efficiency in February, which is found to be 17.07%. Overall, it can be indicated that the average monthly net electricity production and average net monthly overall efficiency for the CPSD-SE/HWT plant are found to be about 4612.97 MWh and 21.45%, respectively.

In addition, Figure 21 shows a typical day in the summertime as an example. According to the fluctuations in the solar and wind energy generation, the Figure shows the variation in load generation across the whole day. From Figure 21, it is observed that the electricity demand surpasses the power generated from the proposed system that is produced during the day, particularly at night. Most of the day, the HWT generated a power range between 380 kWe and 900 kWe. The power generation in the middle of the day is dominated by the solar dish operation. The system starts at 5:20 am and provides 610kWe and ends with 590kWe at 16:30. In fact, throughout the period from 11:00 am to 01:00 pm, the peak power generated by the CPSD-SE is about 1200 kWe, as shown in Figure 21.

Furthermore, it is observed that batteries either produce or consume electricity throughout the day. The battery operates when there is no sun or wind, as shown in Figure 21. In this context, the batteries are discharging and delivering the electricity in the range of 100 to 375 kWe. Overall, it can be seen that the CPSD-SE has the potential to be an attractive system to generate power for residential communities; however, it needs a recovery unit such as a battery, and HWT. Table 8 shows some of the important calculated results under the Mafraq, Jordan operating conditions.

Figure 21. Power generation result related to 24-hour operation for a typical day in summertime in Mafraq, Jordan.

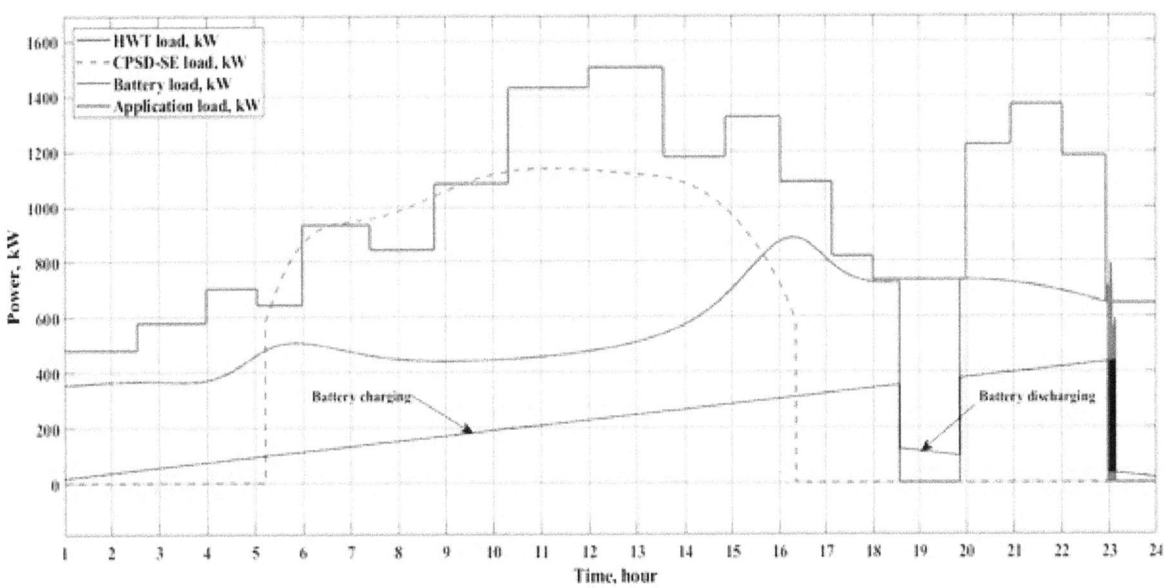

Table 8. Data results according to Mafraq, Jordan

Parameters	Unit	Value
Solar farm output		
Top engine temperature	°C	506.50
Stirling efficiency	%	30.86
Total solar plant efficiency	%	22.75
Stirling engine power	kWe	16.86
Total plant power	kWe	1011.83
Electricity generation	MWh	11.13
Dish concentration ratio, A_d/A_r	-	1344.44
Total plant area	m^2	5702
Dish area	m^2	95
Stirling engine compression ratio	-	7.91
Stirling pressure ratio	-	20.73
Wind farm output		
Rotor swept area	m^2	1564.90
Torque	N.m	61216.83
Power coefficient	-	0.29
Wind power	kW	116.98
Mechanical power	kW	37.26
Net developed power per module	kWe	33.28
Total Farm Power	kWe	499.23
Electricity generation	MWh	11.98
Farm total area	km^2	0.54
Batteries bank output		
Total battery power	kWe	205.69
Electricity generation	MWh	4.94
Microgrid output		
Average overall efficiency	%	23.73
Net electricity production	MWh	17.34

The MATLAB/Simulink® was run to evaluate the hybrid CPSD-SE/HWT productivity throughout the year throughout the 8760 hours. Figure 22 shows the contribution of each sub-system, CPSD-SE, HWT, and batteries for power production. The predicted annual electricity generation of an application load 1500kWe capacity. Also, it demonstrates the dynamic behaviour of the generating electricity for system units during a full year to cover the load demand. The total annual electricity generated by the CPSD-SE, HWT, and battery bank varies from less than 700 kWe, 550 kWe, and 400 kWe to more than 1160 kWe, 900 kWe, and 420 kWe, respectively. As can be seen, despite the large energy output of the proposed system, it is not enough to cater for the current load demand, thus making the use of optimisation techniques an imperative action.

Figure 22. The power produced by the hybrid system over the year for Mafraq, Jordan.

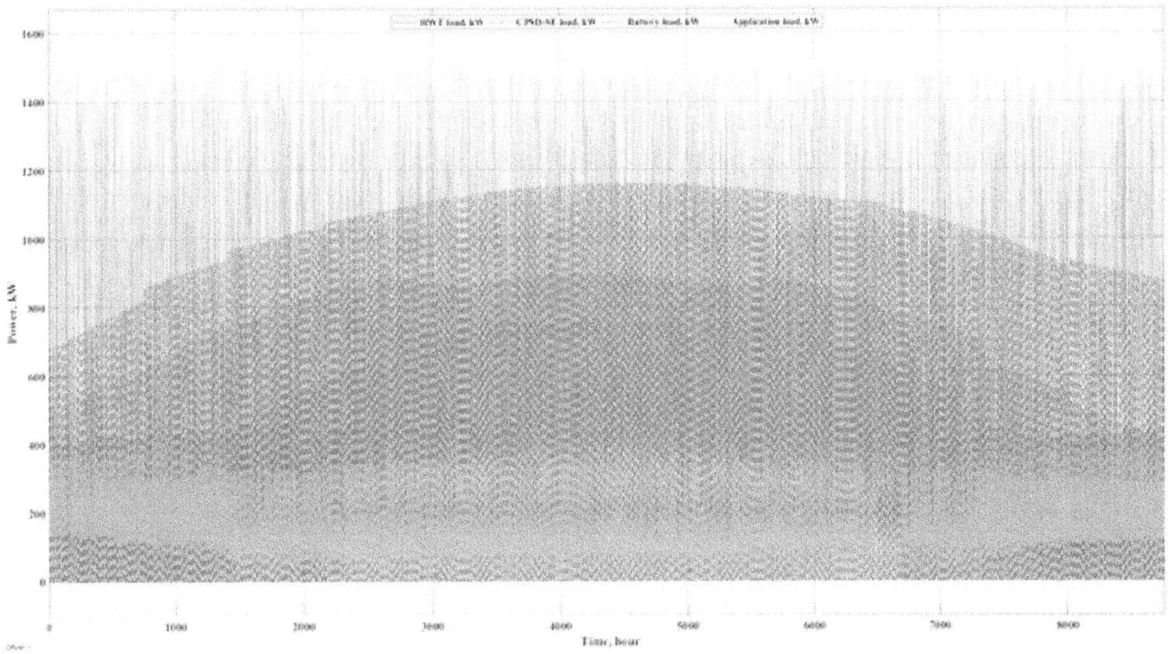

Economic Analysis

The economics of the entire microgrid system includes the total capital cost, total annual cost, total hourly cost, total LCOE, annual electricity savings and payback period. Table 9 presents the cost results for the investigated location (Mafraq). The cost-effectiveness of the hybrid system was examined based on the four economic indices of the whole system, including the total LCOE, $LCOE_{tot}$, *\$/kWh*, total hourly cost, THC_{tot}, \$/h, annual electricity savings, *AES*, *\$* and payback period, *year*. These economic indicators are evaluated from Eqs. (1) to (4).

Overall, the total capital cost of the CPSE-SE/HWT system, including the total direct and indirect cost of the batteries bank, HWT and CPSD-SE is about 912772.84 \$. The average LCOE of the hybrid system is found to be about 0.18 \$/kWh and the average hourly costs are found to be 27.44 \$/h. The estimated payback period for the integrated system is a year. It is evident that the developed system is feasible and competitive.

Table 9. Economic analysis results of the proposed microgrid systems configuration

Parameters	Unit	CPSD-SE	HWT	Batteries bank	Power plant
Total capital cost	$	18784.82	86527.74	950000	912772.84
Total annual cost	$/year	2759.89	10698.34	226875.99	240334.23
Total hourly cost	$/hour	0.32	1.22	25.90	27.44
Total LCOE	$/kWh	-	-	-	0.18
Annual electricity savings	$	-	-	-	828143.54
Payback period	year	-	-	-	1.03

Sensitivity Analysis

In general, the sensitivity analysis would be utilised to find out the optimal system behaviour related to various uncertain parameters. In this investigation, the four input variables are DNI, *W/m*^{2}, wind speed, *m/s,* rotor diameter, *m, and* dish diameter, *m,* that we have been taken to study their effects on the hybrid system results along the year. Table 10 shows the sensitivity variables that are used in the analysis. Each of the examined input variable impacts on the techno-economic performance results of the power plant, which are the $LCOE_{tot}$ and average overall annual efficiency as depicted in Table 10.

Table 10. Effect of the sensitivity variables on the proposed hybrid system

Sensitivity input variables	values	Sensitivity estimated variables
DNI	500-1000	$LCOE_{tot}$
Wind speed	1.5-6	Average overall annual efficiency
Rotor diameter	20-150	
Dish diameter	2-14	

Sensitivity to the Solar Radiation and Wind Speed

The average *DNI* was varied between 500 and 1000 W/m^2 and its effect on the $LCOE_{tot}$, *\$/kWh* and total efficiency, %, may be observed (see Figures 23-a,b). The incident DNI has a moderate effect on the $LCOE_{tot}$ and average annual electricity generation, as shown in Figures 23-a and b. However, the solar radiation has a proportional effect on the average annual efficiency, the higher intensity of solar radiation, and the higher the overall efficiency, as depicted in Figures 23-b. In the case of increasing the direct solar radiation, the overall efficiency increases from 20.7% to 25.6%.

Figure 23.

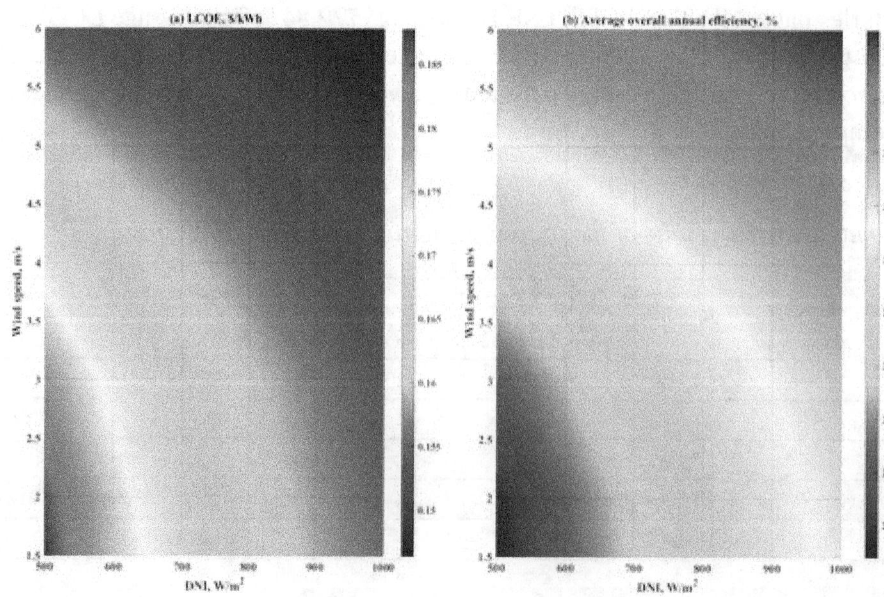

In the case of a variation in the wind speed, it has been observed that the wind speed played a key role in calculating the $LCOE_{tot}$. Accordingly, the wind speed was varied between 1.5 and 6 m/s. By considering Figure 23-a, it can be seen that the cost of electricity decreases with respect to the increase of solar radiation and wind speed, the $LCOE_{tot}$ varies between 0.188 \$/kWh and 0.146 \$/kWh. However, the wind speed also has a significant impact on the average net annual efficiency, the higher the wind speed, the higher is the overall efficiency, as illustrated in Figure 23-b.

Sensitivity of $LCOE_{tot}$ and Overall Annual Efficiency to the Rotor Diameter

The effect of the rotor variation was analysed by MATLAB/Simulink® and in this study, the expected variation in the rotor diameter between 20 m to 150 m was considered. Thus, as the diameter increases then this has a significant change in the $LCOE_{tot}$ and the average overall annual efficiency. Figure 24 depicts the differing diameter variations impact on the economic indices and, as can be seen, the higher the diameter, the lower the LCOE. Hence, the increasing in the rotor diameter decreases the LCOE from 0.5668 \$/kWh to 0.1206 \$/kWh. As a result of comparing the simulation results with the increasing rotor diameter, it was found that the overall efficiency of power plant increases from 20.2% to 24.1%, as shown in Figures 24.

Figure 24.

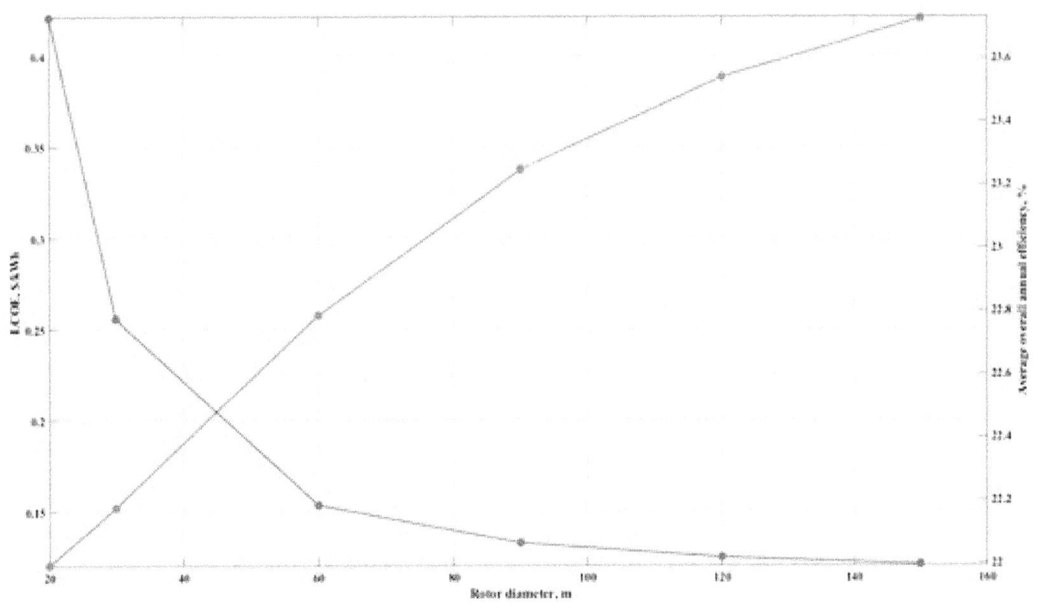

Sensitivity of $LCOE_{tot}$ and Overall Efficiency to the Top Cycle Temperature

The MATLAB/Simulink® is performed to investigate the effect of the high cycle temperature on the system performance and cost. The top cycle temperature is an important parameter when obtaining the most desirable outcome. The expected variation in the top cycle temperature is between 200 °C to 800

°C. As illustrated in Figure 25, it can be seen that a decrease in $LCOE_{tot}$ and an increase in the average overall annual efficiency with the top cycle temperature. Hence, the increasing of high cycle temperature decreases the $LCOE_{tot}$ from 0.1792 \$/kWh to 0.1744 \$/kWh. However, as the top cycle increases, it was found that the overall efficiency increases from 20.1% to 23.7%.

Figure 25.

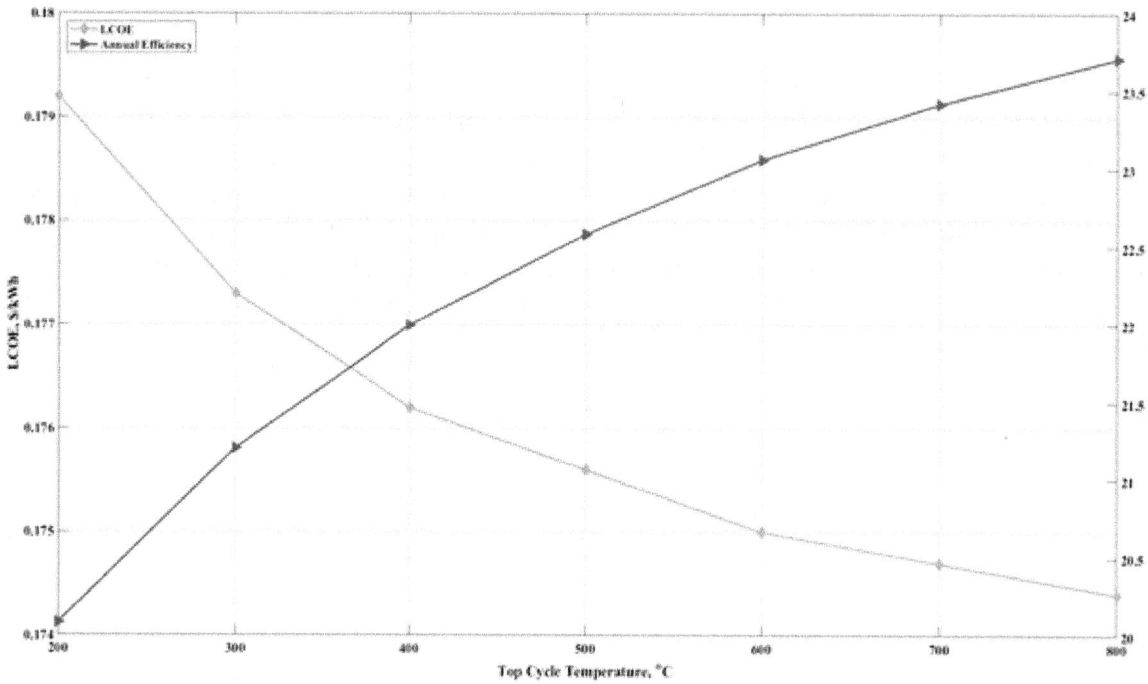

Multi-Objective GA Optimisation Analysis

The developed GA model aims to determine the cost and efficiency range over a wide range of data related to the CPSD-SE/HWT power plant. Therefore, it would be very interesting to obtain optimised data that may help in the design and performance aspects. The optimisation model has been developed for two objective functions: $LCOE_{tot}$ (to be minimised) and plant efficiency (to be maximised).

The optimised data results for the entire CPSD-SE/HWT plant are obtained and analysed according to the input constraints listed in Table 3. To achieve the two objective functions, the performance constraints such as concentrator efficiency, η_c, %, receiver efficiency, η_r, %, and generator efficiency, η_{gen}, % are assumed to have the following values 97%, 80%, and 95%, respectively. These were anticipated to give optimal values related to the total efficiency and this is found to be 30.8% and in this case the $LCOE_{tot}$ is equal to about 0.248 \$/kWh. The obtained results show that the number of dishes should be 54, the top cycle temperature is about 383 °C, the wind turbine power is 1170 kWe, the number of wind turbines is 2 and the rotor diameter is 133 m. These mentioned design values are anticipated to achieve a higher total plant efficiency, η_{tot}, which is 30.8%. In the case of lower $LCOE_{tot}$, the number of dishes, the top cycle temperature, the wind power, the number of turbines, and the rotor diameter are 67, 370°C,

752 kWe, 4 and 125 m, respectively. For the operating conditions, to achieve the maximum of the total plant efficiency and minimum $LCOE_{tot}$, it is assumed that the lower cycle temperature (sink) should be equal to 25 °C. However, solar radiation is assumed to be 875 W/m², which is considered to be high.

It is quite clear that the GA would give a clear decision on the selection between the optimum $LCOE_{tot}$ and the optimum efficiency by specifying the optimum operating conditions and the design aspects needed (constraints).

Figure 26 indicates the Pareto front solution for the hybrid CPSD-SE/HWT with two objectives regarding the $LCOE_{tot}$ and η_{tot}. Each point on this Pareto front depicts a potentially optimum solution for the minimum $LCOE_{tot}$ and maximum η_{tot}. In this regard, the Pareto front assists the decision-maker in selecting a single ideal solution from a group of optimal solutions depending on the decision maker's preferences and criteria for establishing a microgrid CPSD-SE/HWT power plant with higher annual efficiency or a lower $LCOE_{tot}$. It is also indicated the impact of increasing the system efficiency on the Pareto front, dislocating the curves to higher values of efficiency, without significantly changing the values of the $LCOE_{tot}$.

Figure 26.

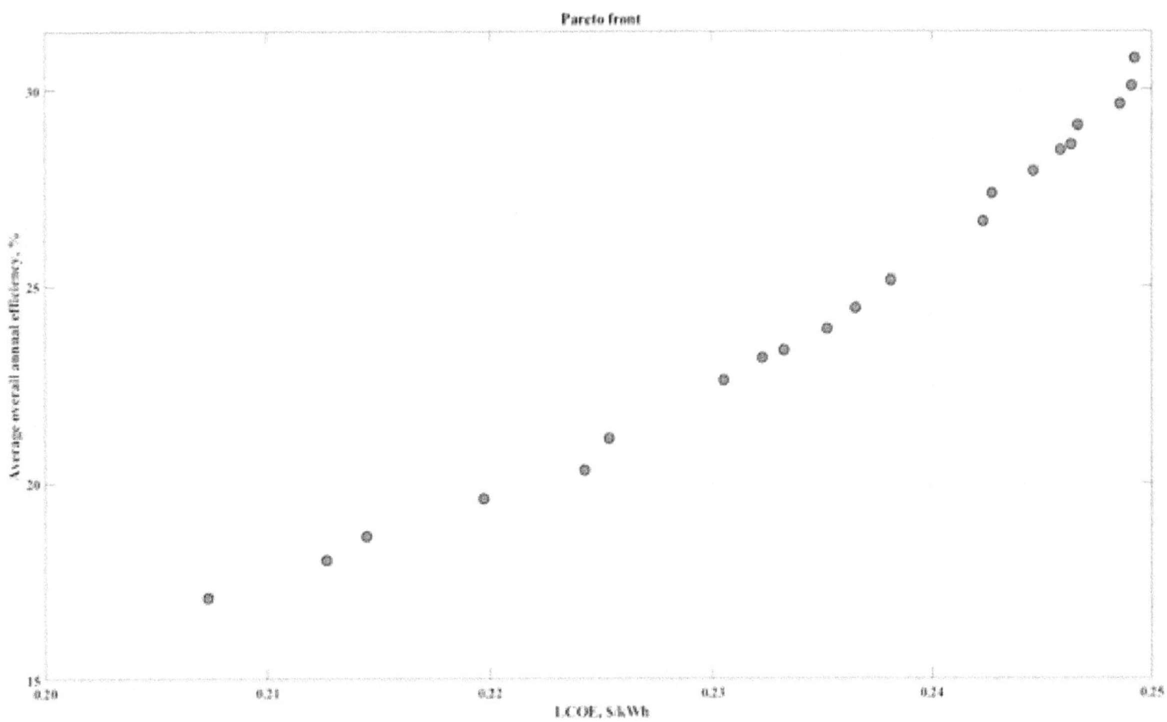

As can be seen in Figure 26, all of the generated solutions occurred between (η_{tot} = 30.8%, $LCOE_{tot}$ = 0.248 $/kWh) and ($\eta_{tot}$ = 17.08%, $LCOE_{tot}$ = 0.207 $/kWh). Hence, it can be indicated that the designer/decision maker can adopt various solutions for this case. From the shape of the curve it appears that solutions with higher efficiency should be preferred as the increase in the $LCOE_{tot}$ is not as sharp as the uplift of the energy efficiency.

CONCLUSION

The aim of this study is to investigate advanced strategies for the integration, performance and optimisation of new hybrid CPSD-SE/HWT for stand-alone microgrid power generation. The system aims to supply power to arid and semi-arid regions and the system depends on the solar and wind energies. The annual simulation of a 1500 kWe hybrid system is performed dynamically for the city of Mafraq based on the operating conditions of the satellite-driven data from Solargis™. The integrated system is validated with the simulated and actual data published in the literature in order to ensure the reliability and accuracy of the developed model.

The results of the dynamic performance are presented. For the solar part, a CPSDE-SE has been considered and for the sun off periods, a HWT and a battery bank have been used. Regarding the CPSD-SE, CO_2 is used as the working fluid instead of air and a total number of 60 dishes are assumed to be implemented while the total number of HWTs is 7. For a typical summer day, during the daylight, the CPSD-SE is dominant and generates about 1500 kWe with a total high efficiency of about 26%. Most of the supplied power throughout the year is dominated by the CPSD-SE with a high cycle temperature that reaches up to 850°C. On the other hand, the HWT generates a limited amount of power. However, it is considered to be extremely helpful to overcome the uncertainties in the solar radiation throughout the day.

A multi-objective GA method that is based on an evolutionary computation algorithm is applied to the CPSD-SE/HWT system for the electricity production to evaluate the optimal design parameters for the system. To fully utilise this, computer modelling techniques using the Multi-objective GA Solver in MATLAB/Simulink® have been employed. The objective functions, design parameters and constraints, and the overall optimisation are elaborated. The results show that the η_{tot} values are raged between 17.08% to 30.8%, where the $LCOE_{tot}$ values are about 0.207 to 0.248 $/kWh, thus, providing a set of optimal solutions for both objective functions to the designer. In general, this novel system could be substantially utilised as a distributed power generation system for a low to medium scale microgrid application.

ACKNOWLEDGMENT

The authors would like to acknowledge Al al-Bayt University and the Oman Ministry of Higher Education for their valuable financial support.

REFERENCES

Agency, I. R. E. (2012). *Renewable Energy Technologies: Cost Analysis Series* (Vol. 1). doi:10.1016/B978-0-12-812959-3.00012-5

Al Badwawi, R., Abusara, M., & Mallick, T. (2015). A Review of Hybrid Solar PV and Wind Energy System. *Smart Science*, *3*(3), 127–138. doi:10.1080/23080477.2015.11665647

Ali, H. H., Al-rub, F. A. A., Shboul, B., & Moumani, H. (2020). *Evaluation of Near-net-zero-energy Building Strategies : A Case Study on Residential Buildings in Jordan.* doi:10.32479/ijeep.10107

Ali, H. M. (2020). Recent advancements in PV cooling and efficiency enhancement integrating phase change materials based systems – A comprehensive review. *Solar Energy, 197*(June), 163–198. doi:10.1016/j.solener.2019.11.075

Bakir, H., & Kulaksiz, A. A. (2020). Modelling and voltage control of the solar-wind hybrid micro-grid with optimized STATCOM using GA and BFA. *Engineering Science and Technology, an International Journal, 23*(3), 576–584. doi:10.1016/j.jestch.2019.07.009

Belfkira, R., Zhang, L., & Barakat, G. (2011). Optimal sizing study of hybrid wind/PV/diesel power generation unit. *Solar Energy, 85*(1), 100–110. doi:10.1016/j.solener.2010.10.018

Bernal-Agustín, J. L., Dufo-López, R., & Rivas-Ascaso, D. M. (2006). Design of isolated hybrid systems minimizing costs and pollutant emissions. *Renewable Energy, 31*(14), 2227–2244. doi:10.1016/j.renene.2005.11.002

Cristóbal-Monreal, I. R., & Dufo-López, R. (2016). Optimisation of photovoltaic-diesel-battery stand-alone systems minimising system weight. *Energy Conversion and Management, 119*, 279–288. doi:10.1016/j.enconman.2016.04.050

Diaf, S., Belhamel, M., Haddadi, M., & Louche, A. (2008). *Technical and economic assessment of hybrid photovoltaic / wind system with battery storage in Corsica island.* doi:10.1016/j.enpol.2007.10.028

Dincer, I., Rosen, M. A., & Ahmadi, P. (2017). Optimization of Energy Systems. doi:10.1002/9781118894484

Ding, Z., Hou, H., Yu, G., Hu, E., Duan, L., & Zhao, J. (2019). Performance analysis of a wind-solar hybrid power generation system. *Energy Conversion and Management, 181*(September), 223–234. doi:10.1016/j.enconman.2018.11.080

Dufo-López, R., Bernal-Agustín, J. L., Yusta-Loyo, J. M., Domínguez-Navarro, J. A., Ramírez-Rosado, I. J., Lujano, J., & Aso, I. (2011). Multi-objective optimization minimizing cost and life cycle emissions of stand-alone PV-wind-diesel systems with batteries storage. *Applied Energy, 88*(11), 4033–4041. doi:10.1016/j.apenergy.2011.04.019

Erdinc, O., & Uzunoglu, M. (2012). Optimum design of hybrid renewable energy systems: Overview of different approaches. In Renewable and Sustainable Energy Reviews (Vol. 16, Issue 3, pp. 1412–1425). Elsevier Ltd. doi:10.1016/j.rser.2011.11.011

Guo, S., He, Y., Pei, H., & Wu, S. (2020). The multi-objective capacity optimization of wind-photovoltaic-thermal energy storage hybrid power system with electric heater. *Solar Energy, 195*(August), 138–149. doi:10.1016/j.solener.2019.11.063

Halabi, L. M., & Mekhilef, S. (2018). Flexible hybrid renewable energy system design for a typical remote village located in tropical climate. *Journal of Cleaner Production, 177*, 908–924. doi:10.1016/j.jclepro.2017.12.248

Helioscsp. (2014). *225 MW from concentrated solar power in Jordan.* Author.

Hirbodi, K., Enjavi-Arsanjani, M., & Yaghoubi, M. (2020). Techno-economic assessment and environmental impact of concentrating solar power plants in Iran. *Renewable & Sustainable Energy Reviews, 120*, 109642. Advance online publication. doi:10.1016/j.rser.2019.109642

Kabalci, E. (n.d.). *Hybrid renewable energy systems and microgrids.* Academic Press.

Kalantar, M., & Mousavi, G. S. M. (2010). Dynamic behavior of a stand-alone hybrid power generation system of wind turbine, microturbine, solar array and battery storage. *Applied Energy, 87*(10), 3051–3064. doi:10.1016/j.apenergy.2010.02.019

Keyif, E., Hornung, M., & Zhu, W. (2020). Optimal configurations and operations of concentrating solar power plants under new market trends. *Applied Energy, 270*(May), 115080. doi:10.1016/j.apenergy.2020.115080

Khan, F. A., Pal, N., & Saeed, S. H. (2018). Review of solar photovoltaic and wind hybrid energy systems for sizing strategies optimization techniques and cost analysis methodologies. *Renewable & Sustainable Energy Reviews, 92*(March), 937–947. doi:10.1016/j.rser.2018.04.107

Koutroulis, E., & Kolokotsa, D. (2010). Design optimization of desalination systems power-supplied by PV and W/G energy sources. *Desalination, 258*(1–3), 171–181. doi:10.1016/j.desal.2010.03.018

Koutroulis, E., Kolokotsa, D., Potirakis, A., & Kalaitzakis, K. (2006). Methodology for optimal sizing of stand-alone photovoltaic/wind-generator systems using genetic algorithms. *Solar Energy, 80*(9), 1072–1088. doi:10.1016/j.solener.2005.11.002

Kumar, R., Gupta, R. A., & Bansal, A. K. (2013). Economic analysis and power management of a stand-alone wind/photovoltaic hybrid energy system using biogeography based optimization algorithm. *Swarm and Evolutionary Computation, 8*, 33–43. doi:10.1016/j.swevo.2012.08.002

Lagorse, J., Paire, D., & Miraoui, A. (2009). Hybrid stand-alone power supply using PEMFC, PV and battery Modelling and optimization. *2009 International Conference on Clean Electrical Power, ICCEP 2009*, 135–140. 10.1109/ICCEP.2009.5212069

Liu, H., Zhai, R., Fu, J., Wang, Y., & Yang, Y. (2019). Optimization study of thermal-storage PV-CSP integrated system based on GA-PSO algorithm. *Solar Energy, 184*(December), 391–409. doi:10.1016/j.solener.2019.04.017

Mohammadi, K., Khanmohammadi, S., Khorasanizadeh, H., & Powell, K. (2020). A comprehensive review of solar only and hybrid solar driven multigeneration systems: Classifications, benefits, design and prospective. *Applied Energy, 268*(March), 114940. Advance online publication. doi:10.1016/j.apenergy.2020.114940

Motevasel, M., Seifi, A. R., & Niknam, T. (2013). Multi-objective energy management of CHP (combined heat and power)-based micro-grid. *Energy, 51*, 123–136. doi:10.1016/j.energy.2012.11.035

Nafey, A. S., Sharaf, M. A., & García-Rodríguez, L. (2010). Thermo-economic analysis of a combined solar organic Rankine cycle-reverse osmosis desalination process with different energy recovery configurations. *Desalination, 261*(1–2), 138–147. Advance online publication. doi:10.1016/j.desal.2010.05.017

Olatomiwa, L., Mekhilef, S., & Ohunakin, O. S. (2016). Hybrid renewable power supply for rural health clinics (RHC) in six geo-political zones of Nigeria. *Sustainable Energy Technologies and Assessments, 13*, 1–12. doi:10.1016/j.seta.2015.11.001

Ould Bilal, B., Sambou, V., Ndiaye, P. A., Kébé, C. M. F., & Ndongo, M. (2010). Optimal design of a hybrid solar-wind-battery system using the minimization of the annualized cost system and the minimization of the loss of power supply probability (LPSP). *Renewable Energy*, *35*(10), 2388–2390. doi:10.1016/j.renene.2010.03.004

Petrescu, S., Petre, C., Costea, M., Malancioiu, O., Boriaru, N., Dobrovicescu, A., Feidt, M., & Harman, C. (2010). A methodology of computation, design and optimization of solar Stirling power plant using hydrogen/oxygen fuel cells. *Energy*, *35*(2), 729–739. doi:10.1016/j.energy.2009.10.036

Rani, D., Jain, S. K., Srivastava, D. K., & Perumal, M. (2013). Genetic Algorithms and Their Applications to Water Resources Systems. In Metaheuristics in Water, Geotechnical and Transport Engineering (pp. 43–78). Elsevier Inc. doi:10.1016/B978-0-12-398296-4.00003-9

Rojas-Zerpa, J. C., & Yusta, J. M. (2015). Application of multicriteria decision methods for electric supply planning in rural and remote areas. In *Renewable and Sustainable Energy Reviews* (Vol. 52, pp. 557–571). Elsevier Ltd., doi:10.1016/j.rser.2015.07.139

Saffari, M., de Gracia, A., Fernández, C., Belusko, M., Boer, D., & Cabeza, L. F. (2018). Optimized demand side management (DSM) of peak electricity demand by coupling low temperature thermal energy storage (TES) and solar PV. *Applied Energy*, *211*, 604–616. doi:10.1016/j.apenergy.2017.11.063

SAM. (2021, February 18). *CSP Cost Data*. SAM.

Sharaf Eldean, M. A., Rafi, K. M., & Soliman, A. M. (2017). Performance analysis of different working gases for concentrated solar gas engines: Stirling & Brayton. *Energy Conversion and Management*, *150*(August), 651–668. doi:10.1016/j.enconman.2017.08.008

Shboul, B., AL-Arfi, I., Michailos, S., Ingham, D., AL-Zoubi, O. H., Ma, L., Hughes, K., & Pourkashanian, M. (2021a). Design and Techno-economic assessment of a new hybrid system of a solar dish Stirling engine instegrated with a horizontal axis wind turbine for microgrid power generation. *Energy Conversion and Management*, *245*, 114587. Advance online publication. doi:10.1016/j.enconman.2021.114587

Shboul, B., AL-Arfi, I., Michailos, S., Ingham, D., Ma, L., Hughes, K. J., & Pourkashanian, M. (2021). A new ANN model for hourly solar radiation and wind speed prediction: A case study over the north & south of the Arabian Peninsula. *Sustainable Energy Technologies and Assessments*, *46*, 101248. Advance online publication. doi:10.1016/j.seta.2021.101248

Singh, S., Singh, M., & Kaushik, S. C. (2016). A review on optimization techniques for sizing of solar-wind hybrid energy systems. In International Journal of Green Energy (Vol. 13, Issue 15, pp. 1564–1578). Taylor & Francis. doi:10.1080/15435075.2016.1207079

Siva Reddy, S. C. K., & S. K. T. (2012). Exergetic analysis and performance evaluation of parabolic dish Stirling engine solar power plant. *Archives of Thermodynamics, 33*(4), 23–40. doi:10.1002/er

Solargis. (n.d.). *Solar resource maps of Jordan*. Retrieved July 3, 2020, from https://solargis.com/maps-and-gis-data/download/jordan

Starke, A. R., Cardemil, J. M., Escobar, R., & Colle, S. (2018). Multi-objective optimization of hybrid CSP+PV system using genetic algorithm. *Energy*, *147*, 490–503. doi:10.1016/j.energy.2017.12.116

Tafreshi, S. M. M., Zamani, H. A., Ezzati, S. M., Baghdadi, M., & Vahedi, H. (2010). Optimal unit sizing of distributed energy resources in microgrid using genetic algorithm. *Proceedings - 2010 18th Iranian Conference on Electrical Engineering, ICEE 2010*, 836–841. 10.1109/IRANIANCEE.2010.5506961

The Wind Power. (n.d.). Retrieved May 10, 2020, from https://www.thewindpower.net/windfarms_list_en.php

Wang, X. Q., Li, X. P., Li, Y. R., & Wu, C. M. (2015). Payback period estimation and parameter optimization of subcritical organic Rankine cycle system for waste heat recovery. *Energy*, *88*, 734–745. https://doi.org/10.1016/j.energy.2015.05.095

Xu, D., Kang, L., Chang, L., & Cao, B. (2005). *Optimal sizing of standalone hybrid wind / PV power systems using genetic algorithms.* Academic Press.

Yang, H., Wei, Z., & Chengzhi, L. (2009). Optimal design and techno-economic analysis of a hybrid solar-wind power generation system. *Applied Energy*, *86*(2), 163–169. https://doi.org/10.1016/j.apenergy.2008.03.008

Yang, H., Zhou, W., Lu, L., & Fang, Z. (2008). Optimal sizing method for stand-alone hybrid solar-wind system with LPSP technology by using genetic algorithm. *Solar Energy*, *82*(4), 354–367. https://doi.org/10.1016/j.solener.2007.08.005

Yang, Y., Guo, S., Liu, D., Li, R., & Chu, Y. (2018). Operation optimization strategy for wind-concentrated solar power hybrid power generation system. *Energy Conversion and Management*, *160*(January), 243–250. https://doi.org/10.1016/j.enconman.2018.01.040

Yunus, A., & Çengel, M. A. B. (2015). *Thermodynamic an Engineering Approach* (8th ed.). McGraw-Hill Education.

Zayed, M. E., Zhao, J., Li, W., Elsheikh, A. H., Zhao, Z., Khalil, A., & Li, H. (2020). Performance prediction and techno-economic analysis of solar dish/stirling system for electricity generation. *Applied Thermal Engineering*, *164*(April), 114427. https://doi.org/10.1016/j.applthermaleng.2019.114427

KEY TERMS AND DEFINITIONS

Concentration Ratio: A ratio of the dish area to the receiver area.

Genetic Algorithm: A heuristic optimisation technique, which is biologically inspired using to find a set of optimal solutions.

Levelised Cost of Electricity: An economic parameter representing a cost per unit of electricity generated.

Microgrid: An autonomous and local energy network integrated with a central control system to generate power in multiple scales.

Power Coefficient: A parameter indicates the efficiency of wind turbine to convert the kinematic energy in the wind into electrical energy.

Rim Angle: A ratio of the focal length to the concentrator diameter.

Solar Dish: A solar thermal device producing high temperature to generate electricity.

APPENDIX

Table 11. Battery bank mathematical model

Parameter	Equation
Single battery Amber hour, AH	$AH_b \times I_b \times t_d$
Single Battery storage, Wh	$E_b = AH_b \times V_b$
Single Battery power, W	$P_b = \frac{E_b \times DOD \times \eta_b}{OH \times NOC}$
Total battery bank power, W	$P_{b,tot} = P_b \times NOB$
Load current, A	$I_l = \frac{P_{b,tot}}{V_l}$

Source: (Shboul et al., 2021a)

Table 12. HWT mathematical model

Parameter	Equation
Air temp based on site elevation, °C	$T_{air} = 15.5 - \frac{19.83 \times H_s}{3048}$
The air density, kg/m³	$\rho = \frac{P_{air} \times 100}{0.287 \times (T_{air} + 273.15)}$
Air density at sea level, kg/m³	$\rho_{air,s} = \rho \times e^{\left(\frac{-1 \times 0.297 \times H_s}{3048}\right)}$
Rotor swept area, m²	$A_r = \pi \times \left(\frac{D_r}{2}\right)^2$
Air mass flow rate, kg/s	$M_{air} = \rho a_{ir,s} \times Ar \times Vw_r$
Wind speed at the blades, m/s	$V_u = \frac{V_{wr}}{V_o / V}$
Power coefficient	$C_p = 4 \times \frac{V_{wr}^2}{V_u^2} \times \left(1 - \frac{V_{wr}}{V_u}\right)$
Required wind power, kW	$P_w = \frac{\left(\frac{1}{2} \times \rho_{air,s} \times A_r \times \left(Vw_r^3\right)\right)}{1000}$
Power delivered by the turbine, kW	$P_{mech} = P_w \times C_p$
Axial force on the turbine wheel, kN	$F_x = \frac{4}{9000} \times \rho_{air,s} \times A_r \times \left(V_{wr}^2\right)$

continued on following page

Table 12. Continued

Parameter	Equation
Rotor speed, rpm	$RPM_r = \frac{60 \times V_u}{\pi \times D_r}$
Rotor speed, *rev/s*	$\omega = \frac{(2 \times \pi \times RPM_r)}{60}$
Rotor torque, N.m	$T_{or} = \frac{(1000 \times P_{mech})}{\omega}$
Power outlet from generator, kW	$P_g = \frac{P_{mech}}{\eta_g}$
Output current, A	$I_g = \frac{1000 \times P_g}{\sqrt{3} \times V \times FP}$
Net power developed, kW	$P_{net} = P_{mech} - (RL \times P_g)$
Generator torque, N.m	$T_g = \frac{1000 \times P_{net}}{w}$
Total Farm power, kW	$P_{form} = P_{net} \times N_{wt}$
Optimum spacing in a row, m	$X_s = 12 \times D_r$
Optimum spacing in cross-wind direction, m	$Y_s = 3 \times D_r$
total land area, km^2	$A_{t,tot} = 2 \times X_s \times Y_s \times N_{wt}$

Source: (Shboul et al., 2021a)

Table 13. CPSD-SE mathematical model

Parameter	Equation
Piston volume, cm^3:	$V_p = Stroke \times \frac{\pi}{4} \times D_p^2$
Dish area, m^2	$A_{dish} = \frac{\pi}{4} \times D_c^2$
Stirling engine efficiency	$\eta_{SE} = 0.5 \times \left[1 - \left(\frac{T_l + 273}{T_h + 273}\right)\right]$
Total efficiency of the module	$\eta o = \eta SE_{\times} \eta gen_{\times} \eta c \times {}_{\eta} r$
Stirling engine compression ratio	$CR_{SE} = e^{\left(\frac{1 \times C_v}{R \times \left(\frac{\left(\frac{1 - \frac{1}{\left(\frac{T_h + 273}{T_l + 273}\right)}}{\eta_{SE}}\right) - 1}{1 - \left(\frac{1}{\left(\frac{T_h + 273}{T_l + 273}\right)}\right)}\right)}\right)}$

continued on following page

Table 13. Continued

Parameter	Equation
Stirling high pressure, kPa	$P_h = 100 \times P_{atm} \times CR_{SE} \times \left(\frac{T_h + 273}{T_1 + 273}\right)$
Stirling pressure ratio	$R_{PSE} = \frac{P_h}{P_{atm} \times 100}$
Maximum specific volumes, m^3/kg	$v_{max} = \frac{R \times (T_1 + 273)}{100 \times P_{atm}}$
Minimum specific volumes, m^3/kg	$v_{min} = \frac{v_{max}}{CR_{SE}}$
Mean pressure, kPa	$P_m = \frac{P_{atm} \times (CR_{SE} + 1) \times \left(\left[\frac{T_h + 273}{T_l + 273}\right] + 1\right)}{4}$
Stirling engine power, kW	$P_{SE} = \frac{\eta_o \times I_s \times A_{dish}}{1000}$
Total plant power of CPSD-SE, kW	$P_{CPSD\text{-}SE} = N_{dishes} \times P_{SE}$
Top cycle temperature, °C	$T_h = \frac{60 \times 10^9 P_{SE} \times (T_h + T_l)}{V_p \times NOC_{SE} \times P_{mean} \times \vec{V}_{SE} \times \pi \times (T_h - T_l)}$
Rim angle ratio, f/D_c	$RAR = 1.003 \times e^{-\left(\frac{\psi_r - 11.28}{13.86}\right)^2} + 2.186 \times e^{-\left(\frac{\psi_r + 100.2}{127.6}\right)^2}$
Focal length, m	$' = \frac{D_c}{4 \times tan\left(\frac{\psi_r}{2}\right)}$
Dish height, m	$H_{dish} = \frac{D_c^2}{16 \times '}$
Receiver area m^2	$A_r = \frac{\pi}{4} \times D_r^2$
Concentration ratio	$CR_{dish} = \frac{A_{dish}}{A_r}$
Total surface area, m^2	$A_{dish,tot} = A_{dish} \times N_{dishes}$

Source: (Shboul et al., 2021a); (Sharaf Eldean et al., 2017)

Table 14. Cost analysis

Economic parameters	
Interest rate, %	$i = 5$
Battery lifetime, year	$LT_b = 5$
Amortization factor, 1/y	$A_f = \frac{i \times (1+i)^{LT_b}}{(1+i)^{LT_b} - 1}$
Plant lifetime, year	$LT_p = 25$
Fixed charge rate	$FCR = 0.098$
Batteries Bank	
Battery cost, $	$C_b = 100$
Variable operating cost of the batteries, $/kWh	$VOC_b = 0.07$ (Charging Electricity Price)
Direct capital cost of the batteries bank, $	$CC_b = 5 \times C_b \times N_b$, where N_b is No. of batteries.
Annual capital cost of the batteries bank, $/yr	$ACC_b = CC_b \times A_f$
Fixed operating cost of the batteries bank, $/y	$FOC_b = 0.05 \times CC_b$
Annual total cost of the batteries bank, $/yr	$ATC_b = ACC_b + FOC_b$
Hourly total cost of the batteries bank, $/hr	$HTC_b = \frac{ATC_b}{365 \times 24}$
Horizontal Axis Wind Turbine (HWT)	
Normalised capital cost, $/kWe	$POC = 1628$
Variable operating cost of the turbines, $/kWh	$VOC_t = 0$
Direct capital cost of the turbines, $	$CC_t = POC \times P_m$
Indirect capital costs of the turbines, $	$ICC_t = WTC + CTC + OCC$ or $ICC_t = 0.86 \times CC_t$
where, the wind turbine cost share: $WTC=0.65 \times CC_t$, construction cost share (Civil works): $CTC=0.16 \times CC_t$, and other capital cost share: $OCC=0.05 \times CC_t$	
Total capital cost of the turbines, $	$TCC_t = CC_t + ICC_t$
Annual capital cost of the turbines, $/yr	$ACC_t = TCC_t \times A_f$
Fixed operating cost of the turbines, $/yr	$FOC_t = FCR \times CC_t$
Annual total cost of the turbines, $/yr	$ATC_t = ACC_t + FOC_t$
Hourly total cost of the turbines, $/hr	$HTC_t = \frac{ATC_t}{24 \times 365}$
Solar Dish Stirling Engine (CPSD-SE)	
SE cost, $/kWe	CSE=370 to 400$/kW
Receiver cost, $/kWe	CCR = 185$/kW
Dish concentrator cost, $/m²	$CDC = 300\$/m^2$
Site cost is 2.2$/m²	$SIC = 2.2\$/m^2$
Variable operating cost of the CPSD-SE, $/kWh	$VOC_{dish} = 0.06$
Direct capital cost of the CPSD-SE, $	$CC_{dish} = (COP \times P_{SE}) + ([CDC + SIC] \times A_C)$ Where $COP = CSE + CCR$
Indirect capital costs of the CPSD-SE, $	$ICC_{dish} = CPEC + CGC + OCC$

continued on following page

Table 14. Continued

Where, the construction, procurement, and engineering cost share are calculated (H. M. Ali, 2020), (Agency, 2012): $CPEC=0.16\times CC_{dish}$, contingency cost share: $CGC=0.10\times CC_{dish}$, other capital cost share: $OCC=0.03\times CC_{dish}$	
Total capital cost of the CPSD-SE, \$	$TCC_{dish}=CC_{dish}+ICC_{dish}$
Annual capital cost of the CPSD-SE, \$/yr	$ACC_{dish}=TCC_{dish}\times A_f$
Fixed operating cost of the CPSD-SE, \$/yr	$FOC_{dish}=FCR\times CC_{dish}$
Annual total cost of the CPSD-SE, \$/yr	$ATC_{dish}=ACC_{dish}+FOC_{dish}$
Hourly total cost of the CPSD-SE, \$/hr	$HTC_{dish}=\frac{ATC_{dish}}{24\times 365}$
Total Plant Cost	
Total capital cost, \$	$TCC_{tot}=TCC_b+TCC_t+TCC_{dish}$
Total annual cost, \$/y	$ATC_{tot}=ATC_b+ATC_t+ATC_{dish}$
Total hourly costs, \$/h	$THC_{tot}=\frac{ATC_{tot}}{8760}$
Total variable operating costs, \$/kWh	$VOC_{tot}=VOC_b+VOC_t+VOC_{dish}$
Total Levelised cost of energy, \$/kWh	$LCOE_{tot}=\frac{THC_{tot}}{Loadfactor\times P_{tot}}+VOC_{tot}$
Annual electricity savings, \$	$Electricity\ cost=LCOE_{tot}\times Annual\ electricity\ generated$
Payback period, year	$Paybackperiod=\frac{TCC_{tot}}{Electricitycost}$

Source: (Shboul et al., 2021a)

Table 15. The developed GA multi-objective functions

GA function code
• The Power coefficient function: $$C_p=\left(4\times\left(\frac{V_{wr}^{\ 2}}{\left(\frac{V_{wr}}{V_o/V}\right)^2}\right)\times\left(1-\left(\frac{V_{wr}}{\frac{V_{wr}}{V_o/V}}\right)\right)\right)$$
• The mechanical power delivered by the turbine, kW: $$P_m=\left(\left(\frac{0.5\times\left(\left(\frac{P_{air}\times 100}{0.287\times\left(\left(15.5-\left(19.83\times\frac{H_s}{3048}\right)\right)+273.15\right)}\right)\times exp\left(-1\times 0.297\times\frac{10}{3048}\right)\right)\times\left(\left(\frac{\pi}{4}\right)\times\left(D_r^{\ 2}\right)\times\left(V_{wr}^{\ 2}\right)\right)}{1}\right)\times C_p\times 1\right)$$

continued on following page

Table 15. Continued

GA function code
• The Stirling engine power, kW: $P_{SE}=\frac{\left(\left(0.5\times\left(1-\left(\frac{T_l+273}{T_h+273}\right)\right)\right)\times\eta_{gen}\times\eta_r\times\eta_c\right)\times DNI\times A_c}{1000}$
• Total levelised cost of energy is obtained as: $LCOE_{tot}=\left(\left(\frac{\left(\frac{(140.5\times NOB)+(0.230\times POC\times P_m)+\left[\left(108.3\times\left(\frac{\left(\left(0.5\times\left(1-\left(\frac{T_l+273}{T_h+273}\right)\right)\right)\times\eta_{gen}\times\eta_r\times\eta_c\right)\times DNI\times A_c}{1000}\right)\right)+(57.3\times A_c)\right]}{8760}\right)}{LF\times(P_{SE}+P_{mech})}\right)+0.13\right)$
• Average overall annual efficiency is calculated as: $\eta_{tot}=\frac{(P_{mech}+P_{SE})\times1000}{(DNI\times A_c\times N_{dishes})+\left(\frac{0.5\times1.25\times\left(\frac{\pi}{4}\times D_r^2\right)\times V_w^3}{1000}\times N_{wt}\right)}\times100$

Table 16. Random numbers within a specific range for load calculation

```
%% Hours:
% n = 8760 during the year.
% n = The number of hours (n=24*number of days in each month)
% n = 744 in Jan, Mar, May, Jul, Aug, Oct, Dec.
% n = 720 in Apr, Jun, Sep, Nov.
% n = 672 in Feb.
```

```
h1 = 1;
h2 = n;
Hour = sort(((h2-h1).*rand(n,1) + h1),'ascend');
%% Power, kW:
p1 = 100;
p2 = 1500;
Power = (p2-p1).*rand(n,1) + p1;
```

Chapter 10
Numerical and Experimental Investigations on a Bio-Inspired Design of Darrieus Vertical Axis Wind Turbine Blades With Leading Edge Tubercles

Nishant Mishra
https://orcid.org/0000-0002-2302-4609
Shiv Nadar University, India

Punit Prakash
https://orcid.org/0000-0002-6892-4504
Shiv Nadar University, India

Anand Sagar Gupta
Shiv Nadar University, India

Jishnav Dawar
Shiv Nadar University, India

Alok Kumar
Shiv Nadar University, India

Santanu Mitra
https://orcid.org/0000-0002-7141-4695
Shiv Nadar University, India

ABSTRACT

Various improvements can be made to Darrieus vertical axis wind turbines (VAWT) for maximum performance in an urban environment. One such improvement is the inclusion of bio-inspired leading-edge tubercles to increase the aerodynamic performance. These structures, found on the flippers of humpback whales, are believed to aid the mammal in quick maneuvering. The objective of the chapter is to investigate and compare the performance of a Darrieus type VAWT with the inclusion of leading edge tubercles. The performance of the turbine with leading-edge tubercles on the blades is compared with the turbine with normal blade, computationally (with computational fluid dynamics using transition SST turbulence model) and experimentally. The focus lies on building an experimental setup to compare the performance of leading-edge tubercles with the baseline turbine.

DOI: 10.4018/978-1-7998-8561-0.ch010

1 INTRODUCTION

Easily available to us in one of the greatest quantities as a source of power, the wind is also one of the most inexhaustible sources of energy. To harness the potential, of this energy, different kinds of turbines are used. Based upon the orientation of their axis of rotation, these turbines can be placed in two clusters – horizontal axis wind turbines (HAWTs) and vertical axis wind turbines (VAWTs). The fact that VAWTs can perform, irrespective of the direction of the winds, is what primarily segregates them from HAWTs. This is what gives the VAWTs an edge over the HAWTs, as the latter requires orientation in the direction of the wind, resulting in a need for additional mechanism, which means enhanced costs, in addition to more chances of failure. What makes VAWTs the most suited choice for application in urban environments is the fact that these turbines are omnidirectional. The impact of the direction of the wind does not affect them and they can support repeated occurrences of changes in wind direction that are violent or unsteady. The current designs are considered to be highly effective in comparison to the older designs, as has been shown by various proposed designs and testing. (Amano, 2017; Gupta, 2015).

Considering the factors that they can be installed near the ground, do not require a strict operating mechanism, and can take-off without high wind speed, are economical, safe for the environment, unaffected by irregular patterns or the direction of the wind, made us choose in VAWTS for our research.

The categories of Darrieus and Savonius in turbines are the two further classifications of VAWTs. Because of the high negative torque produced by the returning blade (Alom & Saha, 2018), the efficiency of the Savonius type turbine is comparatively less. The research is centered on Darrieus type turbines which work using the lift forces generated due to the aerofoil geometry of the blade. To create a cost-effective model design of a straight blade VAWT for small-scale power generation is the focus of the study. Parasitic drag and induced drag are the significant reasons behind the low performance of Darrieus turbines (Islam et al., 2008). Tip vortices play center stage in induced drag.

To perceive and advance efficient HAWTs, major work has been carried out in the past few years, but not much attention had been laid on VAWTs. Work on many criteria that affect the performance of a VAWT were published by Islam et. al (Islam et al., 2008). In-depth research was done on the consequences of the utilization of many numbers of design parameters, which resulted in more cost-effective and efficient turbines. We learn from the additional study by Persico et. al (Amato et al., 2013) that large-scale vortices have a powerful impact on the tip region of the wake shed by the H-shape turbine. As a consequence of the periodic fluctuation of the blade aerodynamic loading, these vortices pulsate notably during the period. This resulted in a great advantage in the execution of the present work to develop an efficient turbine for carrying out experimental studies. As observed in a recent study done on various wingtip devices, namely endplates and winglets, by Mishra et al, it was found that the endplates improved the performance of a Darrieus type VAWT significantly (Mishra et al., 2018).

The evolution of all living organisms to a well-adapted structure and material over all these years has been through natural selection. New technologies inspired by biological solutions at macro and nanoscales have taken off, courtesy of Bio-mimetics. All through the existence of the human race, the species have bought from nature the answers to problems and challenges. Nature has provided solutions to engineering problems such as self-healing capabilities, tolerance, and resistance to harsh environmental and climatic exposures, hydrophobicity, ability to self-assemble and harness solar energy. Scientists, researchers, and academicians have been ever enthralled to study nature and take inspiration from it. In the field of wind turbines, research has been done on bio-inspired materials for better efficiency. Ryoichi S Amano et al introduced a novel concept by using self-healing polymer composites for turbine blades. (Arun et al.,

2015). This helped the turbine blades to self-heal when the cracks were formed due to fatigue. Further investigations into the bio-inspired designs and materials led to another novel idea of self-healing in the eventuality of failures due to bending stresses (Matt, 2017). A special issue for the Second International Conference on 2016 Next Generation of Wind Energy (ICNGWE) also included research based on bio-inspiration (Amano et al., 2017).

Biomimetic adds new avenues into engineering design that may otherwise go invalidated. The different structures may propose a method to increase maneuverability and stealth. Taking inspiration from nature, another notable research that just came up a few years ago by Watts et. al (Watts & Fish, 2001) studies the fins of humpback whales as shown in Figure 1 (Ask Nature, n.d.).

Figure 1. Leading Edge tubercles on a humpback whale (Ask Nature, n.d.)

It was shown that the tubercles on the leading edges of humpback whale flippers augment maneuverability while capturing prey. It is thus suggested that tubercles may be used as adaptations that increase the functionality of the blades of the turbine. With tubercles on the blades, it was found that there was a significant increment in lift and lift to drag ratio while observing a reduction in induced drag. It has been shown that leading-edge modifications of streamlined bodies can offer performance enhancements (Watts & Fish, 2001).

Work done on understanding the effects of various kinds of tubercle shapes (both sinusoidal and spherical) on the performance of an airfoil blade were studied by Aftab *et. al* (Aftab & Ahmad, 2017). It showed that tubercles when added to the leading edge of the blade, reduce the induced drag by shortening the wake region. The paper also provides key insights into the computational fluid analysis of the tubercle blade and also led the authors to choose the right kind of geometry to work on and implement onto the actual turbine to test on.

Some experimental studies do provide the quantification of performance enhancement for different airfoils. Around 5% enhancement in the lift, 11% decrease in induced drag, and 18% increment in the lift to drag ratio for the airfoil NACA634-021 were reported by Watts et. al (Watts & Fish, 2001). Approximately 41% increment for stall angle, 5% increase in maximum lift coefficient and a reduction in total drag in the post-stall regime for airfoil NACA0020 was reported by Miklosovic et. al (Miklosovic et al., 2004). A decrease in the maximum lift coefficient and stall angle pre-stall but a 50% increase in post-stall regime was reported by Johari et.al (Johari et al., 2009). The tubercle profile can be obtained with the formula stated in Equation (1) Seyhan M., and Sarioglu M. (Seyhan et al., 2021)

$$\frac{a\cos(2\pi x)}{\lambda} \tag{1}$$

Where, a is the amplitude and λ is the wavelength. Aerodynamic post stall was studied by Leknys at. al. stated the post stall behavior of leading edge tubercle was significantly improved compared to symmetrical blade at low tip-speed ratio (Leknys et al., 2018; Yan et al., 2021). Arunvinthan et. al. observed lift characteristics increase with increasing turbulence intensity irrespective of variation of an incremental or detrimental AOA values (Arunvinthan et al., 2020). Wang at. al. (Wang et al., 2018; Wang & Zhuang, 2017) uses taguchi method along with CFD to report the improvement using serration on leading edge of a blade, the coefficient of power improved by 18.3%. Suppuration of dynamic stall was observed from 75° to 160° of azimuth angle. Lositano and Danao published work on three bladed Darrieus VAWT with cambered Tubercle blades using CFD with NACA 0025 blades and found the performance to be detrimental but states this effect to be specific for the choosed design and operating condition. Lou et al. (Lositano & Danao, 2019) applied biomimetic design to a high pressure turbine stage and the loos of energy has been reduced.

2 COMPUTATIONAL FLUID DYNAMICS

To create a valid setup to compare our geometries, the setup studied by Aftab et. al (Aftab & Ahmad, 2017) were replicated. Their work was carried out on an airfoil with a NACA4415 blade profile to which sinusoidal and spherical tubercles of various amplitudes and wavelengths were added and compared to the normal turbine. Instead of checking the readings for all of the blades, a clean blade and a tubercle blade were chosen with amplitude .05c and wavelength as .25c, where c is the chord length. Thus, a blade with a NACA4415 blade profile was modeled using SOLIDWORKS. When it comes to tubercles, a sinusoidal profile on the leading edge was created as specified in the literature. The chord length (c) for both the blades was chosen to be 75mm while the span length (L) of the blades was chosen to be 100mm.

2.1 Domain Details

The CAD models created in SolidWorks were imported as STEP files into ANSYS workbench and an enclosing domain was created. This was created across the wing, which had a width equal to the length of the aerofoil using the design modeler. A distance of −1.3 times c and 10.3 times c is maintained from the aerofoil leading edge for the inlet and outlet. To mitigate confinement effects, the domain is elongated twice times c above and below the aerofoil. Here, two zones were created, the inside one close to the aerofoil to get a fine grid as shown in Figure 2 and Figure 3.

Figure 2. Mesh Domain

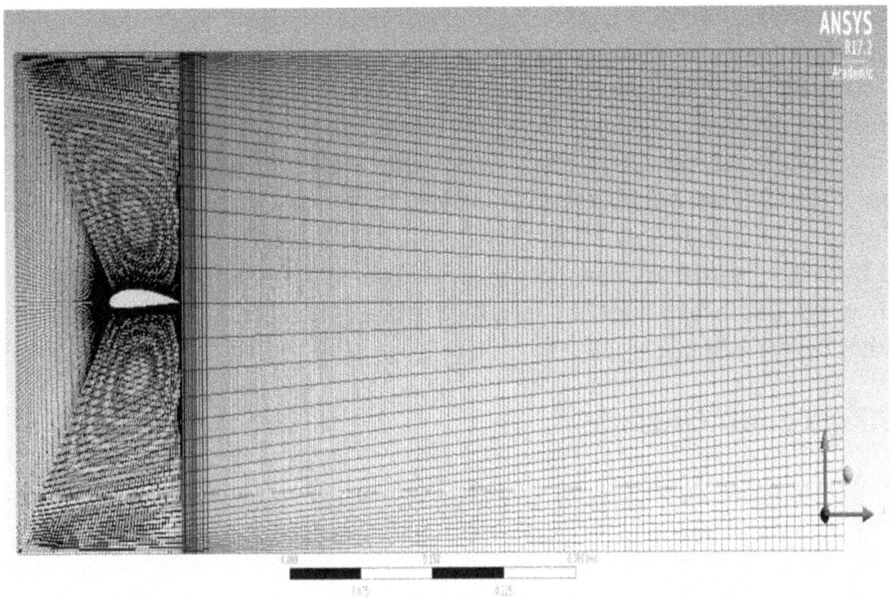

Figure 3. Meshed Domain (3D view)

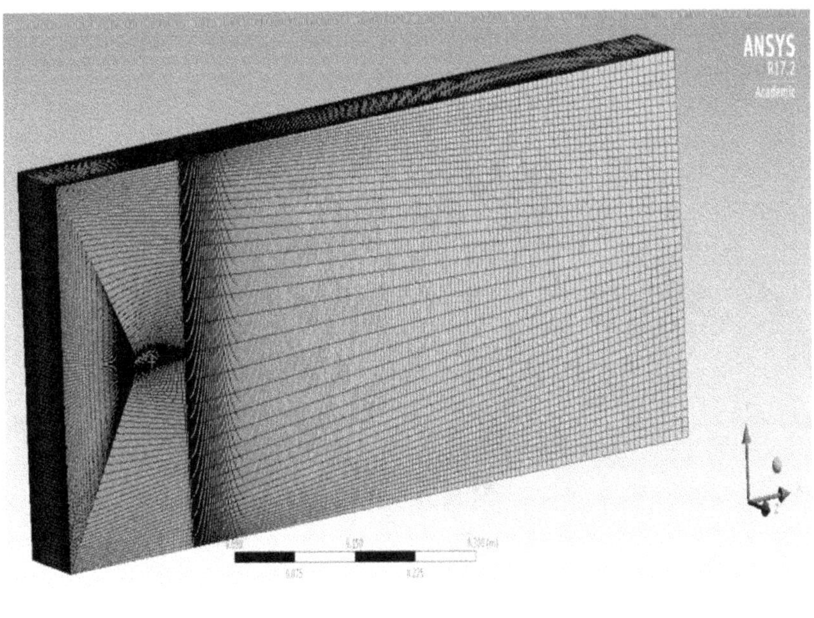

2.2 Computational Aspects

The education license of ANSYS for research in academic institutions was used to carry out the analysis of the steady-state simulations. The transition SST turbulence model was considered for the analysis. SST turbulence model is a well-established model to provide design solutions for aerodynamic applications with low Reynolds numbers.

Four transport equations are used in Transition SST to model the transition behavior. The computation load is substantially is less and the model is found more precise and accurate. Implementation of SIMPLE pressure coupling was done, and convergence criteria were set to 10-6 while analysis was done for accuracy of second order. The Angle of Attack (AoA) of 0° till 18° was chosen for the analysis. Conditions at sea level were deliberated for the input criterion e.g., pressure, density, and viscosity. The input velocity was maintained at 1.8 m/s for a chord based on a Reynolds number of 1.2e5. All the data obtained from the simulations were post-processed using Microsoft Excel. A comparison is done between our readings and the readings obtained by S. M. A. Aftab et. al (Matt, 2017) and was found to be similar within the error limits.

2.3 Simulations on NACA0018

Now, as the setup was ready, simulations were run on the chosen blade profile i.e. NACA0018 for clean and tubercles blade turbine. The only difference is in the leading edge profile of the tubercles. Since the blades were to be manufactured and tested on an actual turbine, getting a sinusoidal leading edge would require manufacturing techniques beyond the scope of those available to the authors. Thus, a simpler triangular wave geometry was used and the blade was designed using SOLIDWORKS as shown in Figure 4, Figure 5 and Figure 6 below.

Figure 4. Blade profile

Figure 5. Front view

Figure 6. Side view

A similar domain as used in the previous case was generated around both the blades as shown in Figure 7 and Figure 8.

Figure 7. Blade Meshing

Figure 8. Tubercle Blade Meshing

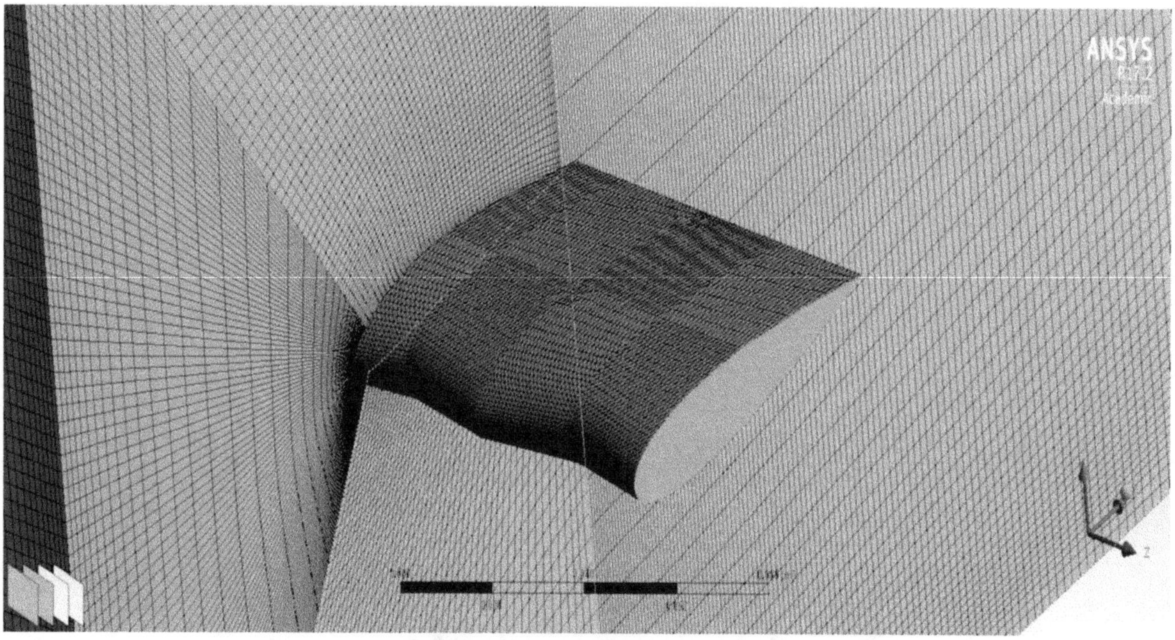

Again, the Transition SST turbulence model was used here. The simulation was done at inlet speeds varying from 4 m/s to 20 m/s. The drag and lift forces and drag and lift coefficients were calculated and compared which is shown in the graphs below as shown in Figure 9, Figure 10 and Figure 11 respectively.

These simulations provide us with enough data on NACA0018 to take this study further into the experimentation stage. As the graphs suggest, an average of 1.44 percent and 8.81 percent increase in average lift force and maximum lift force was found out respectively. Also, it was found that using tubercles on an airfoil blade decreased the average drag force by 2.64 percent, and the maximum reduction was found out to be 5.24 percent. It is worth noting that the Tubercle effect came into effect after a certain inlet velocity threshold was crossed.

3 EXPERIMENTATION

The designing of the blades, shaft, and connecting rod of the turbine was done in SolidWorks. The designed files were 3D printed. The total time taken for the printing of each of the base turbine and tubercles turbine was about 14 hours. The endplates were printed similarly. PLA material was used to print the turbine parts. The blades, shaft, and connecting rod were fixed to each other and the turbine was finally ready for testing, as shown in Figure 12, Figure 13, Figure 14 and Figure 15.

The 3D printed turbine was then made to test in the test section of the experimental setup as shown in Figure 16. A shaft was made to pass through the hollow shaft of the turbine supported by bearing at both ends. The upper part of the free shaft was connected to a dynamo to measure the voltage developed due to the rotation of the turbine at different inlet wind speeds. The RPM at different wind speeds was measured using a tachometer. The wind inlet speed was varied using an autotransformer. The base turbine, tubercles turbine, and tubercles turbine with endplates were tested simultaneously in the test section and the corresponding observations were recorded.

The graphs in Figure 17 and Figure 18 show that the Tubercle geometry out-performs the normal geometry and even more so in the case when end plates are attached to it.

Figure 9. C_L v/s Inlet Velocity

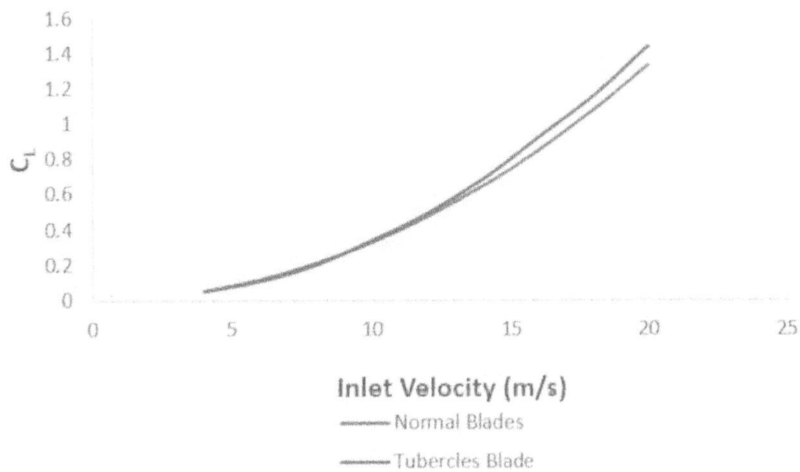

Figure 10. C_D v/s Inlet Velocity

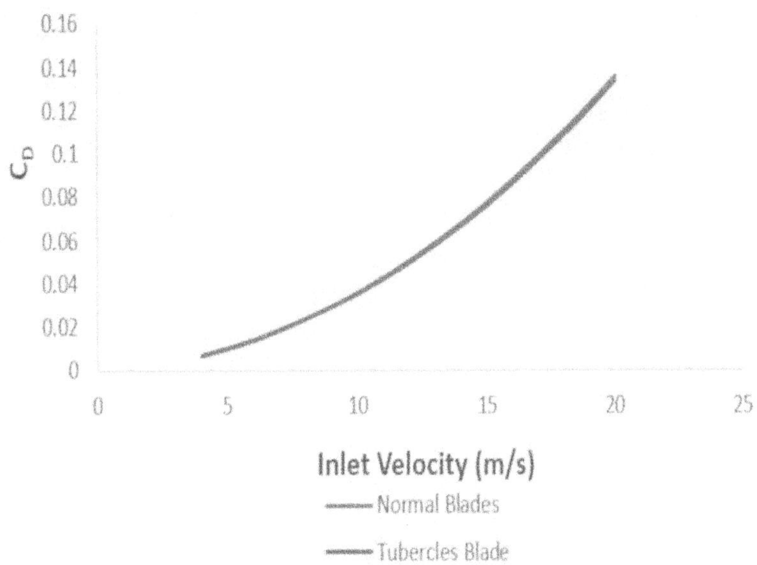

Figure 11. Lift to Drag ratio v/s Inlet Velocity

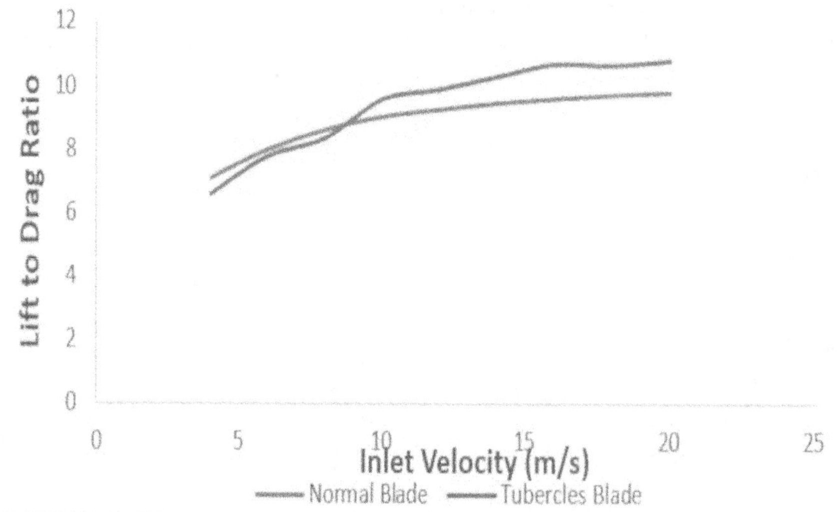

Figure 12. 3D printed turbine

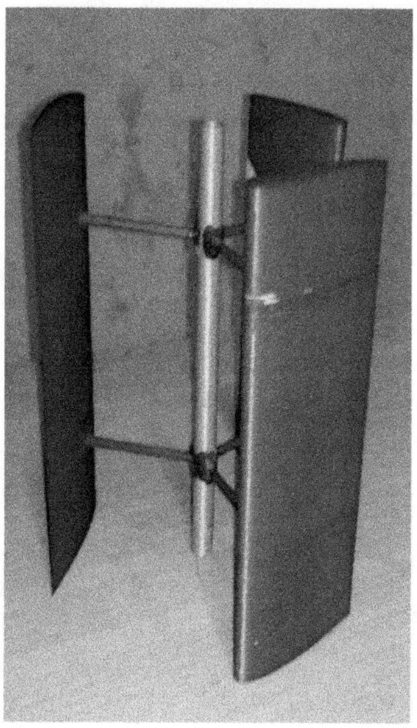

Figure 13. 3D printed tubercle turbine

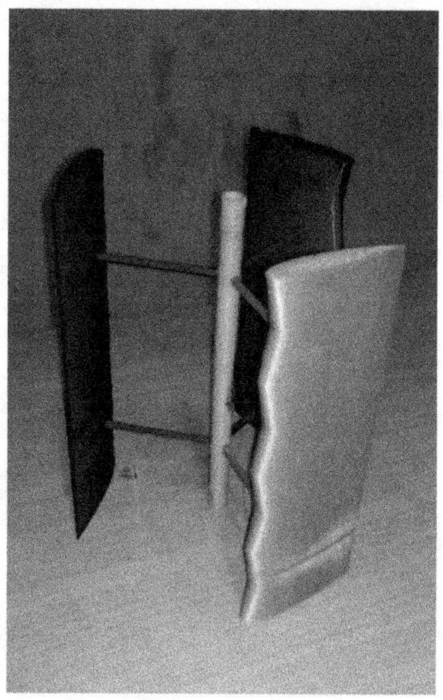

Figure 14. 3D Printed tubercle turbine with endplates

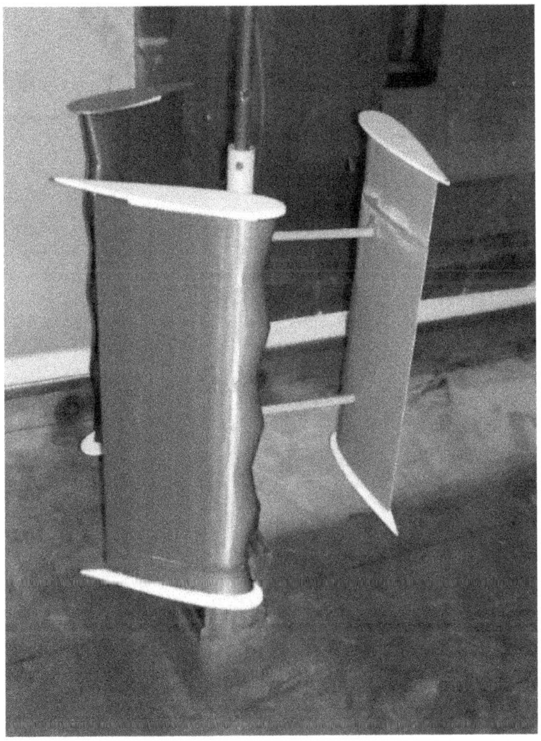

Figure 15. 3D Printed turbine with endplates

Figure 16. Experimental setup for testing of 3D printed turbine

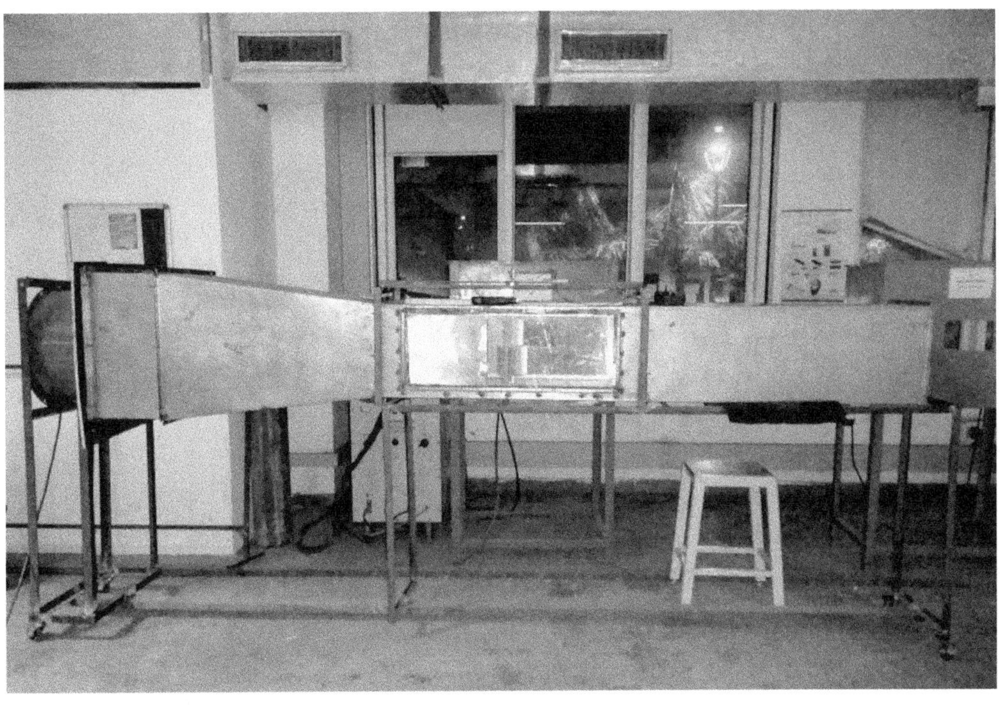

Figure 17. RPM vs V (Inlet) -Comparative performance of all the models of the turbine

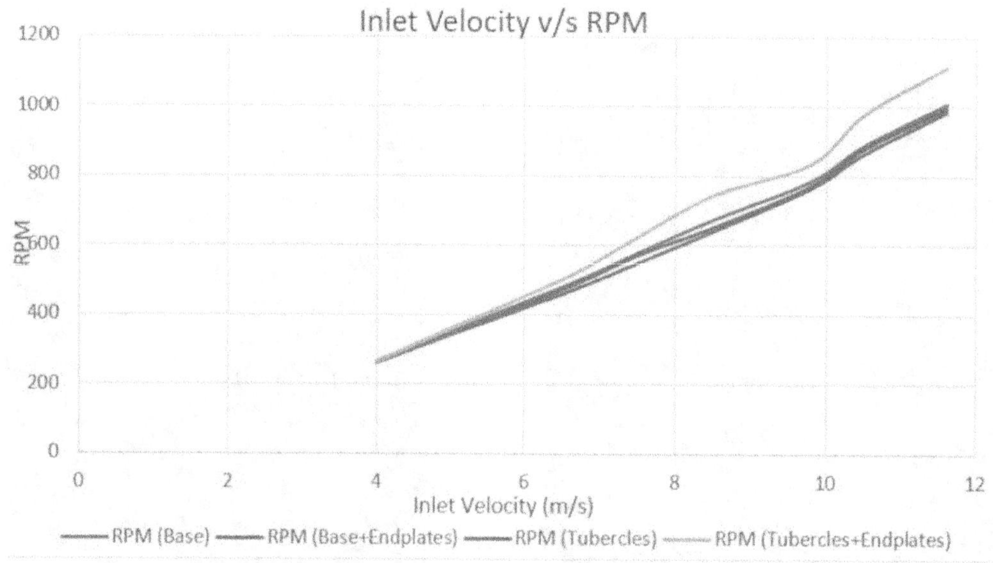

Figure 18. C_p vs V (Inlet) -Comparative performance of all the models of the turbine

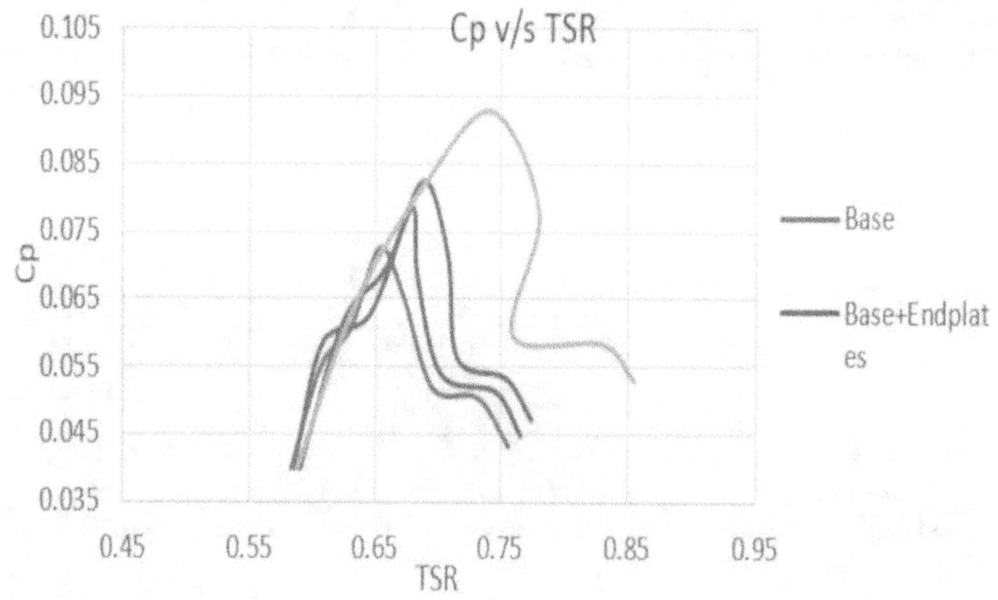

4 RESULTS

- Leading-edge tubercles show enhancement in efficiency but at lower speeds that enhancement is not significant.
- A combination of both, leading-edge tubercles and end plates show a significant increment in the turbine efficiency.

- For all configurations, efficiency increases up to a certain point only and after that, it starts to decrease. So to run the turbine most efficiently, that particular point needs to be achieved i.e. that particular value of TSR needs to be maintained to have the maximum efficiency.

ACKNOWLEDGMENT

The authors would like to thank Anand Sagar Gupta, Jishnav Dawar and Alok Kumar, former undergraduate students, Mahipal Singh, Sunil Gupta, Sanjay Singh and Amit Kumar, Lab Assistants, Shiv Nadar University for their help in establishing the experimental setup.

REFERENCES

Aftab, S. M. A., & Ahmad, K. A. (2017). CFD study on NACA 4415 airfoil implementing spherical and sinusoidal Tubercle Leading Edge. *PLoS One*, *12*(8), e0183456. doi:10.1371/journal.pone.0183456 PMID:28850622

Alom, N., & Saha, U. K. (2018). Four Decades of Research into the Augmentation Techniques of Savonius Wind Turbine Rotor. *Journal of Energy Resources Technology*, *140*(5), 050801. doi:10.1115/1.4038785

Amano, R., Sunden, B., & Gupta, A. (2017). Special Issue for the Second International Conference on 2016 Next Generation of Wind Energy (ICNGWE). *ASME Journal of Energy Resources Technology*. 10.1115/1.4037711

Amano, S. (2017). Review of Wind Turbine Research in 21st Century. *Journal of Energy Resources Technology*, *139*(5), 050801. doi:10.1115/1.4037757

Amato, Bedon, Castelli, & Benini. (2013). Numerical Analysis of the Influence of Tip Devices on the power coefficient of a VAWT. *International Journal of Mechanical, Aerospace, Industrial, Mechatronic and Manufacturing Engineering, 7.*

Arun, K. K. M., Strong, S., ElGammal, T., & Amano, R. S. (2015). Development of Novel Self-Healing Polymer Composites for Use in Wind Turbine Blades. *Journal of Energy Resources Technology*.

Arunvinthan, S., Pillai, S. N., & Cao, A. (2020). Aerodynamic characteristics of variously modified leading-edge potuberanced (LEP) wind turbine under various turbulent intensities. *Journal of Wind Engineering and Industrial Aerodynamics*, *202*, 104188. doi:10.1016/j.jweia.2020.104188

Ask Nature. (n.d.). Available: https://asknature.org/strategy/flippers-provide-lift-reduce-drag/

Gupta, A. K. (2015). Efficient Wind Energy Conversion: Evolution to Modern Design. *Journal of Energy Resources Technology*, *137*(5), 051201. doi:10.1115/1.4030109

Islam, M., Fartaj, A., & Carriveau, R. (2008). Analysis of the Design Parameters related to a fixed pitch straight bladed vertical axis wind turbine. Wind Engineering, 32. doi:10.1260/030952408786411903

Johari, H., Henoch, C., Custodia, D., & Levshin, A. (2009). Effects of leading edge protuberances on hydrofoil performance. *ExHFT-7.*

Leknys, R. R., Arjomandi, M., Kelso, R. M., & Birzer, C. H. (2018). Thin airfoil load control during post-stall and large pitch angles using leading-edge trips. *Journal of Wind Engineering and Industrial Aerodynamics, 179*, 80–91. doi:10.1016/j.jweia.2018.05.009

Lositano, I. C. M., & Danao, L. A. M. (2019). Steady wind performance of a 5 kW three-bladed H-rotor Darrieus Vertical Axis Wind Turbine (VAWT) with cambered tubercle leading edge (TLE) blades. *Energy, 175*, 278–291. doi:10.1016/j.energy.2019.03.033

Luo, Y., Wen, F., Wang, S., Hou, R., Wang, S., & Wang, Z. (2021). Numerical study on biomimetic trailing edge in the environment of a high pressure turbine stage. *Aerospace Science and Technology, 115*(August), 106770. doi:10.1016/j.ast.2021.106770

Matt. (2017). *Self-Healing of Wind Turbine Blades Using Microscale Vascular Vessels. ASME Journal of Energy Resources Technology.*

Miklosovic, Murray, Howle, & Fish. (2004). *Leading-edge tubercles delay stall on humpback whale (Megaptera novaeangliae) flippers.* Academic Press.

Mishra, N., Gupta, A. S., Dawar, J., Kumar, A., & Mitra, S. (2018). Numerical and Experimental Study on Performance Enhancement of Darrieus Vertical Axis Wind Turbine with Wingtip Devices. *ASME Journal of Energy Resources Technology, 140*(12), 121-201 doi:10.1115/1.4040506

Seyhan, M., Sarioglu, M., & Akansu, Y. E. (2021, October). Influence of Leading-Edge Tubercle with Amplitude Modulationon NACA 0015 Airfoil. *AIAA Journal, 59*(10), 3965–3978. doi:10.2514/1.J060180

Wang, Z., Wang, Y., & Zhuang, M. (2018). Improvement of the aerodynamic performance of vertical axis wind turbines with leading-edge serrations and helical blades using CFD and Taguchi method. *Energy Conversion and Management, 177*, 107-121.

Wang, Z., & Zhuang, M. (2017). Leading –edge serrations for performance improvement on a vertical axis wind turbine at low tip-speed-ratio. *Applied Energy, 208*, 1184-119.

Watts, P., & Fish, F. E. (2001). The influence of passive, leading edge tubercles on wing performance. *12th International Symposium on Unmanned Untethered Submersible Technology.*

Yan, Y., Avital, E., Williams, J., & Cui, J. (2021). *Aerodynamic performance improvements of a vertical axis wind turbine by leading-edge protuberance.* Academic Press.

Chapter 11
Computational Analysis of Symmetric and Cambered Blade Darrieus Vertical Axis Wind Turbine With Bio-Mimicked Blade Design

Punit Prakash
https://orcid.org/0000-0002-6892-4504
Shiv Nadar University, India

Nishant Mishra
https://orcid.org/0000-0002-2302-4609
Shiv Nadar University, India

Praveen Laws
New York University, Abu Dhabi, UAE

Santanu Mitra
https://orcid.org/0000-0002-7141-4695
Shiv Nadar University, India

ABSTRACT

Vertical axis wind turbine suffers from low performance, and the need for improvement is a challenge. This work addresses this problem by using computational fluid dynamics. This chapter aims to analyze and compare symmetric and cambered Darrieus turbine. These analyses are usually carried for straight leading-edge blades, and cambered resembles more the natural shape of the wing of birds and other aquatic mammals, which helps them generate extra lift during movement. Moreover, recent studies suggest better performance was observed for NACA0018 symmetric aerofoil blades, and a similar trend has been observed for NACA2412 cambered aerofoil profiles. Turbine models having symmetric NACA0018 and cambered NACA2412 profiles have been studied. By comparing the symmetric model with cambered blade models, differences in coefficient of torque have been presented. OpenFOAM is used for performing the 2D simulation with dynamicOverset-FvMesh for motion solver with overset mesh method. Meshed geometry was constructed with GMSH codes and the simulation uses overPimpleDyMFoam algorithm as a solver.

DOI: 10.4018/978-1-7998-8561-0.ch011

1 INTRODUCTION

The wind energy sector is one of the leading sectors of growth for sustainable development across the globe. According to the latest global wind report 2021, the industry showed a 53% increase in installation globally in 2020 (Global Wind Energy Council, 2021). Wind energy turbines are categorized as Horizontal Axis Wind Turbines (HAWT) and Vertical Axis Wind Turbines (VAWT). HAWTs are the traditional three-bladed lift-based turbines whose rotational axis is parallel to the ground and has higher acceptability. Vertical axis turbines are less used and have both drag-based like Savonius and lift-based like Darrieus. The axis of rotation is perpendicular to the ground for vertical axis wind turbines. However, Vertical Axis Wind Turbine suffers from low performance, and the need for improvement is a challenge. The VAWTs are considered to have lesser efficiency as compared to HAWTs; however, they have a wide range of operations. Many works have been published to improve the efficiency of these turbines with new design considerations (Ahmad et al., 2016; Mishra et al., 2018; Seeni et al., 2018). Hand et al. (Hand et al., 2021) have presented aerodynamic design parameters that influence the lift type VAWT. Elsakka et al. (Elsakka et al., 2019) pointed out how the Angle of Attack (AoA) variation can enhance performance for Darrieus turbines with a variable pitch by comparing it with the fixed pitch model. Bio-inspiration is one of the techniques of mimicking nature to improve the aerodynamics of the present models. Fish (Fish, 2020) introduced tubercles to the blade's leading edge as a bio-inspired design inspired by gigantic humpback whale flippers. Yadav (Yadav et al., 2021) implemented a bio-inspired nose inspired by cetacean marine mammals for flow separation on NACA airfoil for the 4 and 6 series. Work states for a lower angle of attack, larger perturbation of the nose leads towards multiple accelerations on the extended surface, creating a forward-facing step and thus improving the aerodynamic efficiency. But, at a high angle of attack, it is ineffective due to early flow separation. Lositano & Danao (Lositano & Danao, 2019) studied cambered tubercle leading-edge blades for VAWT and stated a detrimental effect compared to cambered NACA0025 airfoil. Rezaeiha et al. (Rezaeiha et al., 2018) published a work on VAWT CFD study that comprises of tip speed ratio, solidity azimuthal increment and simulation convergence. Bai et al. (Wang et al., 2015) worked on computational analysis of VAWt with tubercle leading edge on NACA0015 and stated the thrust is lower than the straight blade turbine. Howell et al. (Howell et al., 2010) showed 2D results overestimate 3D results and the reason for such drop in result in 3D case is due to end tip vortices. Kjellin at el. (Kjellin et al., 2011) conducted experimental study for 12 KW VAWT and got coefficient of performance 0.29 at tip speed ratio of 3.3 using NACA0025 airfoil. Mohamed (Mohamed et al., 2015) stated the zero pitch angle gives best performance while comparing 25 airfoil profile, LS(1)-0413 airofoil gives 10% extra power compared to NACA0018 and NACA 63-415 gives a wider operating range than all other airfoil models. Jones(Jones et al., 2018) published a computational work for low Reynolds no flow for morphing airfoil the effect of dynamic surface around airfoil was explored computationally. Razaeiha et al. (Rezaeiha et al., 2017) performed experimental analysis and for increasing the pitch angel to -2° the C_p increases for 6.6% in comparison to zero degree pitch angle they stated the change in configuration changes the strength of shed vortices and can decrease the wake generation for VAWTs. Rostamzadeh et al. (Rostamzadeh et al., 2013) has published an experimental and computational for a new design for tubercle was introduced the result for computational and experiment are in good agreements with each other. Khaleghinia et al. (Khaleghinia et al., 2021) published a study for NACA0018 with cavity layouts and claimed the performance increased by the addition of these cavities, a computational study was performed for shape position size and no. of cavities. The greatest value of torque was achieved with a single cavity and a dual cavity result was detrimental. Saha and Siddhart (Saha & Jain, 2020) worked on dynamic stall of H-blade Darrieus rotors variation of C_l and C_d with different phases for dynamic stall is discusses. They observed three different stall leading mixed and trailing edge stalls. Mohamed et al. (Mohamed et al., 2019) published a paper with different blade shapes effect on turbine performance and stated the J-shaped design with wind lens enhance the performance by a factor of 2.24. they also stated using these wind lens is noisy still it enhances the performance. Sengupta et al. (Sengupta et al., 2019) worked on high solidity and low Reynolds no. with a cambered blade profile, they suggested a better desigh of airforil shape with higher blade curvature on inner surface around moment center. Blade profile S815 and EN0005 blades are considered better design. Chenet al. (Chen et al., 2015) published a study of novel Darrieus VAWT, the effect of opening ratio was analyzed and resulted in increasing torque and decreasing the power, the opening ratio of 0.48 and 0.60 was desirable for this design. Dessoky et al. (Dessoky et al., 2019) worked on wind lens technology to enhance the aerodynamic performance, they also studied the noise generated during the study. They used a DDES approach after 2D URANS computational study. The study improves the performance by 82% by using a cycloid diffuser. Scungio et al. (Scungio et al., 2016) studied a wind tunnel testing of an auxiliary straight blade Darrieus VAWT and static and dynamic power and torque were validated with standard results. Castelli et al. (Castelli et al., 2011) present a new CFD code that combines blade elemental theory and dynamic quantities like rotor torque and blade tangential and normal force. They run simulation for NACA 0021 three rotor blades and validate with standard results. Shukla (Shukla & Kaviti, 2017) studies the effect of gurney flap and dimples independently and combined effect on various straight blade profile and find the dimple and Gurney flay combined gives best result for NACA0015. In the study NACA0012, NACA0015, NACA0018 and NACA0021 were studied. Bangga et al. (Bangga et al., 2020) published a work that used prediction approaches to study three VAWTs at various solidity, various models like DMS, IDMS, UBEM, Vortex model and CFD method has been used in the study. different rotor solidity 0.23, 0.53 and 1.325 are used and observed CFD predictions have discrepancy in showing consistent power prediction still it manages to predict accurate thrust values. Giorgetti et al. (Giorgetti et al., 2015) published a work for setting VAWT in close proximity, the turbine selected is micro turbine and has low solidity and low tip speed ratio. The study observed the accelerated free stream helps in extraction of more power in closed proximity arrangement.

Wind turbines work on the principle of Betz's law, which states no turbine can surpass the maximum wind efficiency of 0.529. C_p–λ curve is used to measure the performance of a turbine, where Cp is the coefficient of power and (λ=T*SR(T*ip Speed Ratio)) (Giorgetti et al., 2015). VAWTs have a better performance as compared to HAWTs at lower wind speed regimes. Cp $_i$s stated as the efficiency of the wind turbines, representing the ratio of the power being captured by the turbine blades to the power available in wind hitting the blades (Jain, 2016).

$$C_p = \frac{P_{out}}{P_{in}} = \frac{T\omega}{0.5\rho A v^3} = \frac{T}{0.5\rho A v^2 R} \times \frac{\omega r}{v} = c_m \times \lambda \tag{1}$$

The study comprises of computational flow solution for a wind turbine and uses OpenFoam an open source CFD (computational fluid dynamics) solver.

2 PROBLEM STATEMENT

A comparative parametric study has been performed between symmetric and cambered airfoil turbines, symbolized by model A and model B, respectively, as shown in Figure 1 and Figure 2. Model A represents a straight blade VAWT with a NACA0018 profile. Model B represents the cambered blade profile with the NACA2412 profile. The chord length has been kept constant for both the airfoil. The specifications have been presented in Table 1.

Figure 1. Schematic Overview

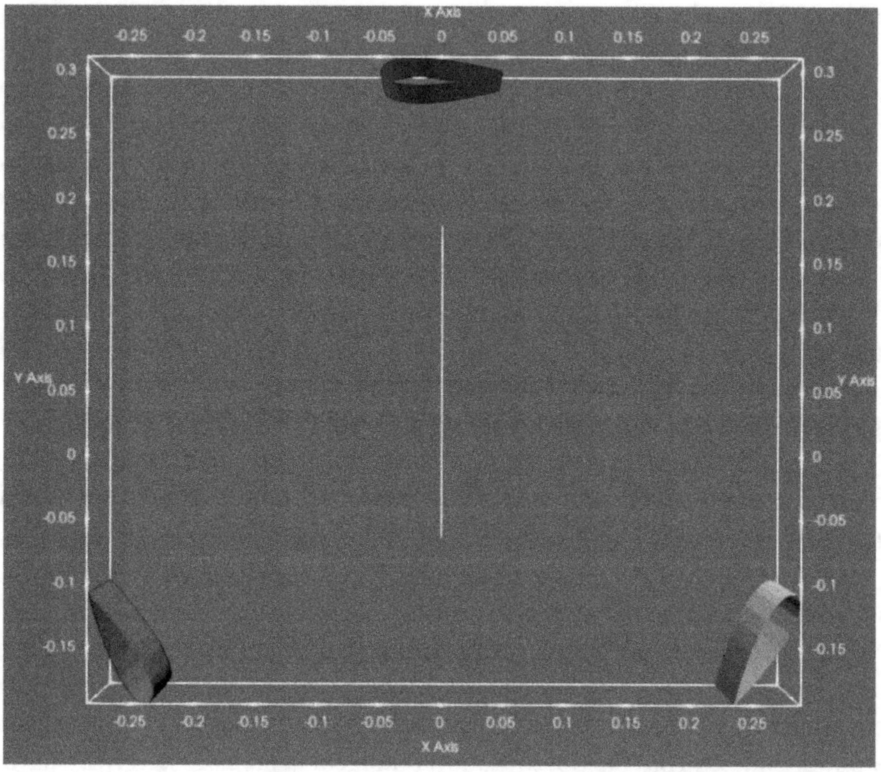

Table 1. Specification for rotor

Parameters	Symbols	Dimension
Chord length (B)		100 mm
Chord length (A)		100 mm
Length	L	0.1 m
Diameter of turbine	D	600 mm
NACA profile A		0018
NACA profile B		2412

Figure 2. Schematic Overview

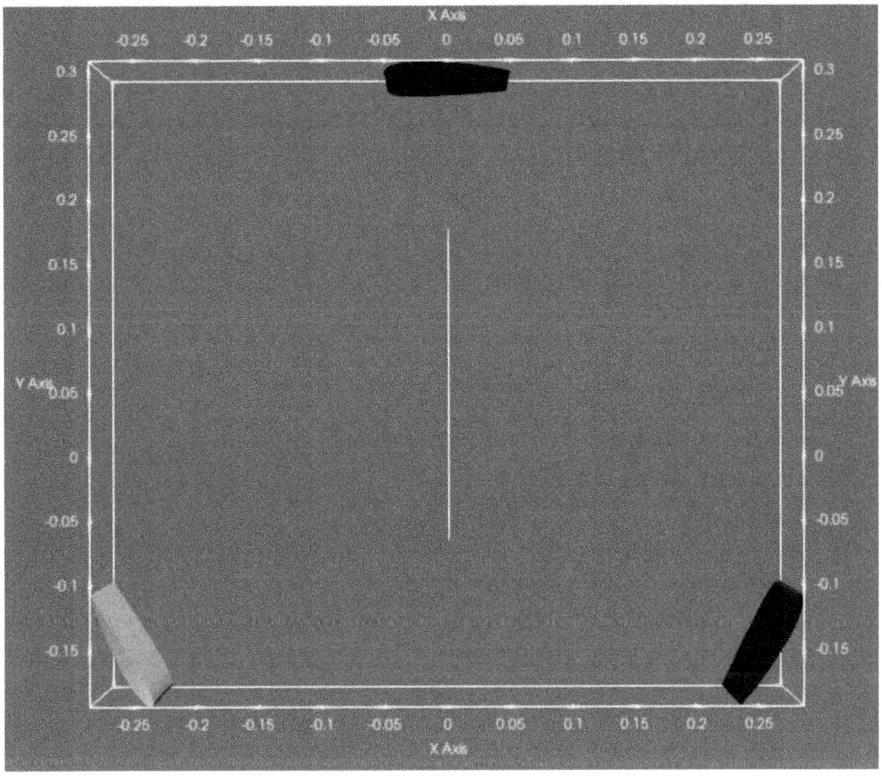

3 GOVERNING PHYSICS AND SOLVER DETAILS

This study was performed using OpenFoam a CFD toolbox for computational analysis on a system having Intel Xeon Gold-5222 CPU @8 processors with 64 GB 2666 MHz DDR4 RAM. A meshing code from Gmsh an Opensource codes have been used, and Open foam is used as a solver for describing the governing physics. The geometry was divided into two sets, Rotor (overset) and Domain. Overset serves as a boundary for the transfer of cell information between the domain and blades. Domain is the stationary portion for the fluid flow interaction. This computation focus on fluid flow analysis, where a solid body is treated as a wall with no-slip boundary condition, and only fluid equations need to be solved. An angular momentum ω is provided to the solid body, and change in the fluid is computed using overPimpleDyM algorithm. The script uses K–ωSS*T to* solve Reynolds averaged Navier-Stokes equations as a turbulence model, the transport model as Newtonian, and the time scheme as Euler. A detailed description has been tabulated in Table 2. The overset method, also known as Chimera mesh, is being used for the study. An overlapping of mesh is required for this scheme. The cell information of the domain passes to the outer boundary of the blade zones through an overset. The outer cells of the blade containing the mesh zone need to be identical to the background domain mesh, as shown in Figure 3. A detailed description of the overset method can be found in (Jain, 2016). For simulation of the external sub-sonic flow boundary conditions comprise of three velocity components and pressure value at the outlet, as shown in Table 2. The inlet is placed at a distance of 3D, and the outlet is at 6D from the rotor blades.

Table 2. Computational details

Constant			
Transport model	Newtonian	Dynamic viscosity	$1.42E^{-5}$
Simulation type	RAS	RAS model	$K\text{-}\omega SST$
Dynamic meshsolver	dynamicOversetFvMesh	Solver	multisolidBodyMotionSolver
Systems			
Fvscheme		FvSolution	
ddt scheme	Euler	Residuals	10^{-6}
gradScheme	Gauss Linear		
Div scheme	Limited linear		
Laplacian Scheme	Gauss linear corrected		
Interpolation Scheme	linear		
Boundary condition			
Inlet	(5,0,0)m/s	Patch = zero Gradient	
Outlet	Pr = 0 bar	Patch = zero Gradient	

4 MESHING

Meshing is the most important aspect of a CFD problem. The domain for fluid flow is shown in Figure 3. Overset meshing schemes have been used with all hexahedra cells for meshing. This method is effective for dynamic mesh where cell information passes through the overset surface to the zone where a motion has been defined. A y^+ Value of 0.1 has been used in these simulations, and the domain is kept constant along with the no. of cells and only parameter that has changed is the airfoil shape from symmetric to cambered, as shown in Figure 4 and Figure 5. The first layer thickness have been kept constant for both case as $8.54E^{-6}$ as shown in Figure 6 and Fig 7. Mesh statistics have been provided in Table 3, and Table 4 states domain specifications.

Table 3. Mesh information

Model	Cells
Model A	5928
Model B	5928

Figure 3. Domain for analysis

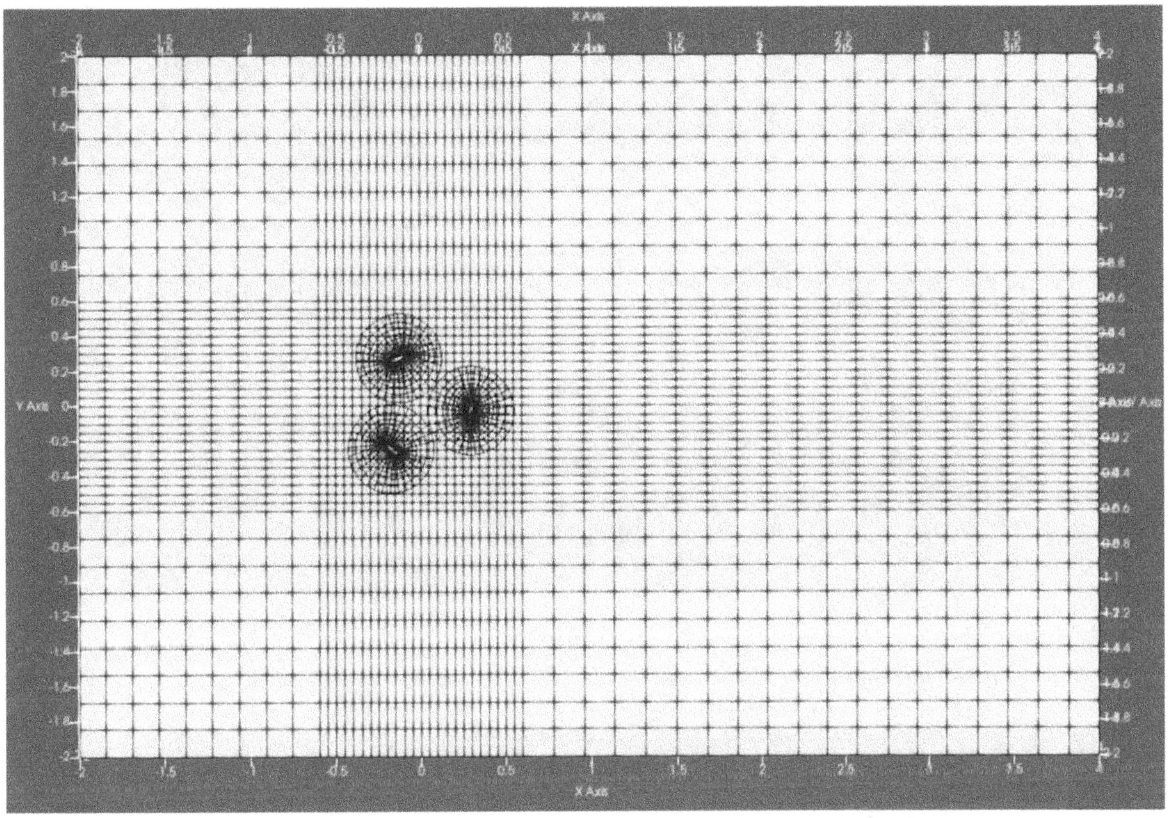

Figure 4. Meshed blade geometry

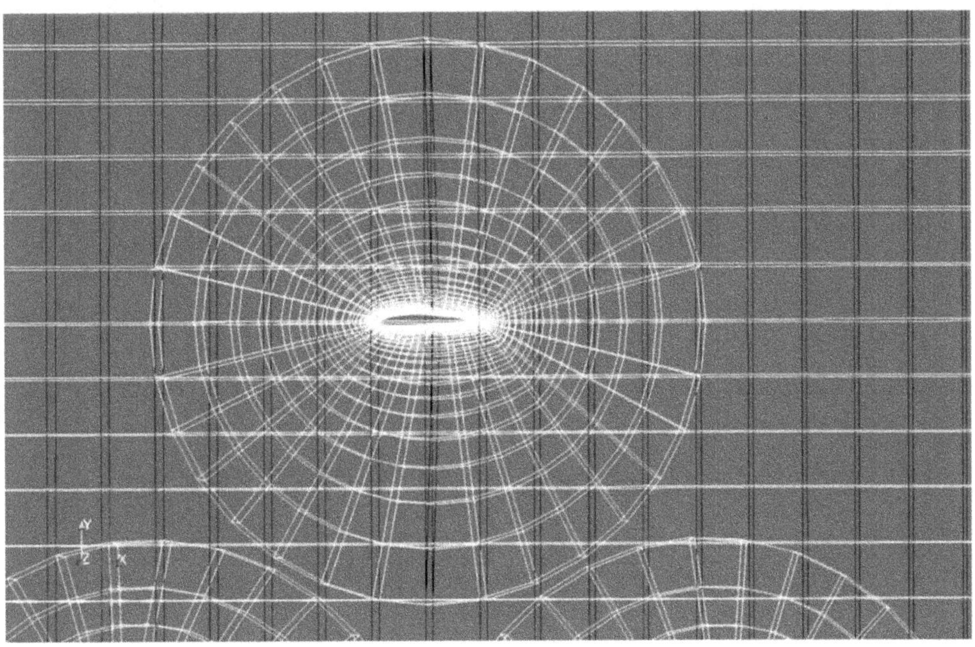

Figure 5. Meshed blade geometry

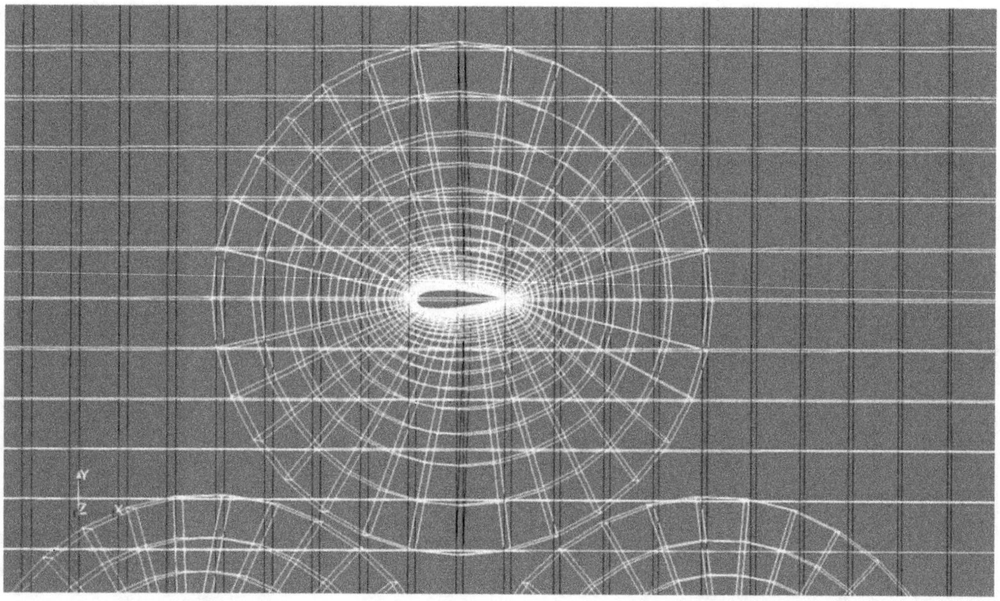

Figure 6. First layer thickness

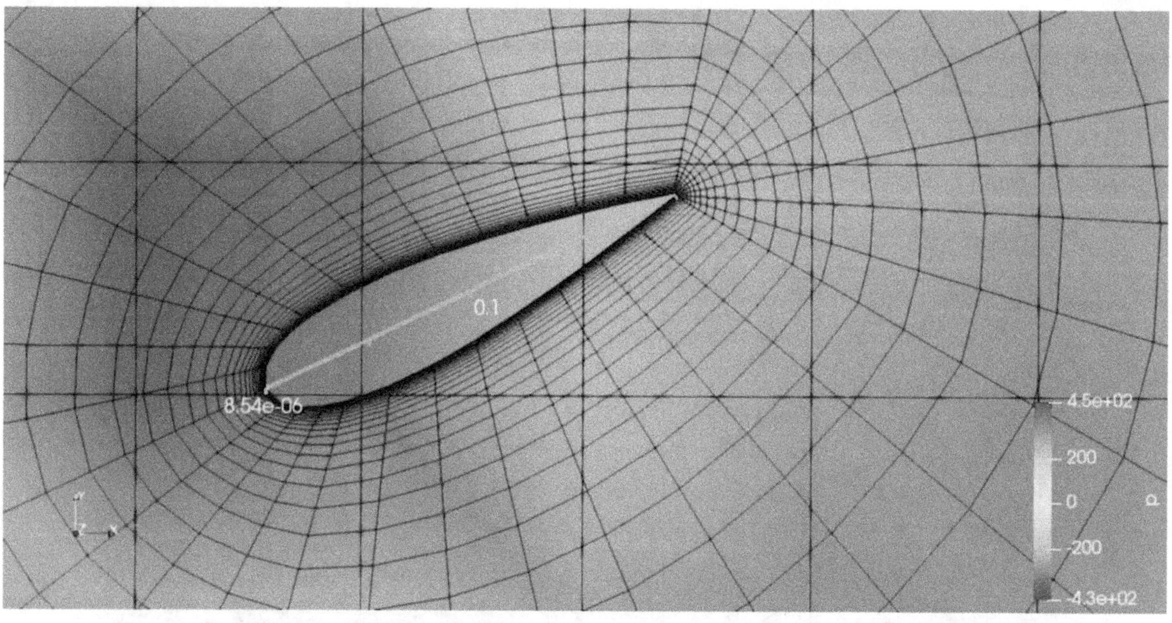

Figure 7. First layer thickness

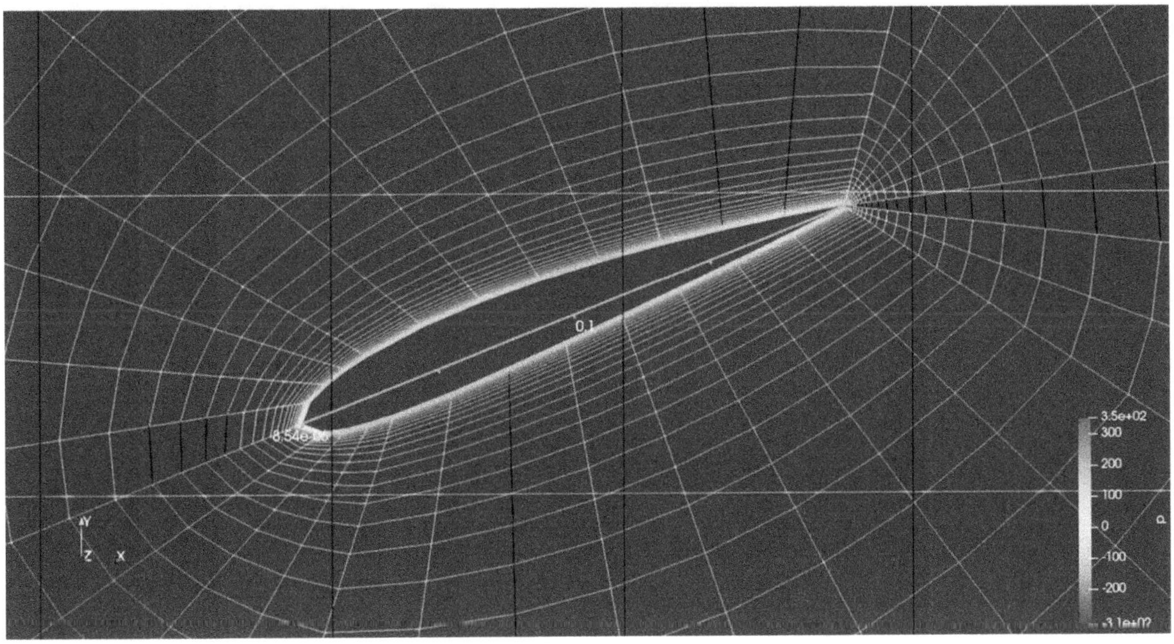

Table 4. Specifications for the domain

Symbol	Dimension
Overset radius	0.180 m
Domain size	(-2,-2,0) (4,2,0.1) m
First layer thickness	$8.54E^{-5}$

5 RESULTS

The Airfoils have been studied, and corresponding pressure and velocity contour at time 1.79s have been shown for a uniform time period for both the models. Model A shows a higher pressure profile at the tail end of the airfoil than Model B, as shown in Figure 8 and Figure 9. Both airfoil has been studied for a transient simulation and the timestep 1.79s has been selected for comparison in as a steady state has been reached as can be seen from the coefficient of torque plot. The velocity distribution appears to be uniform at this time step, as shown in Figure 10 and Figure 11. The residual plot containing the Ux Uy, Omega, k, and P values at each time step for a simulation has been shown in Figure 12 for the simulation of NACA0018 at a 2.02 tip speed ratio. Its coefficient of torque is shown in Figure 13.

Figure 8. Pressure contour

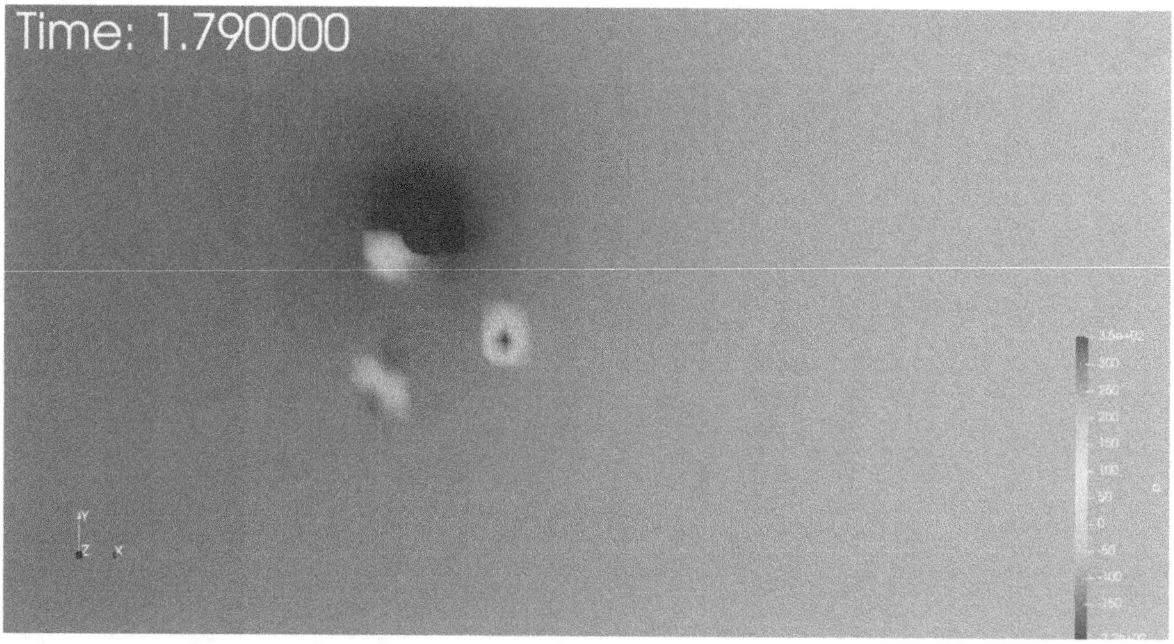

Figure 9. Pressure contour

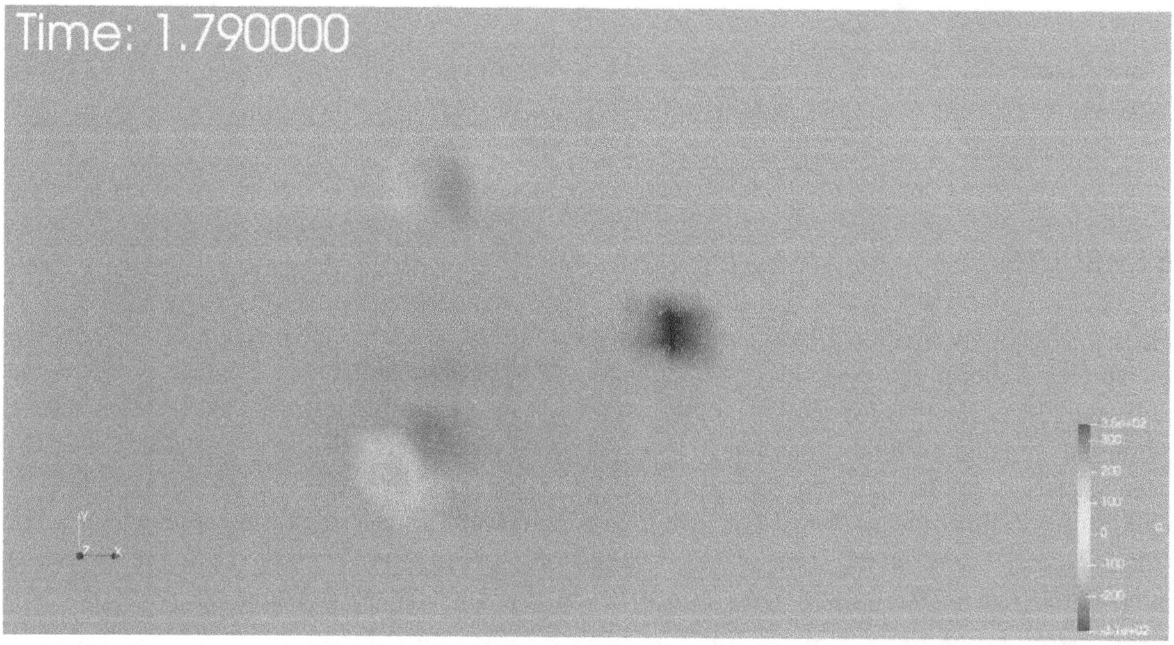

Figure 10. Velocity contour

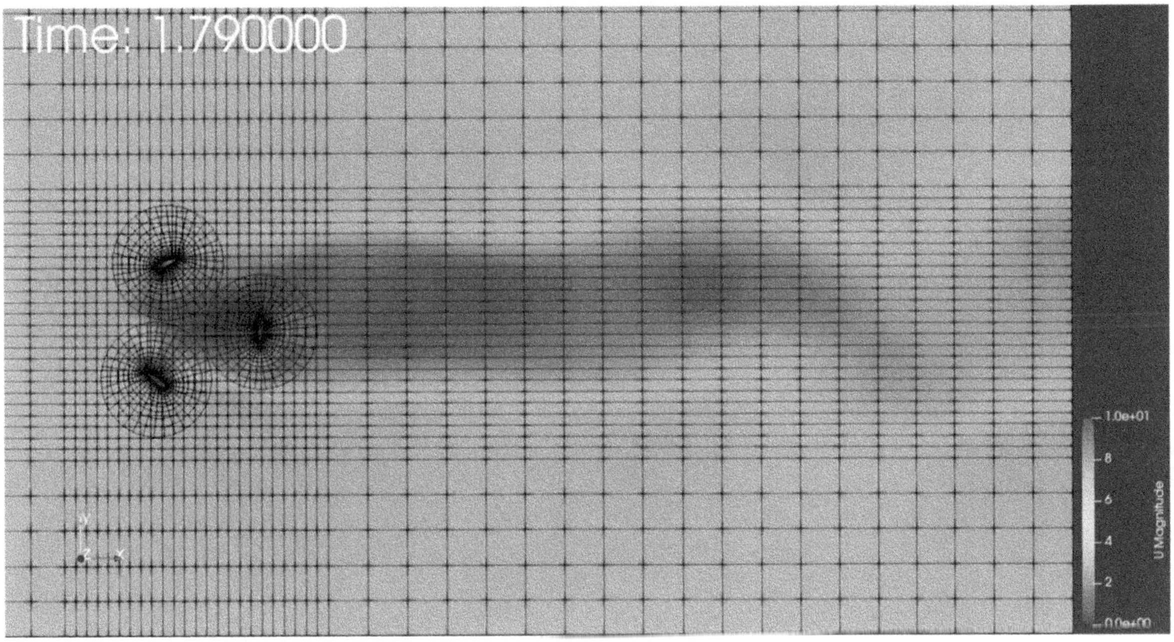

Figure 11. Velocity contour

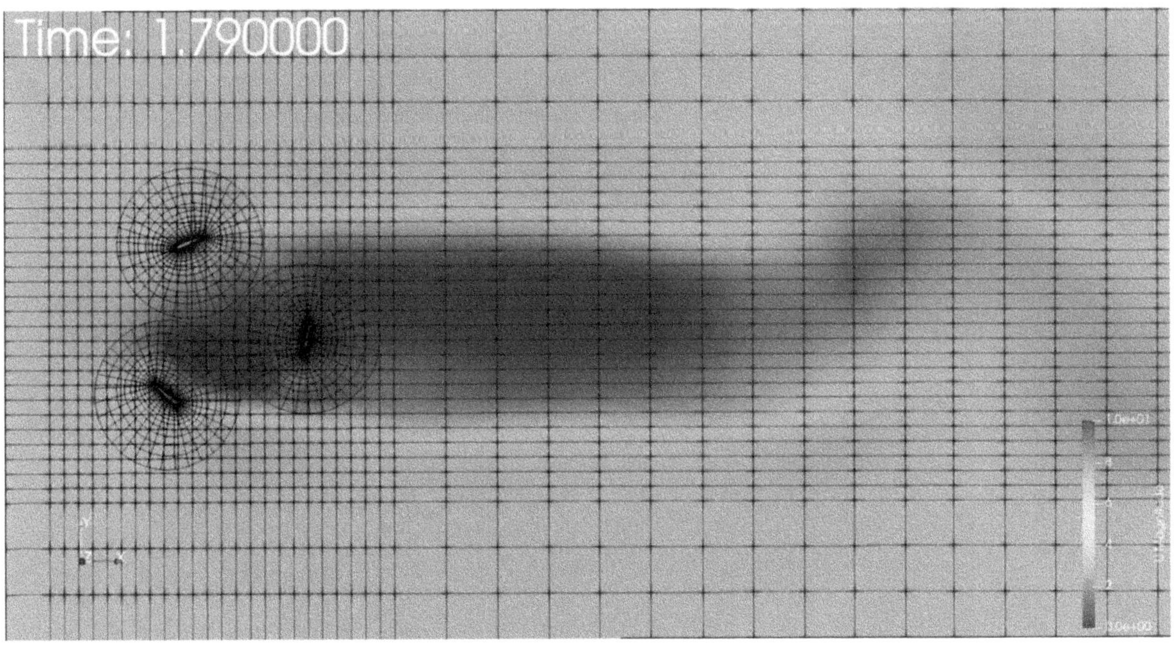

The simulations are carried out for different values of TSR, and the corresponding coefficients of torque vs TSR are shown in Figure 14. Using Eqn. (1) the coefficient of performance values can be easily found out. The corresponding C_p vs *TSR* curve for performance characteristics has been shown in Figure 15.

Figure 12. Residuals for the simulations

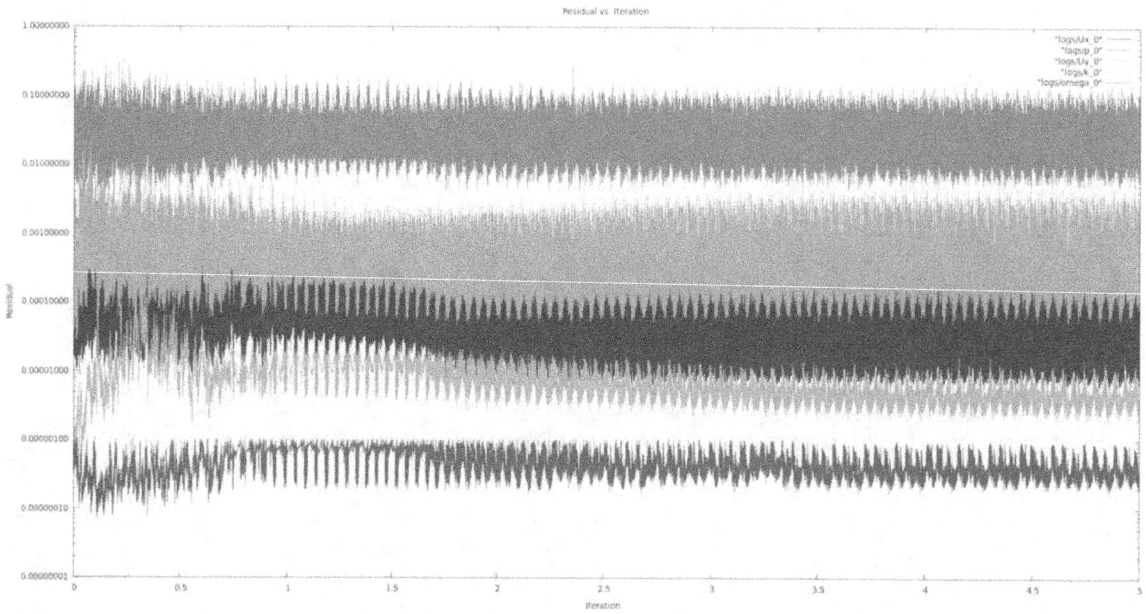

Figure 13. Coefficient of Torque vs Time for TSR 2.02 for NACA0018

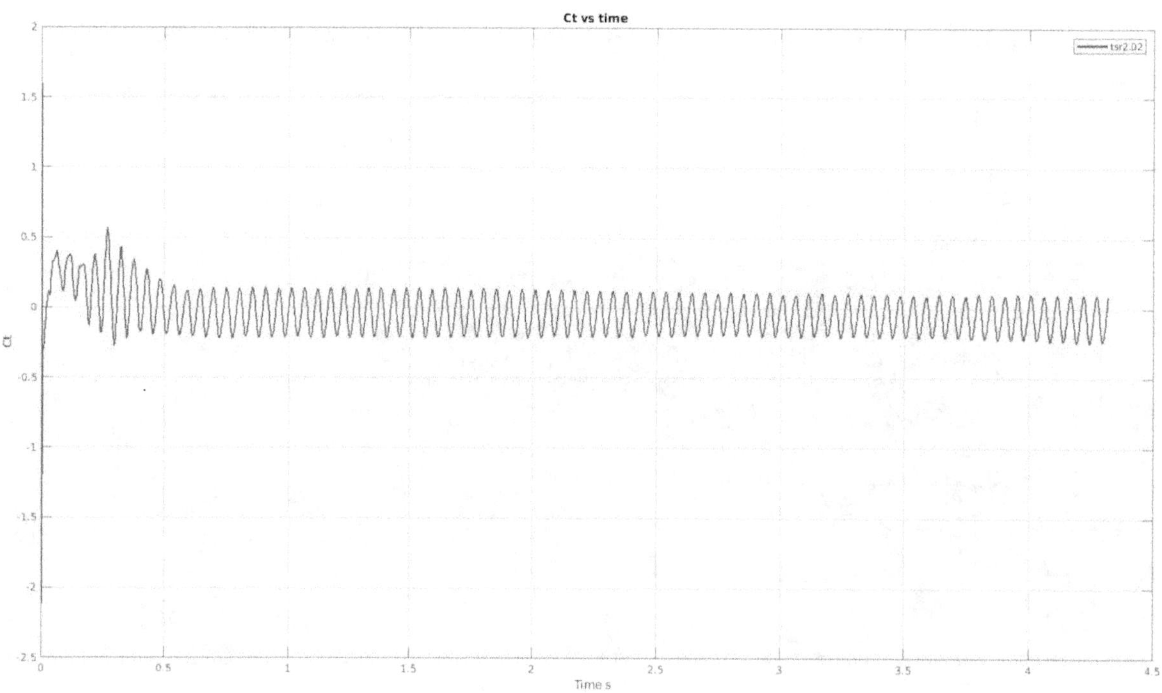

Figure 14. Comparison of C_p vs TSR

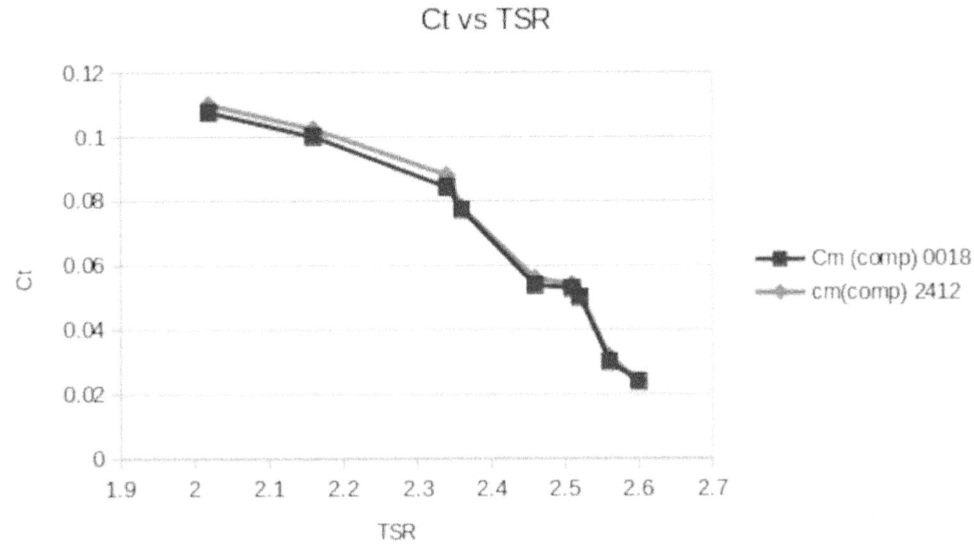

Figure 15. Comparison of C_p vs TSR

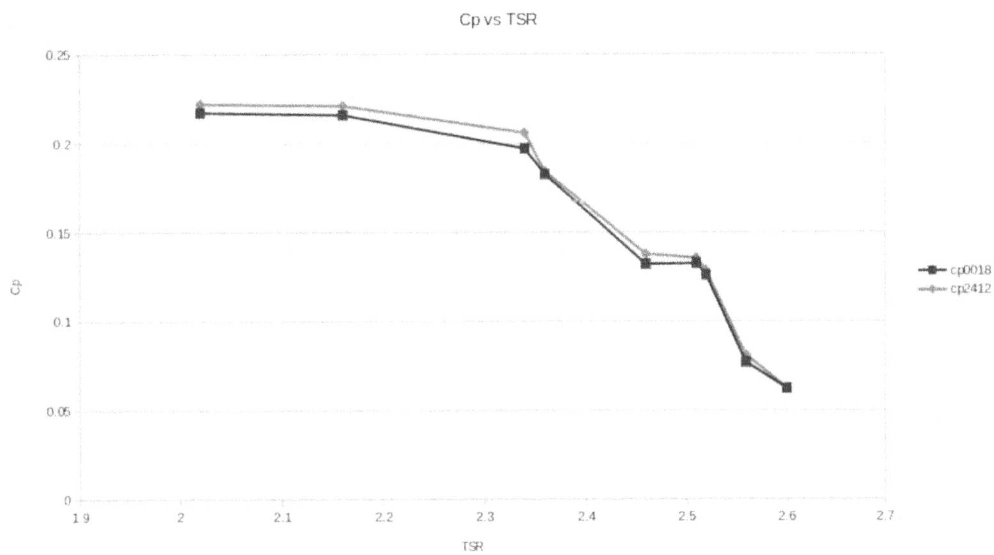

6 CONCLUSION

Two turbine configurations represented as Model A and B are compared with uniform chord length. A 2D CFD numerical study has been performed with different TSR values and has been shown. A Transient steady case with the k–ω SST turbulence model has a prerequisite of 10 complete VAWT rotations to attain converged results. The simulation is allowed to run for 5 seconds to eliminate the transient phase, and a stable result can thus be obtained. Coefficients of lift, drag and moment on blades were independently examined during the study and converged using a MATLAB script to smoothen the curve to get a $Ct_{(}$averaged) value to be presented.

The results clearly indicate a better performance of the cambered blade profile in comparison with the symmetric blades. There is still a future scope of 3D study to verify the results obtained in the present study.

REFERENCES

Ahmad, K. A., Aftab, S. M. A., Razak, N. A., & Rafie, A. S. M. (2016). Mimicking the humpback whale: An aerodynamic perspective. *Progress in Aerospace Sciences*, *84*, 48–69. doi:10.1016/j.paerosci.2016.03.002

Bangga, G., Dessoky, A., Wu, Z., Rogowski, K., & Hansen, M. O. L. (2020, September). Accuracy and consistency of CFD and engineering models for simulating vertical axis wind turbine loads. *Energy*, *206*(1), 118087. doi:10.1016/j.energy.2020.118087

Castelli, M.R., Englaro, A., & Benini, E. (2011). The Darrieus wind turbine: proposal for a new performance predication model based on CFD. *Energy*, *36*(8), 4919-4934.

Chen, J., Yang, H., Yang, M., & Xu, H. (2015). The effect of opening ratio and location on the performance of a novel vertical axis Darrieus turbine. *Energy*, *89*(September), 819–834. doi:10.1016/j.energy.2015.05.136

Dessoky, A., Bangga, G., Lutz, T., & Kramer, E. (2019, May). Aerodynamic and aeroacoustic performance assessment of H- rotor Darrieus VAWT equipped with wind lens technology. *Energy*, *175*(15), 76–97. doi:10.1016/j.energy.2019.03.066

Elsakka, M. M., Ingham, D. B., Ma, L., & Pourkashanian, M. (2019). CFD analysis of the angle of attack for a vertical axis wind turbine blade. *Energy Conversion and Management*, *182*, 154–165. doi:10.1016/j.enconman.2018.12.054

Fish, F. E. (2020). *Biomimetics and the Application of the Leading-Edge Tubercles of the Humpback Whale Flipper* (1st ed.). Springer. doi:10.1007/978-3-030-23792-9_1

Giorgetti, S., Pellegrini, G., & Zanforlin, S. (2015). CFD Investigation on the Aerodynamic Interferences between Medium-solidity Darrieus Vertical Axis wind Turbines. *Energy Procedia*, *81*(December), 227–239. doi:10.1016/j.egypro.2015.12.089

Global Wind Energy Council. (2021). *Global Wind Report 2021*. GWEC. http://www.indiaenvironmentportal.org.in/files/file/global%20wind%20report%202021.pdf

Hand, B., Kelly, G., & Cashman, A. (2021). Aerodynamic design and performance parameters of a lift-type vertical axis wind turbine: A comprehensive review. *Renewable and Sustainable Energy Reviews*, *139*. doi:10.1016/j.rser.2020.110699

Howell, R., Oin, N., Edwards, J., & Durrani, N. (2010). Wind tunnel and numerical study of a small vertical axis wind turbine. *Renewable Energy*, *35*(2), 412–422. doi:10.1016/j.renene.2009.07.025

Jain, P. (2016). Wind Energy Engineering (2nd ed.). McGraw-Hill Education.

Jones, G., Santer, M., & Papadakis, G. (2018). Control of low Reynolds number flow around an airfoil using periodic surface morphing: A numerical study. *Journal of Fluids and Structures*, *76*(January), 95–115. doi:10.1016/j.jfluidstructs.2017.09.009

Khaleghinia, J., Roshan, M. Y., Nimvari, M. E., & Salarian, H. (2021, October). Performance improvement of Darrieus wind turbine using different cavity layouts. *Energy Conversion and Management*, *246*(15), 114693.

Kjellin, J., Bulow, F., Eriksson, S., Deglaire, P., Leijon, M., & Bernhoff, H. (2011). Power coefficient measurement on a 12 kW straight bladed vertical axis wind. *Renewable Energy*, *36*(11), 3050–3053. doi:10.1016/j.renene.2011.03.031

Laws, P., Saini, J. S., & Kumar, A. (2019). A *Study on OpenFOAM's Overset Mesh Support Using Flow Past NACA 0018 Airfoil*. doi:10.20944/preprints201907.0217.v1

Lositano, I. C. M., & Danao, L. A. M. (2019). Steady wind performance of a 5 kW three-bladed H-rotor Darrieus Vertical Axis Wind Turbine (VAWT) with cambered tubercle leading edge (TLE) blades. *Energy*, *175*, 289–291. doi:10.1016/j.energy.2019.03.033

Mishra, N., Gupta, A. S., Dawar, J., Kumar, A., & Mitra, S. (2018). Numerical and Experimental Study on Performance Enhancement of Darrieus Vertical Axis Wind Turbine with Wingtip Devices. *Journal of Energy Resources Technology*, *140*(12), 121–201. doi:10.1115/1.4040506

Mohamed, M.H., Ali, A.M., & Hafiz, A.A. (2015). CFD analysis for H-rotor Darrieus turbine as a low speed wind energy converter. *Energy Science and Technology*, *18*(1), 1-13.

Mohamed, M. H., Ghazalla, R. A., & Hafiz, A. A. (2019, December). Synergistic analysis of Darrieus wind turbine using computational fluid dynamics. *Energy*, *189*(15), 116214.

Rezaeiha, A., Kalkman, I., & Bert, B. (2017). *Effect of Pitch angle on power performance and aerodynamics of a vertical axis wind turbine*. Academic Press.

Rezaeiha, A., Montazeri, H., & Blocken, B. (2018). Towards accurate CFD simulations of vertical axis wind turbines at different tip speed ratios and solidities: Guidelines for azimuthal increment, domainsize and convergence. *Energy Conversion and Management*, *156*, 301–316. doi:10.1016/j.enconman.2017.11.026

Rostamzadeh, N., Kelso, R.M., Daily, B.B., & Hansen, K.L. (2013). *The effect of Undulating leading-edge modifications on NACA 0021 airfoil characteristics*. Academic Press.

Saha, U., & Jain, S. (2020). On the influence of blade thickness-to-chord ratio on dynamic stall phenomenon in H-type Darrieus wind rotors. *Energy Conversion and Management, 218.*

Scungio, M., Arpino, F., Focanti, V., Profili, M., & Rotondi, M. (2016, December). Wind tunnel testing of scaled model of a newly developed Darrieus-style vertical axis wind turbine with auxiliary straight blades. *Energy Conversion and Management*, *130*(15), 60–70. doi:10.1016/j.enconman.2016.10.033

Seeni, P., Rajendran, P., & Kutty, H. A. (2018). A Critical Review on Tubercles Design for Propellers. *IOP Conf. Series: Materials Science and Engineering, 370.* 10.1088/1757-899X/370/1/012015

Sengupta A.R., Biswas, A., & Gupta, R. (2019). Comparison of low wind speed aerodynamics of unsymmetrical blade H-Darrieus rotor-blade camber and curvature signature for performance improvement. *Renewable Energy, 139*, 1412-1427.

Shukla, V., & Kaviti, A.K. (2017). *Performance evaluation of profilr modification on straight- bladed vertical axis wind turbine by energy and Spalart Allmaras models*. Academic Press.

Wang, W. C., Bai, C. J., Lin, Y. Y., & Lin, S. Y. (2015). Computational fluid dynamics analysis of the vertical axis wind turbine blade with tubercle leading edge. *Journal of Renewable and Sustainable Energy*, *7*(3), 033124. doi:10.1063/1.4922192

Yadav, R., Mohamed, M. A. R., & Guven, U. (2021). Flow Separation Control using a Bio-Inspired Nose for NACA 4 and 6 Series Airfoils. *Aircraft Engineering and Aerospace Technology*, *93*(2), 251–266. doi:10.1108/AEAT-08-2019-0170

Chapter 12
Study on Reliability Design of the Domestic Compressor Subjected to Repetitive Internal Stresses by Parametric Accelerated Life Testing

Seongwoo Woo
Ethiopian Technical University, Ethiopia

Dennis L. O'Neal
Baylor University, USA

Yimer Mohammed Hassen
Ethiopian Technical University, Ethiopia

ABSTRACT

This chapter explains the parametric accelerated life testing (ALT) to recognize design defects in mechanical products. A life-stress model and a sample size formulation are suggested. A compressor is used to demonstrate this method. Compressors were failing in the field. At the first ALT, the compressor failed due to a fractured suction reed valve. The failure modes were similar to those valves returned from the field. The fatigue of the suction reed valves came from an overlap between the suction reed valve and the valve plate. The problematic design was modified by the trespan dimensions, tumbling process, a ball peening, and brushing process for the valve plate. At the second ALT, the compressor locked due to the intrusion between the crankshaft and thrust washer. The corrective action plan performed the heat treatment to the exterior of the crankshaft made of cast iron. After the design modifications, there were no troubles during the third ALT. The lifetime of compressor was secured to have a B1 life 10 years.

DOI: 10.4018/978-1-7998-8561-0.ch012

1. INTRODUCTION

Because of demands in the global market, refrigerators must be designed to have low energy usage and reliability. For those aimed functions, the compressor in the refrigerator often needs to be redesigned to enhance its comprehensive energy efficiency on a yearly basis. It involves recently designed features for the system, which should be rapidly brought to the end-users. With insufficient testing or no assumption of how the new attributes may be utilized, their launching may increase product failures in the marketplace and negatively influence the manufacturer's brand name. These attached features for the product should be totally assessed in the design phase before being released into the marketplace. Therefore, reliability quantitative (RQ) specifications utilizing an established system of method should be presented (CMMI Product Team, 2018; Woo et al., 2021).

A compressor is designed to increase the refrigerant pressure in a refrigeration cycle by several mechanical compression mechanisms. One of the key components of the compressor is the crankshaft that converts the rotational motion into reciprocating motion by a crank mechanism. A compressor is subjected to repeated stresses due to internal pressure loadings over the course of its lifetime. New rotary compressors introduced in 1987 in refrigerators were experiencing large recalls due to the locking of the compressor in the field (Magaziner & Patinkin, 1989). Oil metal sludge formed during normal refrigeration operations was separating from the sintering crankshaft and plugging the capillary tubes, which forced the refrigerator to no longer function. To stop a compressor recall, any flawed components needed to be identified and altered using testing methodology such as parametric ALT, which can produce reliability quantitative (RQ) specifications before the system launches.

The procedures of robust design, such as the Taguchi approach (Chowdhury & Taguchi, 2016; Rosa et al., 2009) and design of experiments (DOE)(Allen, 2020), were developed to help identify the most advantageous designs for products. Especially, Taguchi's method employs design parameters to put it in the right location where ''noise" parameters do not have any effect on the output. As a result, the right designs of mechanical products can be selected. However, without identifying failure mechanisms such as fatigue, this methodology can only pursue system optimization. If there is a design fault, the product may fail during its lifetime as loads are exerted on it. Many parameters must be considered to identify an optimal design of a mechanical structure. However, the large number of parameters may require huge computations unless some options are neglected to reduce computational requirements.

Material faults, such as extremely small voids and contacts when subjected to repeated loads, may begin to fail because of fatigue. Fatigue is the chief source of destruction in metallic elements, explaining roughly 80%–95% of all constructional failures. Fatigue in ductile metals appears itself in the shape of cracks that grow in stress accumulation, such as holes, grooves, etc. Those failures can affect the reliability of mechanical systems such as moving automobiles, airplane wings, marine ships, turbo engines, and atomic reactors. The fatigue procedure covers three fluctuating stress/time modes: 1) symmetrical about zero stresses, 2) asymmetrical about zero stresses, and 3) random stress cycles. The fatigue may also rely on the parameters such as the cyclic stress amplitude, mean stress or stress ratio, R $(=\sigma_{min}/\sigma_{max})$, which can be defined as s the proportion of the minimum cyclic stress to the maximum cyclic stress (Campbell & Fatigue, 2008). In other words, in periodical shapes, the peaks on both the maximum (high side) and the minimum (low side) are crucial. When employing an elevated load which can be stated as an accelerated factor (AF), accelerated life testing (ALT) can be examined to discover the design defects such as stress raiser in the structure.

The ALT combined on the reliability block diagram was investigated as another method (Modarres et al., 2016). It included a test plan for the system, identifying failure mechanics such as fatigue, and using sample size equation, accelerated loads, etc. Elsayed (Elsayed, 2012) categorized statistical, physics/statistics, and physics/experimental-established prototypes for examination. Meeker (Hahn & Meeker, 2004) suggested numerous practical ways to organize an ALT. Carrying out an ALT (McPherson, 1989; McPherson, 2010) necessitates numerous notions such as the BX life for the system test scheme, a simplified life-stress description, sample size formulation, and fracture mechanics because failure may happen suddenly from the fragile components in a product. Contemporary test techniques might be hard to replicate the design flaws of components in a multi-module system because those procedures evaluate insufficient component samples and do not identify the fatigue(s) which actually occur in the marketplace.

To attain the sound design of a mechanical product, designers have employed traditional techniques such as fracture mechanics and strength of materials (Hertzberg et al., 2020). As quantum as one of the branches of mechanics has developed, engineers identified which failures came from micro-void co-alescence (MVC) and discovered many metallic alloys or some engineering plastics. To find the fatigue source of a mechanical product, a life-stress model can be incorporated with conventional design methods and relevant techniques to help identify the failure of electronic parts due to present material flaws or small cracks when the parts are subjected to (mechanical) stresses. Finite element methods (FEMs) may not identify the source of failure (Zupančič et al., 2021).

To show the effectiveness for identifying and altering the design defects of a mechanical product, this research proposes using a parametric ALT as a systematic reliability method that can create the RQ specifications—mission cycles. It covers: (1) a system BX lifetime created on the ALT scheme, (2) a load examination for ALT, (3) tailored ALTs with the alterations, and (4) an estimation of whether the product design(s) accomplishes the objective BX lifetime. The derivation of the sample size equation, time-to-failure, and BX lifetime is provided. To confirm ALT results in real life and compare the current design with previous one, it would then be necessary to monitor the introduction of the new design in the market to ensure it reached the targeted lifetime and failure rate. A newly designed compressor in a domestic refrigerator subjected to repeated pressure impact loading is used as an example of this methodology.

2. PARAMETRIC ALT FOR MECHANICAL PRODUCT

2.1. BX Lifetime of a Mechanical System

To perform an ALT, the BX life, L_B, is needed as a measure of the product lifetime. It can identify the accumulative failure rate and meet the market needs for reliability requirements of the product. BX life is the elapsed time for which X% of a population of a product fails. Thus, a “BX life Y years” is a way for expressing the system lifetime. For example, if a system lifetime has a B20 life of 10 years, then 20% of the population will fail during 10 years of operation. On the other hand, the mean time to failure (MTTF)—the B60 lifetime—is not acceptable for identifying the system lifetime because it takes too long for 60% of the population of the product to fail. The BX life is a more suitable measure of a lifetime.

In a mechanical product, movements of power for acquiring mechanical advantages are utilized in functions that require forces and movement by accommodating certain system mechanisms. As products are subjected to repeated stresses, they should have a proper lifetime for those stresses. For instance, to maintain the freshness of food products, a refrigerator is designed to provide cooled air from a heat exchanger such as an evaporator to the refrigerator (or freezer) sections. The refrigerator is made up of different modules: the cabinet, door, internal fixtures (shelves and drawers), controls and instruments, electric motor, compressor, heat exchangers (evaporator and condenser), water supply device, and numerous dissimilar parts. A refrigerator has approximately 2,000 components (Figure 1a). A target of the refrigerator's lifetime is set to have a B20 life 10 years. If a refrigerator consists of 20 units and each unit has 100 components, the lifetime of each unit should be designed to have a B1 life 10 years. The required system lifetime of a refrigerator relies on the many components, including the compressor. If the compressor has design faults, it impacts the potential life of the refrigerator (Figure 1b).

*Figure 1. Product lifetime decided by newly designed module (**a**) categorization of multi-module refrigerator; (**b**) product lifetime L_B and failure rate λ_s.*

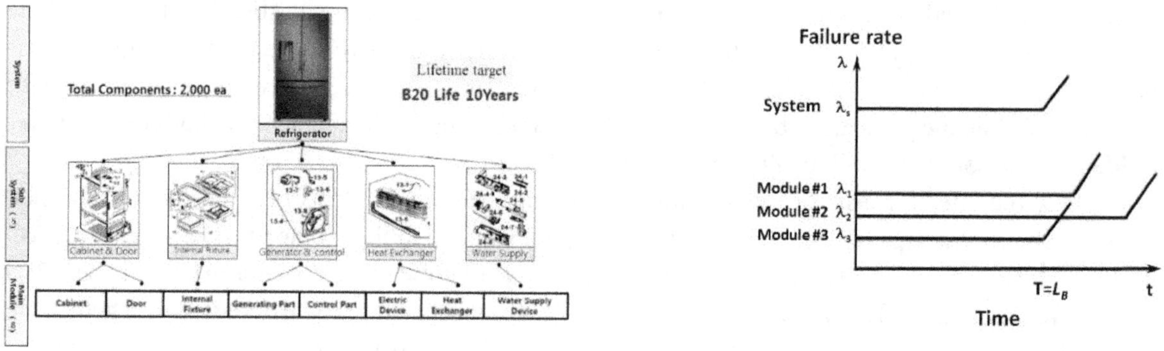

(a) (b)

2.2. Placing an Entire Parametric ALT Scheme

For a stated period of time, the reliability for a compressor can be explained as the capacity needed to continue performing the intended function under specified environmental/operational conditions (IEEE Standard Glossary of Software Engineering Terminology, 2002). It can be explained using the traditional "bathtub curve", which is composed of three regions, as seen in the top curve of Figure 2. In the first region, during the early product life, there is some reduction in the failure rate. In the second region, during its middle life, there is a relatively constant failure rate. In the third region, there is a growing failure rate until the final life of the product. If a product follows this classic pattern, it might not be successful in the marketplace due to the tall initial failure rates and the relatively short lifetime of the product due to built-in design flaws. Companies need to upgrade the system design by setting its reliability targets to (1) remove premature failures, (2) reduce random failures over the product lifetime, and (3) enlarge product lifetime. As the design of a mechanical product improves, its failure rate in the field decreases, and the system lifetime can be increased. For such situations, the traditional bathtub curve might be altered to the bottom curve in Figure 2, where failure rates are low throughout the life of the product except toward the end of its expected life.

Figure 2. Bathtub curve and straight line.

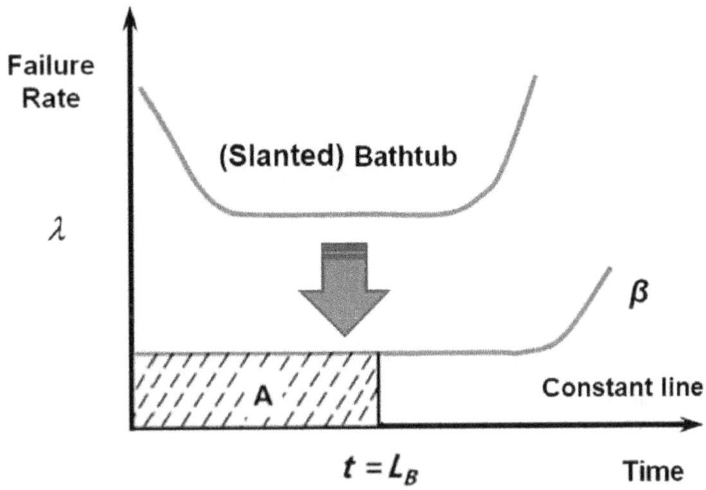

The failure rate on the bathtub in Figure 2 can be stated as

$$\lambda = \frac{f}{R} = \frac{dF/dt}{R} = \frac{(1-R)'}{R} = \frac{-R'}{R} \tag{1}$$

where λ is the failure rate, f is the failure density function, R is the reliability, and F is the unreliability of the product.
If Equation (1) is integrated over time, the X% cumulative failure $F(L_B)$ at BX life, $L_{B,}$ can be obtained:

$$\int \lambda dt = -\ln R \tag{2}$$

That is, it can be stated as:

$$A = \langle \lambda \rangle \cdot L_B = \int_0^{L_B} \lambda(t) \cdot dt = -\ln R(L_B) = -\ln(1-F) \cong F(L_B)(=X) \tag{3}$$

As a product has a lower failure rate of the straight line that follows an exponential distribution, the reliability of a mechanical system could be stated as the product lifetime L_B and failure rate λ:

$$R(L_B) = 1 - F(L_B) = e^{-\lambda L_B} \cong 1 - \lambda L_B \tag{4}$$

This relationship is suitable below about 20% of the cumulative failure (Kreyszig, 2011). After targeting the lifetime, L_B, and failure rate, λ, the mechanical product can be redesigned by identifying the troublesome structures and altering them by a parametric ALT (Figure 3).

Figure 3. Parameter diagram of compressor (example).

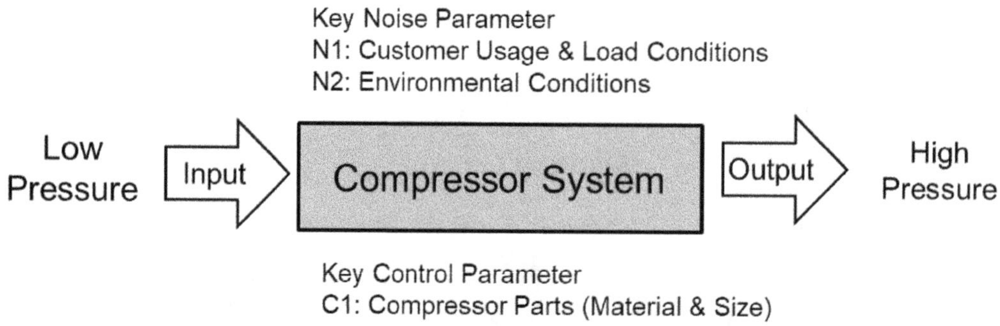

To target the lifetime of a mechanical system through a parametric ALT, there are three possible system modules: (1) modified independent units, (2) newly designed independent units, and (3) similar independent units to the earlier design base on request in the field. The modified compressor in the refrigerator inspected here is used as a case study. It was originally redesigned to enhance its energy efficiency and make the refrigerator more competitive in the marketplace. However, the redesigned compressor had design faults that needed to be altered because of failures in the field.

The altered independent units D from the field data manifested in Table 1 had a failure rate of 0.31% per year and a B1 life of 3.2 years. The lifetime of compressors, based on data from the field, indicated they had an expected B1 life of 1.6 years because they had a failure rate of 0.62% per year. To satisfy consumer demands, a new lifetime target for the mechanical system such as compressor was put to have a B1 life of 10 years with 0.1 percent per year.

Table 1. Whole ALT scheme of mechanical modules in a refrigerator.

Modules	Field Data		Expected Reliability				Aimed Reliability	
	Failure Rate Per Year, %/Year	BX Life, Year	Failure Rate Per Year, %/Year			BX Life, Year	Failure Rate Per Year, %/Year	BX Life, Year
A	0.35	2.9	Similar	×1	0.35	2.9	0.10	10(BX = 1.0)
B	0.24	4.2	New	×5	1.20	0.83	0.10	10(BX = 1.0)
C	0.30	3.3	Similar	×1	0.30	3.33	0.10	10(BX = 1.0)
D	0.31	3.2	Altered	×2	0.62	1.61	0.10	10(BX = 1.0)
E	0.15	6.7	Altered	×2	0.30	3.33	0.10	10(BX = 1.0)
Others	0.50	10.0	Similar	×1	0.50	10.0	0.50	10(BX = 5.0)
Product	1.9	2.9	-	-	3.27	0.83	1.00	10(BX = 10)

2.3. Failure Mechanics and Parametric ALT for Redesign

As mentioned in Section 2.1, mechanical products usually transfer power (or energy) from one position to another by adapting proper mechanisms. For example, the compressor in a refrigerator, by utilizing the vapor-compression cycle, increases the refrigerant pressure using a crankshaft mechanism. It is subjected to repeated stresses because of pressure loading. If there is a design defect in the structure that creates an insufficient strength (or stiffness) when the loads are applied, the mechanical product may abruptly collapse before its expected lifetime.

Metal fatigue is the familiar word that can be used to express the unanticipated failure of product components that develop fractures in their lifetime. It is connected to the number of stress cycles experienced by a component and the amount of stress exerted on the component. The S-N curve shows that an infinite life is possible for a metal component if the stresses in the component remain below a clearly defined lower stress boundary. Metal fatigue grows if there is a stress raiser such as notches in a component. An engineer should identify these design flaws using appropriate analysis or testing. There is also a correlation between a metal component's ultimate tensile strength and hardness and its capability to endure fatigue loads. The loftier the tensile strength and hardness, the more probably it will fatigue if subjected to higher fluctuating loads.

Fatigue failure due to design defects can be characterized by two components: (1) the stress due to loads on the constructional system and (2) the type of materials (or shape) utilized in the structure. In recognizing the product failure by a parametric ALT, an engineer could optimally design parts with good forms and proper materials. The system could bear repeated loads over its lifetime so that it could fulfill the reliability target (Figure 4).

Figure 4. Fatigue failure on the structure created by repetitive loading and design flaws.

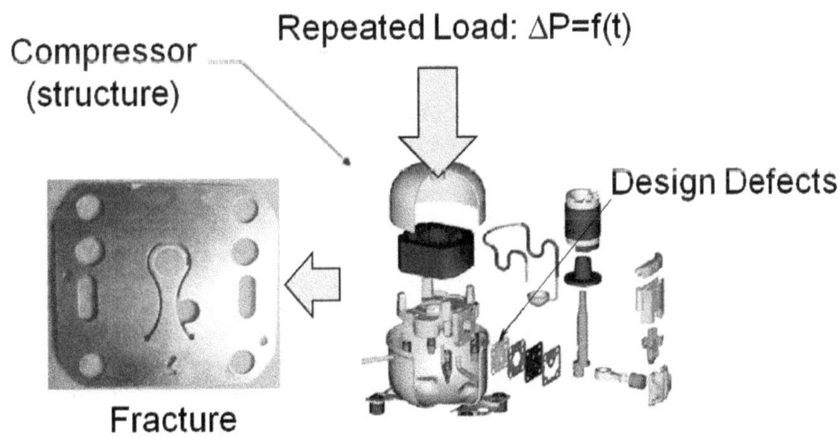

The main problem for reliability testing is to determine how quickly the potential failure mode can be identified. To complete it, it is necessary to successfully formulate a simplified failure description and determine the correct coefficients for the model. A life-stress (LS) model can be developed, which incorporates stresses and reaction parameters. This model can explain various mechanical failures such as fatigue in the structure. Fatigue failures can occur not due to stresses in a perfect part but rather due to the present defects or very small cracks on the exterior of a component.

As fatigue originates from material flaws created on a macro or microscopic scale level, the life-stress model can be defined from such a point of view. For instance, we can utilize the procedures for solid-state diffusion of impurities in silicon, which is commonly used as semi-conduct material, such as in the following procedure: electro-migration-induced voiding, growth of chloride ions, and catching of electrons or holes. As an electric magneto-motive force, ξ, is exerted, the impurities such as voids in materials, created by electronic movement, are effortlessly migrated because the barriers of junction energy are dropped and distorted/phase-shifted. For solid-state diffusion of impurities in silicon, the junction equation J might be expressed as follows (Grove, 1967; Minges, 1989) (Figure 5):

$$J = \left[aC(x-a)\right] \cdot \exp\left[-\frac{q}{kT}\left(w - \frac{1}{2}a\xi\right)\right] \cdot v \tag{5}$$

[Density/Area]·[Jump Probability]·[Jump Frequency]

$$\begin{aligned} &= -\left[a^2 v e^{-qw/kT}\right] \cdot \cosh\frac{qa\xi}{2kT}\frac{\partial C}{\partial x} + \left[2ave^{-qw/kT}\right] C \sinh\frac{qa\xi}{2kT} \\ &= \Phi(x,t,T)\sinh(a\xi)\exp\left(-\frac{Q}{kT}\right) = B\sinh(a\xi)\exp\left(-\frac{Q}{kT}\right) \end{aligned} \tag{6}$$

where C is the concentration, q is the magnitude of electric charge, ν is the frequency, Φ() is a constant, B is a constant, a is the interval between atoms, ξ is the applied field, k is the Boltzmann's constant, T is the temperature, and Q is the energy.

Figure 5. Potential exchange in material such as silicon as electrical field is exerted.

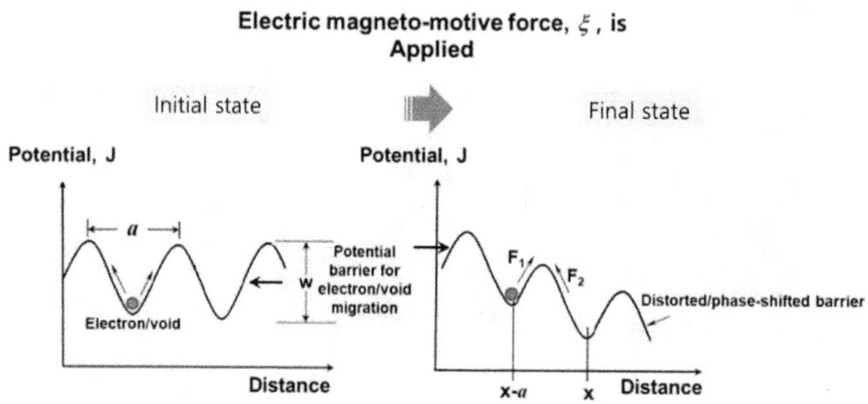

The reaction process, which depends on speed, could be stated as

$$\begin{aligned} K &= K^+ - K^- = a\frac{kT}{h}e^{-\frac{\Delta E - aS}{kT}} - a\frac{kT}{h}e^{-\frac{\Delta E + aS}{kT}} \\ &= a\frac{kT}{h}e^{-\frac{\Delta E}{kT}}\sinh(\frac{aS}{kT}) = B\sinh(aS)\exp\left(-\frac{\Delta E}{kT}\right) \end{aligned}$$

where K is the reaction rate, S is the (chemical) effect, T is the temperature, k is Boltzmann's constant, E is the (activation) energy, and Δ is the difference.
The reaction rate K from Equations (5) and (6) could be simplified as

$$K = B\sinh(aS)\exp\left(-\frac{E_a}{kT}\right) \tag{7}$$

If Equation (7) takes an inverse function, the life-stress (LS) model might be stated as

$$TF = A\left[\sinh(aS)\right]^{-1}\exp\left(\frac{E_a}{kT}\right) \tag{8}$$

The sine hyperbolic expression $[\sin h(aS)]^{-1}$ in Equation (8) can be stated as (Figure 6):

1. $(S)^{-1}$ at first has a little linear result;
2. $(S)^{-n}$ has what is regarded as a medium result;
3. $(e^{aS})^{-1}$ at end is large.

Figure 6. Hyperbolic sine stress term versus S-N curve from a standpoint of stress range.

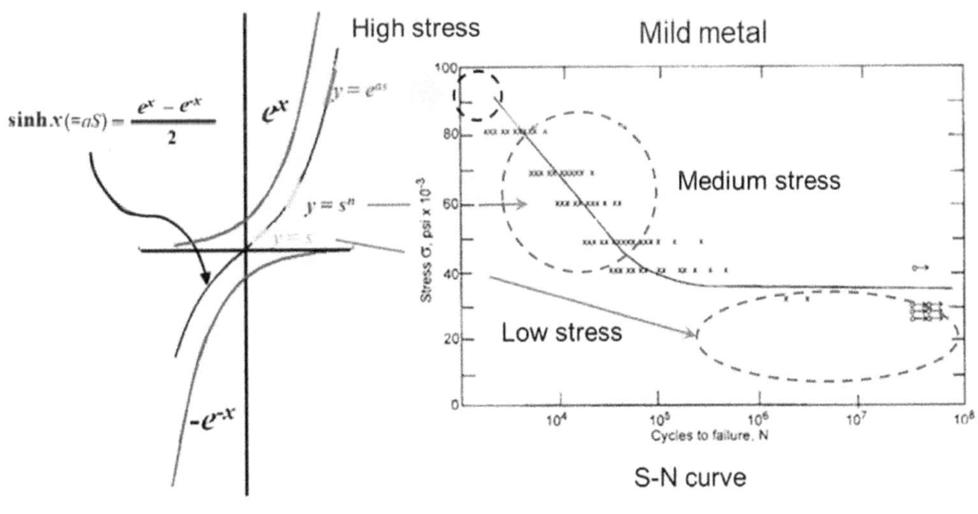

An ALT is usually conducted in the medium scope, and Equation (5) might be stated as

$$TF = A(S)^{-n}\exp\left(\frac{E_a}{kT}\right) \tag{9}$$

Because the stress level in a mechanical system may be hard to be computed during ALT, Equation (9) must be restated. As the power is stated as the product of flows and effort, stresses may originate from effort in a multi-port product (Table 2) (Karnopp et al., 2012).

Table 2. Power concept in a multi-port product expressed as effort and flow.

System	Effort, $e(t)$	Flow, $f(t)$
Translation system	Force, $F(t)$	Velocity, $V(t)$
Rotation system	Torque, $\tau(t)$	Angular velocity, $\omega(t)$
Pump, compressor	Pressure difference, $\Delta P(t)$	Volume flow rate, $Q(t)$
Electric system	Voltage, $V(t)$	Current, $i(t)$
Magnetic	Magneto-motive force, e_m	Magnetic flux, φ

Stress is a material quantity that specifies the inner forces which adjoining particles of a continuum material exert on each other. For a mechanical system, because stress originates from effort, Equation (9) might be stated as

$$TF = A(S)^{-n} \exp\left(\frac{E_a}{kT}\right) = B(e)^{-\lambda} \exp\left(\frac{E_a}{kT}\right) \tag{10}$$

where A and B are constants.

To obtain the acceleration factor (AF), which can mostly affect the evaluation of fatigue strength in a product, AF might be stated as the ratio between the proper elevated stress levels and normal working conditions. AF might be modified to integrate the effort notion:

$$AF = \left(\frac{S_1}{S_0}\right)^n \left[\frac{E_a}{k}\left(\frac{1}{T_0} - \frac{1}{T_1}\right)\right] = \left(\frac{e_1}{e_0}\right)^\lambda \left[\frac{E_a}{k}\left(\frac{1}{T_0} - \frac{1}{T_1}\right)\right] \tag{11}$$

2.4. Parametric ALT for Mechanical Systems

To acquire the mission cycle of parametric ALTs from the targeted BX lifetime on the test plan in Table 1, the sample size equation combined with the AF should be attained (Woo et al., 2021) (see Appendix A).

$$n \geq (r+1) \times \frac{1}{x} \times \left(\frac{L_B^*}{AF \cdot h_a}\right)^\beta + r \tag{12}$$

If the lifetime of a mechanical system, such as the domestic compressor, is targeted to have a B1 life of 10 years, the mission cycles might be acquired for an assigned set of samples subjected to (impact) loading. In parametric ALTs, the design defects of the new product might be identified to help satisfy the lifetime target.

2.5. Case Study—Design of a Recently Designed Domestic Compressor in a Refrigerator

Operating with unspecified consumer usage conditions, the suction reed valves in compressors used in the field were unsuccessful. The fractured suction reed valve caused the domestic compressor to lock and quit functioning. As the primary purpose of the refrigerator, including the compressor, was lost, consumers would solicit to have the product exchanged. To address the problem, it was crucial to reproduce the failure mode(s) of the compressor in simulated conditions in a room or building equipped for engineering experiments. Seemingly, the troublesome compressors that came from the field had two evident design flaws: (1) the suction reed valve had a quantity of overlap with the valve plate, and (2) the valve plate had a sharp edge. As the suction reed valve impacted the valve plate, it would fail before its expected lifetime (Figure 7).

*Figure 7. Failed suction reed valve: (**a**) suction reed valve and valve plate in a compressor. (**b**) Fractured suction reed valve from the marketplace or parametric ALT.*

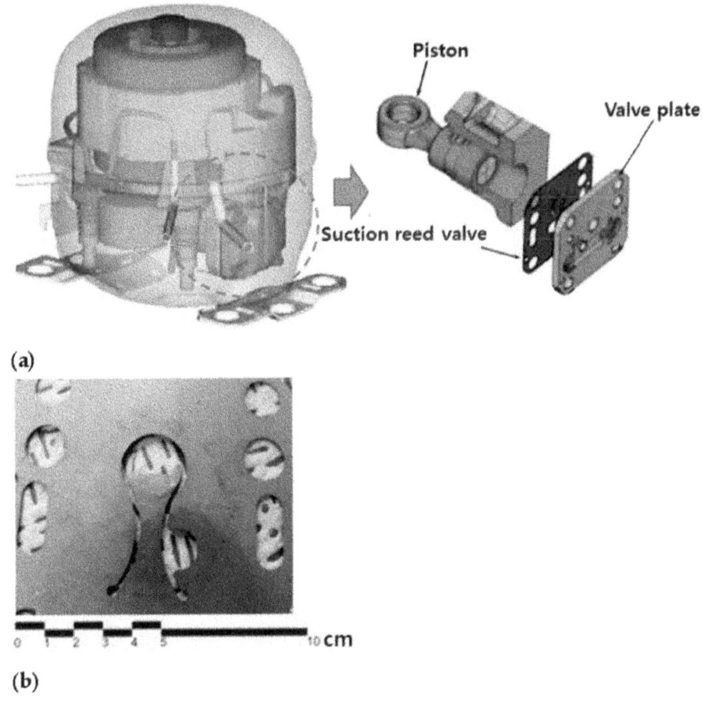

To chill the stored products in a refrigerator, the refrigerator supplies cooled air from the evaporator heat exchanger to the freezer (or refrigerator) departments. The vapor-compression cycle in a refrigerator covers a compressor, condenser, capillary tube, and evaporator. Electrical energy in the compressor motor is converted to work energy in the compressor that is used to elevate the refrigerant pressure. With refrigerant flowing in the system, heat energy absorbed by the evaporator is moved to the condenser, where it is rejected to the surrounding air. A capillary tube manages the refrigerant flow and controls the flow of refrigerant from the high-pressure refrigerant in the condenser to the low-pressure refrigerant in the evaporator. In an ideal vapor-compression refrigeration cycle, the refrigerant enters the compressor as a saturated vapor (Processes 1–2) and is cooled to the saturated liquid state in the condenser (Processes 2–3). It then drops in pressure as it goes through the expansion device to the lower pressure evaporator (Processes 3–4). In the evaporator, the refrigerant is vaporized as it absorbs heat from the refrigerated space (Processes 4–1) (Figure 8).

*Figure 8. A domestic compressor in the refrigeration cycle: (**a**) vapor-compression refrigerator cycle, (**b**) P h diagram.*

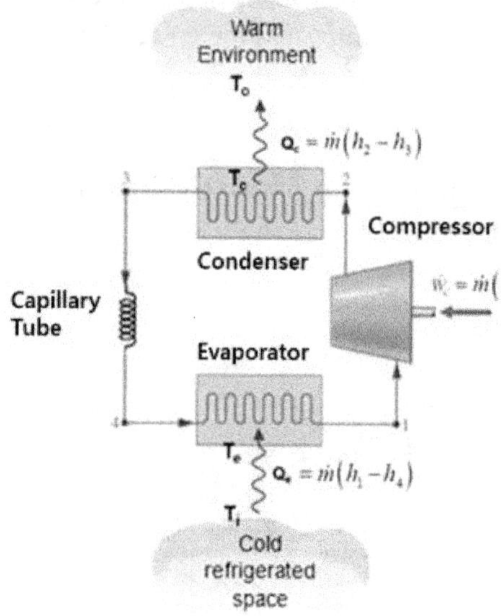

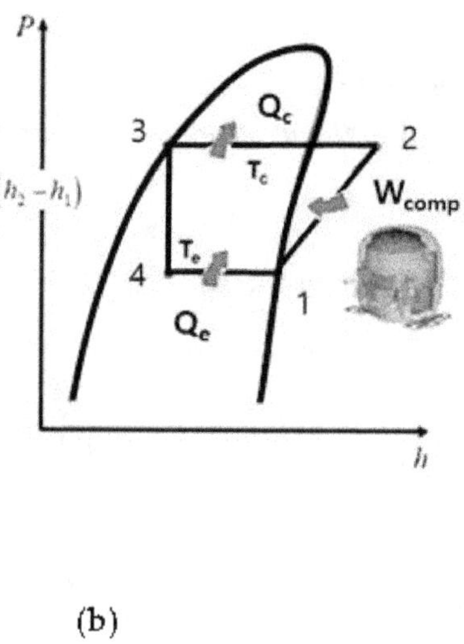

During normal operation, refrigerator compressors are subjected to repetitive stresses due to pressure loads differences in the compressor. If there is a design fault in a component, such as an insufficiency of strength, when the loads are applied in the compressor, the component may abruptly fail and not meet its expected lifetime. By identifying the system failure by a parametric ALT, an engineer can select proper material and redesign the component using the best or most favorable way so the compressor can support repeated loads and its lifetime can be increased. It is necessary to analyze the pressure loads in the compressor.

In a representative refrigeration cycle, to assess its design, it was required to determine both the condensing temperature, T_c, and evaporating temperature, T_e. The mass flow rate of refrigerant in a compressor can be modeled as

$$\dot{m} = PD \times \frac{\eta_v}{v_{suc}} \tag{13}$$

where PD is the piston displacement, η_v is the compressor's volumetric efficiency, and ν_{suc} is the specific volume at compressor suction port.

The mass flow rate of the refrigerant at the capillary tube can be modeled as (Whitesel, 1957).

$$\dot{m}_{cap} = A\left[\frac{-\int_{P_3}^{P_4} \rho dP}{\frac{2}{D} f_m \Delta L + \ln\left(\frac{\rho_3}{\rho_4}\right)}\right]^{0.5} \tag{14}$$

By conservation of mass, the mass flow rate can be stated as:

$$\dot{m} = \dot{m}_{cap} \tag{15}$$

The energy conservation at the condenser can be stated as

$$Q_c = \dot{m}(h_2 - h_3) = (T_c - T_o) / R_c \tag{16}$$

The energy conservation at the evaporator can be stated as

$$Q_e = \dot{m}(h_1 \quad h_4) = (T_i - T_e) / R_e \tag{17}$$

Using Equations (15) through (17), it is possible to obtain estimates of the mass flow rate, $\dot{m}$, evaporator temperature, T_e, and condenser temperature, T_c. Because the saturation pressure, P_{sat}, is a function of temperature, we can obtain the evaporator pressure, P_e (or condenser pressure P_c).

$$P_e = f(T_e) \text{ or } P_c = f(T_c) \tag{18}$$

Both the condensing pressure, P_c, and evaporating pressure, P_e, are important when examining the load on the compressor. These pressures depend on the environmental conditions, heat exchanger size, and customer usage conditions in the design stage.

During normal compressor operations, it will be subjected to repeated stresses due to the pressure differences between the suction and discharge. The internal stress of the compressor relies on the pressure difference between suction pressure, P_{suc}, and discharge pressure, P_{dis}.

$$\Delta P = P_{dish} - P_{suc} \cong P_c - P_e \tag{19}$$

Under accelerated conditions, the life-stress model (LS model) in Equation (10) can be stated as:

$$TF = A(S)^{-n} \exp\left(\frac{E_a}{kT}\right) = A(\Delta P)^{-\lambda} \exp\left(\frac{E_a}{kT}\right) \tag{20}$$

where A is constant, k is Boltzman's constant, E_a is the activation energy, T is the absolute temperature, n is the quotient, and λ is the cumulative damage exponent in Palmgren–Miner's rule.
So the acceleration factor (AF) might be stated as:

$$AF = \left(\frac{S_1}{S_0}\right)^n \left[\frac{E_a}{k}\left(\frac{1}{T_0} - \frac{1}{T_1}\right)\right] = \left(\frac{\Delta P_1}{\Delta P_0}\right)^\lambda \left[\frac{E_a}{k}\left(\frac{1}{T_0} - \frac{1}{T_1}\right)\right] \tag{21}$$

where S_1 (or P_1) is mechanical stress (or pressure difference) under accelerated conditions, and S_0 (or P_0) is mechanical stress (or pressure difference) under representative conditions.

For a domestic compressor in a refrigerator, the typical operating conditions were for a customer span from 0 °C to 43 °C with a humidity fluctuating from 0% to 95%. Vibration conditions presumed for working the compressor were subjected to 0.2 to 0.24g of acceleration.

When the compressor operates, the suction reed valve opens to allow refrigerant to stream into the compressor. A compressor is expected to cycle on and off 22 cycles per day. A worst-case scenario was also duplicated with on and off of 98 cycles per day. Under the worst cases, the compressor working for 10 years may incur approximately 357,700 usage cycles.

The compressor is usually made of (carbon, cast, stainless, alloy, etc.) steel. The allowable stresses are stated as a function of the yield stress (F_y) or tensile stress (F_u) of the structural material. For steel, the ranges of yield strength, F_y, and ultimate or tensile strength, F_u, ordinarily used are 248–345 MPa and 400–483 MPa, separately. Refrigerant R134a is the refrigerant utilized in the refrigerant cycle. It uses synthetic refrigeration compressor oils that have a high Viscosity Index (VI). They have slight viscosity changes in relation to temperature changes. Therefore, the slope of the viscosity of a synthetic lubricant with a high VI is flatter with respect to temperature. The viscosity thus remains stable across a wide temperature usage range.

In the worst case, the pressure difference was 1.27 MPa, and the compressor dome temperature was 90 °C. For a parametric ALT, the pressure difference was elevated to be 2.94 MPa, and the compressor dome temperature was also elevated to be 120 °C, with an accumulative damage exponent, λ, of 2. The total AF computed from Equation (21) was 7.3 (Table 3).

Table 3. Compressor ALT conditions.

System states		Worst Case	ALT	AF
Pressure (MPa)	High-side Low-side ΔP	1.27 0.0 1.27	2.94 0.0 2.94	5.36 ① - -
Temperature (°C)	Dome	90	120	1.37 ②
Total AF (=(① × ②))	-	-	-	7.32

The test cycles of the ALTs calculated from Equation (12) were 49,000 cycles for 100 samples if the shape parameter, β, was presumed to be 2.0. The parametric ALT was designed to reassure a lifetime target—B1 life of 10 years—with an approximate 60% level of confidence that it would fail less than once during 49,000 cycles. We also exerted the duty cycles of the pressure difference between suction pressure, P_{suc}, and discharge pressure, P_{dis}.

To assess the design of the compressor, a straightforward refrigeration cycle was established. It contained a compressor, condenser, capillary tube, and evaporator. The temperature in the enveloped fiberglass package is controlled by two 60 W lamps and a fan. A thermal switch attached to the cover of the compressor managed a 51 m^3/h axial fan. The test conditions and their borders were set up on the control board. As the test started, the high-side and low-side pressures were displayed on the pressure gauges or apparatus screen (Figure 9).

Figure 9. Equipment for the ALTs.

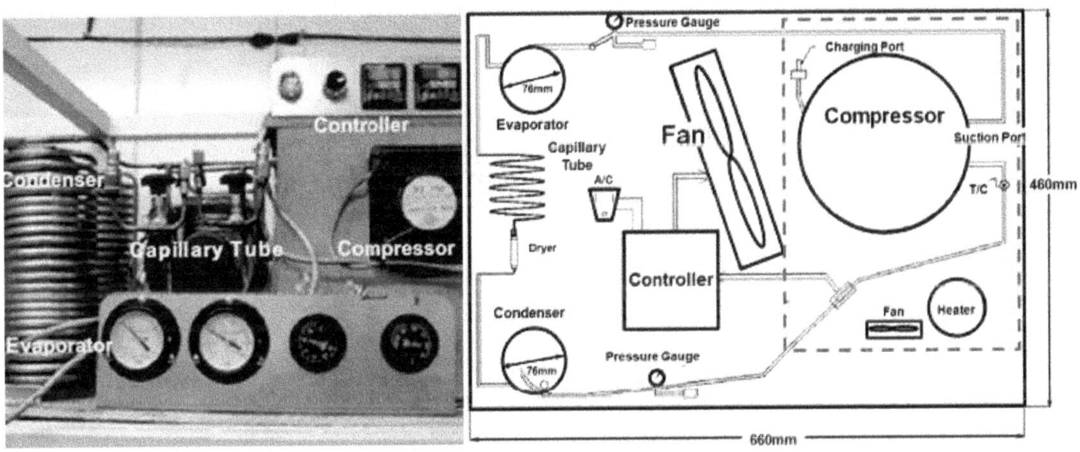

3. RESULTS AND DISCUSSION

A sample in the first ALT (n = 100) locked at 3,500 cycles. Another sample continued working but had a partially broken suction reed at 7,500 cycles (Figure 10). The shape parameter, β, formed from the first ALT, was 2.0 (see Figure 11). The forms and positions of the failure in the samples obtained from the first ALT and the field were alike (Figure 10). The fractured suction reed valve originated from three design faults: (1) it had a quantity of overlap with the valve plate, (2) it had a sharp edge on the valve plate, and (3) it used an insufficiently strong material (see Figures 6 and 12).

When the suction reed valve frequently striked the valve plate, it eventually fractured before its expected lifetime. The main failure mode of the compressor was locking due to the failed suction reed valve. The ALT methodology was well-suited for recognizing the fatigue failure attained from the units in the field. First, the shape of the failed suction reed valves from the marketplace and those in the first ALT were very similar in appearance.

*Figure 10. Failed suction reed valves from the field and first ALT: (**a**) failed products from the field, (**b**) outcome after first parametric ALT, (**c**) partially broken suction reed valve from the first ALT, (**d**) shattered valve plate.*

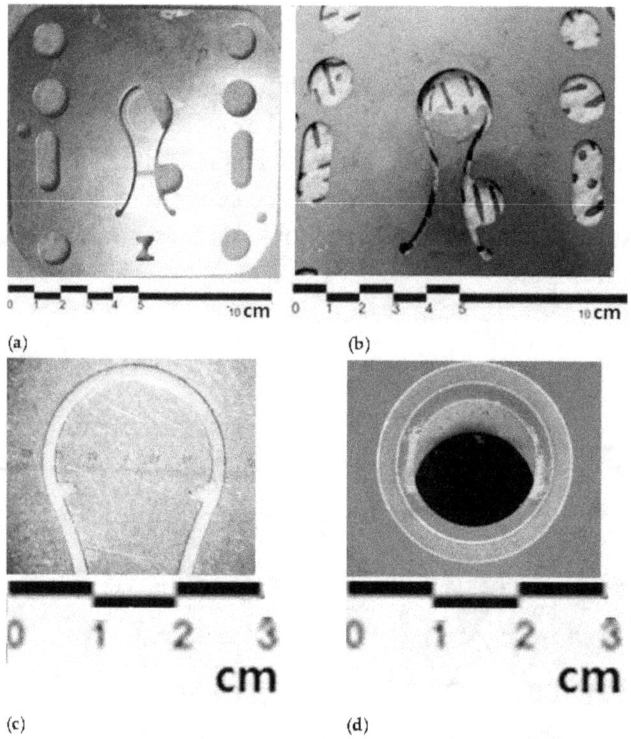

Figure 11. Market statistics and consequences of ALT on Weibull chart.

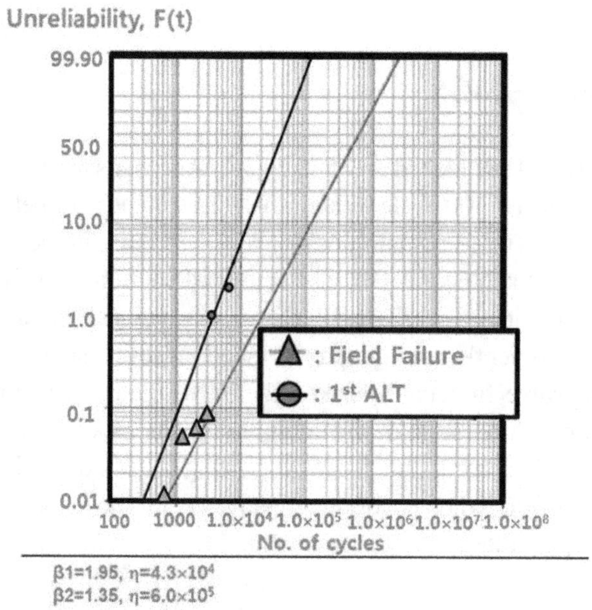

The first ALT failure and market failure data manifested a similar pattern on a Weibull plot (Figure 11). As the data for the two models had similar gradients on the plot, each loading condition of the first ALT and that from the field over the product lifetime were similar. Therefore, it might be anticipated that the test samples in the laboratory would break in a similar way to those in the field. For the shape parameter, β, the last shape parameter from the chart was confirmed to be 2.0, compared with the approximated value—2.0. Based on both test outcomes in the Weibull plot, the parametric ALT was successful because it recognized the design defects which were responsible for the market failures. In other words, as proven by two things—the visual likeness in both photos and similar slopes in the Weibull plot—these ALTs were justified in recognizing the design flaws that were judged as the failures from the market. These failures determined the product lifetime.

*Figure 12. Structural design defects of suction reed and valve plate in a compressor: (**a**) overlapped suction reed valve and valve plate with sharp edge; (**b**) impact load in combination of design defects when compressor is worked.*

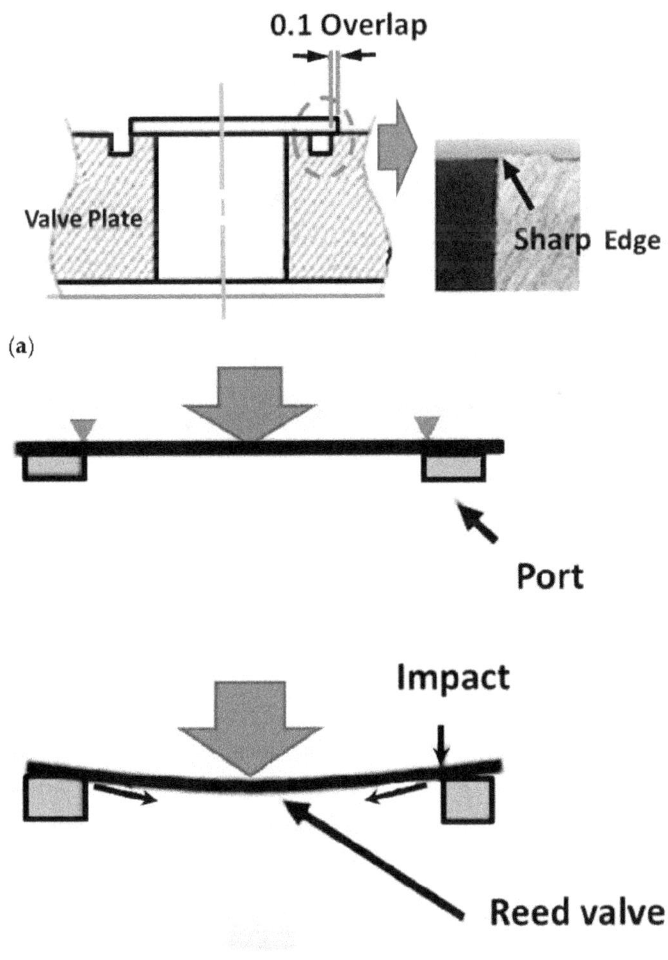

(**a**)

(**b**)

Refrigerators that came from the field had a primary failure mode with no cooling because the compressor did not function. Field data suggested that the troublesome compressors may have had design flaws. Due to these defects, the repetitive pressure loads could produce unanticipated stresses on the suction reed valve, causing it to crack and propagate the crack to its end. To reproduce the dominant failure mode of the compressor, parametric ALTs were carried out. Based on the first ALT and market data, we realized that the AF and β values were 7.3 and 2.0 (see Figure 11). For assigned test samples, the test cycles were computed in Equation (12) if the product lifetime was secured to have a B1 life of 10 years.

To investigate a compressor that failed at 3,500 cycles in the first ALT, the troublesome compressors came from the marketplace and the first ALT were compared to determine the potential design defects. The mode of the compressor failure in the first ALT was very similar to the ones from the field. The suction reed valves failed in areas where they were partly covered by the valve plate. The tests stated that the compressor was improperly designed to operate with this suction reed valve.

The failed suction reed valve originated from the unsuitable design flaws: (1) an quantity of overlap with the valve plate; (2) a sharp edge on the valve plate; and (3) insufficiently strong material (0.178t) utilized in the design of suction reed valve. These defects might cause the compressor system to fracture abruptly when subjected to repetitive pressure loads (see Figures 7 and 12).

*Figure 13. Redesigned valve plate: (**a**) trespan in valve plate; (**b**) ball peening process.*

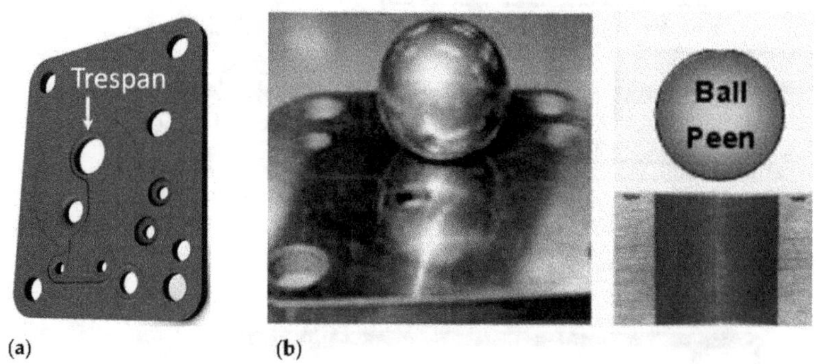

To fix the failed suction reed valve failure that was due to the repetitive pressure stresses in the compressor's lifetime, the valve plate and suction reed valve was modified as follows: (1) trespan size from 0.73 mm to 1.25 mm; (2) adding ball peening and brush process (see Figure 13); (3) thickening the suction reed valve from 0.178 t to 0.203 t (4) expanding tumbling process.

For the second ALT, three samples locked near 17,000 cycles. The problematic compressor system came as follows: (1) erosion of the crankshaft and (2) the intrusion between crank shaft and thrust washer. The design modification was provided to the heat treatment on the surface of crank shaft that can modify the physical, and sometimes chemical, properties of a material to achieve the desired result, such as hardening of a weak crankshaft material (Ductile Iron FCD450 – $C \geq 2.5$ wt%, $S \leq 0.02$ wt%, $Mg \leq 0.09$ wt%).

Table 4. Compressor ALT outcomes.

Parametric ALT	First ALT	Second ALT	Third ALT
	Original Design	Design	Final Design
In 49,000 cycles, there are no issues in the compressor	3,500 cycles: 1/100 locking 7,500 cycles: 1/100 broken suction reed valve	17,000 cycles: 3/100 locking	49,000 cycles: 100/100 OK
Structure	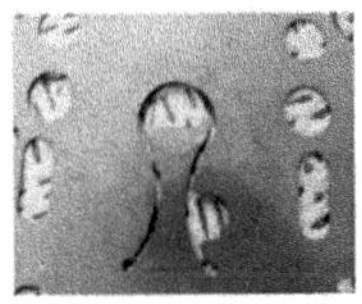	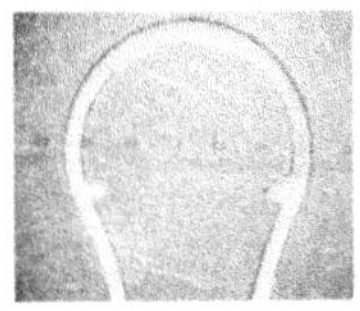 	
Action plans	C1: Trepan size: 0.73 mm→1.25 mm C2: attaching ball peening and brush process C3: SANDVIK 20C: 0.178 t→0.203 t C4: expanding tumbling: 4 h→14 h	C5: FCD500 + no heat treatment → FCD500 + heat treatment on the crankshaft	

In the third ALT, there were no design problems in the compressor until the parametric ALT was carried out to 49,000 cycles. We therefore concluded that the design modifications attained from the first and second ALTs were effective. Table 4 summarizes the parametric ALT results. With the modified design parameters, the compressor samples were ensured to attain the lifetime target—B1 life of 10 years with about a 60% confidence level.

4. CONCLUSION

To increase the lifespan of a new domestic compressor in a refrigerator, the following reliability methodology was developed: (1) the system BX lifetime created the total parametric ALT scheme, (2) the parametric ALT with design modifications, and (3) determine if the product design attains the mission cycles. To show the effectiveness of this methodology for identifying and altering the design defects of a mechanical product, we derived the sample size equation, time-to-failure equation and acceleration method, and BX lifetime. As a test case, we inspected the redesign of a compressor that came from the marketplace.

- In the marketplace and the first ALT, we found that the failed suction reed valve in a compressor originated from the following design flaws: (1) a small overlap between the valve with the valve plate, (2) a sharp edge on the valve plate, and (3) inadequate width of the suction reed valve (SANDVIK20 C0.178t). As design modifications, the trepan on the valve plate was extended from 0.73 mm to 1.25 mm, and a ball peening technique was secured to eliminate the sharp edge.
- For the second ALT, we found that three samples failed near 17,000 cycles because of the intrusion between the crankshaft and thrust washer. As an action plan, the crankshaft was heat-treated.
- During the third ALT, no problems were found. The compressor systems should attain the lifetime target—B1 life of 10 years with about a 60% confidence level.

- By examining problematic products returned from the marketplace and performing parametric ALTs with design modifications, it was possible to improve the expected lifetime of the compressor.

This reliability methodology can be applied to other mechanical products such as airplane, automobiles, and construction machines. To use this methodology, designers should understand why products fail during their lifetime. That is to say, if there are design flaws in the structure that are subjected to repetitive loads, the system will fail before its expected lifetime. Engineers would need to identify the load characteristics of a mechanical system so that the parametric ALT can be carried out until the required mission cycles under accelerated stress conditions. Finally, engineers can use parametric ALT to identify and alter the design problems of a mechanical product.

Abbreviations

A Cross-sectional area of the capillary tube: cm^2
BX Time which is a cumulated failure rate of $X\%$: durability index
E_a Activation energy, eV
e Effort
f Flow
F Impact force, kN
$F(t)$ Unreliability
h Testing cycles (or cycles)
h^* Non-dimensional testing cycles, $h^*=h/L_B \geq 1$
J Junction function equation
K Boltzmann's constant, 8.62×10^{-5} eV/deg
ΔL Capillary tube distance in the two-phase
L_B Target BX life and $x = 0.01X$, on the condition that $x \leq 0.2$
n Number of test samples
ΔP Pressure difference between the condenser and evaporator, MPa
PD Volume flow rate in compressor, m^3/s
P_c Pressure in the condenser, MPa
P_e Pressure in the evaporator, MPa
P_{suc} Pressure at compressor suction, MPa
P_{dis} Pressure at compressor discharge, MPa
Q Amount of energy absorbed or released during the reaction. For the semiconductor, total number of dopants per unit area
Q_c Heat transfer by temperature difference in the condenser, kW
Q_e Heat transfer by temperature difference in the evaporator, kW
R Ratio for minimum stress to maximum stress in stress cycle, $\sigma_{min}/\sigma_{max}$
r Failed numbers
R_c Thermal resistance in the condenser, K/kW
R_e Thermal resistance in the evaporator, K/kW
S Stress
T Temperature, K
T_c Absolute temperature in the condenser, K

T_e Absolute temperature in the evaporator, K
t_i Test time for each sample
TF Time to failure
X Accumulated failure rate, %
x $x = 0.01X$, on condition that $x \leq 0.2$.
W_c Compressor power, kW
Greek symbols
ξ Electrical field exerted
η Characteristic life
λ Cumulative damage exponent in Palmgren–Miner's rule
χ^2 Chi-square distribution
α Confidence level
ν_{suc} Specific volume of refrigerant at compressor suction, m^3/kg
ρ Refrigerant density, kg/m^3
η_v Volumetric efficiency
ω Angular velocity, rad/s
Superscripts
β Shape parameter in Weibull distribution
n Stress dependence, $n = -\left[\frac{\partial \ln(T_f)}{\partial \ln(S)}\right]_T$
Subscripts
0 Normal stress conditions
1 Accelerated stress conditions

ACKNOWLEDGMENT

Author Contributions: S.W. managed the concept forming, methodology, examination and testing, and wrote the manuscript. D.L.O. examined the analysis and writing for the original manuscript and edited it. Both authors have read and agreed to the published version of the manuscript.

Funding: This research accepted no external finance.

Institutional Review Board Statement: Not applicable.

Informed Consent Statement: Not applicable.

Data Availability Statement: The data provided in this research can be obtained on request from the corresponding author.

Conflicts of Interest: The authors express no conflicts of interest.

REFERENCES

Abernethy, R. B. (2000). *The New Weibull Handbook*. Reliability Analysis Center.

Akama, M., & Kiuchi, A. (2019). Fatigue Crack Growth under Non-Proportional Mixed Mode Loading in Rail and Wheel Steel Part 1: Sequential Mode I and Mode II Loading. *Applied Sciences (Basel, Switzerland)*, *9*(14), 2866. doi:10.3390/app9142866

Allen, P. (2020). *Design of Experiments for 21st Century Engineers* (1st ed.). Lulu Press.

Anderson, T. L. (2017). *Fracture Mechanics—Fundamentals and Applications* (3rd ed.). CRC. doi:10.1201/9781315370293

ASTM E399. *Standard Test Method for Linear-Elastic Plane-Strain Fracture Toughness of Metallic Materials*; ASTM International: West Conshohocken, PA, USA, 2020.

ASTM E606/E606M. *Standard Test Method for Strain-Controlled Fatigue Testing*; ASTM International: West Conshohocken, PA, USA, 2019.

ASTM E647. *Standard Test Method for Measurement of Fatigue Crack Growth Rates;* ASTM International: West Conshohocken, PA, USA, 2015.

ASTM E739-10. *Standard Practice for Statistical Analysis of Linear or Linearized Stress-Life (S-N) and Strain-Life (ε-N) Fatigue Data*; ASTM International: West Conshohocken, PA, USA, 2015.

Braco, R., Prates, P., Costa, J. D. M., & Berto, F. (2018). New methodology of fatigue life evaluation for multiaxially loaded notched components based on two uniaxial strain-controlled tests. *International Journal of Fatigue*, *111*, 308–320. doi:10.1016/j.ijfatigue.2018.02.027

Chowdhury, S., & Taguchi, S. (2016). *Robust Optimization: World's Best Practices for Developing Winning Vehicles* (1st ed.). John Wiley & Sons Inc. doi:10.1002/9781119212096

CMMI Product Team. (2018). *Capability Maturity Model Integration (CMMI) Version 2.0, Continuous Representation*. Report CMU/SEI-2002-TR-011; Software Engineering Institute.

Duga, J. J., Fisher, W. H., Buxaum, R. W., Rosenfield, A. R., Buhr, A. R., Honton, E. J., & McMillan, S. C. (1982). *The Economic Effects of Fracture in the United States*. Final Report. NBS Special Publication 647-2. Battelle Laboratories.

Elsayed, E. A. (2012). *Reliability Engineering*. John Wiley & Sons.

Fatigue. (2008). In F. C. Campbell (Ed.), *Elements of Metallurgy and Engineering Alloys*. ASM International.

Griffith, A. A. (1921). The phenomena of rupture and flow in solids. *Philos. Trans. R Soc. Lond. A*, *221*(582-593), 163–198. doi:10.1098/rsta.1921.0006

Grove, A. (1967). Physics and Technology of Semiconductor Device. Wiley International Edition.

Hahn, G. J., & Meeker, W. Q. (2004). *How to Plan an Accelerated Life Test (E-Book)*. ASQ Quality Press.

Hertzberg, R. W., Vinci, R. P., & Hertzberg, J. L. (2020). *Deformation and Fracture Mechanics of Engineering Materials* (6th ed.). John Wiley and Sons Inc.

IEEE Standard Glossary of Software Engineering Terminology. (2002). *IEEE STD 610.12-1990. Standards Coordinating Committee of the Computer Society of IEEE.* Available online: https://ieeexplore.ieee.org/document/159342

Irwin, G. (1957). Analysis of stresses and strains near the end of a crack traversing a plate. *Journal of Applied Mechanics, 24*(3), 361–364. doi:10.1115/1.4011547

Karnopp, D. C., Margolis, D. L., & Rosenberg, R. C. (2012). *System Dynamics: Modeling, Simulation, and Control of Mechatronic Systems* (6th ed.). John Wiley & Sons. doi:10.1002/9781118152812

Kreyszig, E. (2011). *Advanced Engineering Mathematics* (10th ed.). John Wiley and Son.

Li, Q., & Xie, L. (2020). Analysis and Optimization of Tooth Surface Contact Stress of Gears with Tooth Profile Deviations, Meshing Errors and Lead Crowning Modifications Based on Finite Element Method and Taguchi Method. *Metals, 10*(10), 1370. doi:10.3390/met10101370

Magaziner, I. C., & Patinkin, M. (1989). Cold competition: GE wages the refrigerator war. *Harvard Business Review, 89*, 114–124.

McPherson, J. (1989). Accelerated testing. In Electronic Materials Handbook Volume 1: Packaging; ASM International Publishing.

McPherson, J. (2010). *Reliability Physics and Engineering: Time-to-Failure Modeling*. Springer. doi:10.1007/978-1-4419-6348-2

Minges, M. L. (1989). *Electronic Materials Handbook* (Vol. 1). ASM International.

Modarres, M., Kaminskiy, M., & Krivtsov, V. (2016). *Reliability Engineering and Risk Analysis: A Practical Guide* (3rd ed.). CRC Press. doi:10.1201/9781315382425

Rosa, J. L., Robin, A., Silva, M. B., Baldan, C. A., & Peres, M. P. (2009). Electrodeposition of copper on titanium wires: Taguchi experimental design approach. *Journal of Materials Processing Technology, 209*(3), 1181–1188. doi:10.1016/j.jmatprotec.2008.03.021

Sánchez, M., Cicero, S., Arroyo, B., & Álvarez, J. A. (2020). Coupling Finite Element Analysis and the Theory of Critical Distances to Estimate Critical Loads in Al6060-T66 Tubular Beams Containing Notches. *Metals, 10*(10), 1395. doi:10.3390/met10101395

Toribio, J., Kharin, V., Ayaso, F. J., González, B., Matos, J. C., Vergara, D., & Lorenzoc, M. (2010). Failure analysis of a lifting platform for tree pruning. *Engineering Failure Analysis, 17*(4), 739–747. doi:10.1016/j.engfailanal.2009.08.017

Weingart, R.G., & Timoshenko, S. P. (2007, Aug.). Father of Engineering Mechanics in the U.S. *Structure Magazine*.

Whitesel, H. A. (1957). Capillary two-phase flow Part II. *Refrig. Eng., 65*, 35–40.

Woo, S., O'Neal, D., Woldemichael, D. E., Atnaw, S. M., & Tulu, M. M. (2021). Improving the Fatigue of Newly Designed Mechanical System Subjected to Repeated Impact Loading. *Metals*, *11*(1), 139. doi:10.3390/met11010139

Zupančič, B., Prokop, Y., & Nikonov, A. (2021). FEM analysis of dispersive elastic waves in three-layered composite plates with high contrast properties. *Finite Elements in Analysis and Design*, *193*, 103553. doi:10.1016/j.finel.2021.103553

APPENDIX

Derivation of Sample Size Equation for Parametric ALT

Presently, many methods have been proposed to resolve issues on sample size. The Weibayes model by using Weibull examination is a widely acknowledged method of inspecting reliability data. However, it is hard to be straightly used due to the complexity of the mathematical formulation. The entire instances as failures ($r \geq 1$) and no failures ($r = 0$) need to be split. As a result, it is feasible to attain a predictable sample size equation which might supply the mission cycle after correct presumptions.

In selecting the model parameters to maximize the likelihood function, the maximum likelihood estimation (MLE) statistic is a general way of approximating the parameters of a model. The characteristic life h_{MLE} would be stated as:

$$\eta_{MLE}^{\beta} = \sum_{i=1}^{n} \frac{t_i^{\beta}}{r} \tag{A1}$$

where ηM_{LE} is the maximum likelihood estimate for the characteristic life, n is the entire number of samples, ti is the experimental time for each sample, and r is the number of failures.

If the failure numbers, r, greater than or equal to 1 and the confidence level is 100 ($1-\alpha$), then the characteristic life, η_{α}, can be estimated from Equation (A1).

$$\eta_{\alpha}^{\beta} = \frac{2r}{\chi_{\alpha}^{2}(2r+2)} \times \eta_{MLE}^{\beta} = \frac{2}{\chi_{\alpha}^{2}(2r+2)} \times \sum_{i=1}^{n} t_i^{\beta} \quad \text{for } r \geq 1 \tag{A2}$$

where $\chi_{\alpha}^{2}()$ is the chi-square distribution when the p-value is α.

Presuming there are no failure numbers, ln ($1/\alpha$) is similar to the chi-square value, $\frac{\chi^2(2)}{2}$ (Abernethy, 2000).

$$p-value: \alpha = \int_{\chi_{\alpha}^{2}(2)}^{\infty} \left(\frac{e^{-\frac{x}{2}} x^{\frac{v}{2}-1}}{2^{\frac{v}{2}} \Gamma\left(\frac{v}{2}\right)} \right) dx = \int_{2\ln\alpha^{-1}}^{\infty} \left(\frac{e^{-\frac{x}{2}} x^{\frac{v}{2}-1}}{2^{\frac{v}{2}} \Gamma\left(\frac{v}{2}\right)} \right) dx \quad \text{for } x \geq 0 \tag{A3}$$

where ν is the shape parameter and Γ is the gamma function.
For $r = 0$, the characteristic life η_{α} from Equation (A2) might be stated as:

$$\eta_{\alpha}^{\beta} = \frac{2}{\chi_{\alpha}^{2}(2)} \times \sum_{i=1}^{n} t_i^{\beta} = \frac{1}{\ln\frac{1}{\alpha}} \times \sum_{i=1}^{n} t_i^{\beta} \tag{A4}$$

As Equation (A.2) is demonstrated for all cases $r \geq 0$, characteristic life, η_{α}, might be stated as:

$$\eta_\alpha^\beta = \frac{2}{\chi_\alpha^2(2r+2)} \times \sum_{i=1}^{n} t_i^\beta \quad \text{for } r \geq 0 \tag{A5}$$

If the logarithm in the Weibull distribution is used, the relationship between characteristic life and BX life, L_B, might be stated as:

$$L_B^\beta = \left(\ln\frac{1}{1-x}\right) \times \eta_\alpha^\beta \tag{A6}$$

If the estimated characteristic life of the p-value α, η_α, in Equation (A5), is substituted into Equation (A6), we obtain the BX life equation:

$$L_B^\beta = \left(\ln\frac{1}{1-x}\right) \times \frac{2}{\chi_\alpha^2(2r+2)} \times \sum_{i=1}^{n} t_i^\beta \tag{A7}$$

As whole reliability testing has insufficient sample numbers to assess the lifetime for the designated failures, which might be less than that of the sample size, the experiment plan can be stated as:

$$nh^\beta \geq \sum t_i^\beta \geq (n-r) \times h^\beta \tag{A8}$$

If Equation (A8) is changed with Equation (A7), the BX life equation might be restated as:

$$L_B^\beta \cong \left(\ln\frac{1}{1-x}\right) \times \frac{2}{\chi_\alpha^2(2r+2)} \cdot nh^\beta \geq \left(\ln\frac{1}{1-x}\right) \times \frac{2}{\chi_\alpha^2(2r+2)} \times (n-r)h^\beta \geq L_B^{*\beta} \tag{A9}$$

If Equation (A9) is reorganized, the sample size equation with the failures can be stated as:

$$n \geq \frac{\chi_\alpha^2(2r+2)}{2} \times \frac{1}{\left(\ln\frac{1}{1-x}\right)} \times \left(\frac{L_B^*}{h}\right)^\beta + r \tag{A10}$$

Because $\frac{\chi_\alpha^2(2r+2)}{2} \cong (r+1)$ for α = 0.6 and $\ln(1-x)^{-1} = x + \frac{x^2}{2} + \frac{x^3}{3} + \cdots \cong x$, the sample size equation (A10) can be straightforwardly adjacent to:

$$n \geq (r+1) \times \frac{1}{x} \times \left(\frac{L_B^*}{h}\right)^\beta + r \tag{A11}$$

where the sample size equation can be stated as n ~ (failure numbers + 1)×(1/cumulative failure rate)×((target lifetime/(plan testing time)) ^ β + r.

Chapter 13
Energy Resources and Their Consumption

Harpreet Kaur Channi
Chandigarh University, India

ABSTRACT

Power is a significant cause of economic growth and crucial to the sustainability of the economy. Energy consumption is an indicator of a nation's economic growth. Economic growth is focused, among other aspects, on the long-term acquisition of affordable, existing resources, and their use does not pollute the environment. Industrialization serves economic growth and consumes energy. In 2018, 68% of total capital power was consumed by largest energy-intensive areas. When fossil fuel is the primary source of energy, energy consumption is positively correlated with ecosystem cleanliness. Fossil fuels account for more than 70% of the decent energy expectations of India and other economies. In this chapter, problems related to non-renewable energy sources are discussed, and emphasis is given to use more renewable sources.

1. INTRODUCTION

Electricity is the strength to do work and is crucial for operations of life. An alternative energy source can create heat, power life, move objects, or generate power. It is titled as fuel which carries energy. Intake of human energy has steadily grown throughout human history. Early humans required limited energy, mainly food and fuel to cook and keep warm. (Ministry of New and Renewable Energy, n.d.).

Human absorbs as much as 110 times as much power per person as early humans in today's age. Much of the energy we are using today comes from fossil fuels (solar power stored). But fossil fuels have a limitation because on a human time scale they are non-renewable and cause other possibly damaging environmental consequences (Ren 21, n.d.).

1.1 Classification of Energy Resources

Energy resources are classified as shown in figure 1:

DOI: 10.4018/978-1-7998-8561-0.ch013

Figure 1. Classification of energy resources(Ren 21, n.d.)

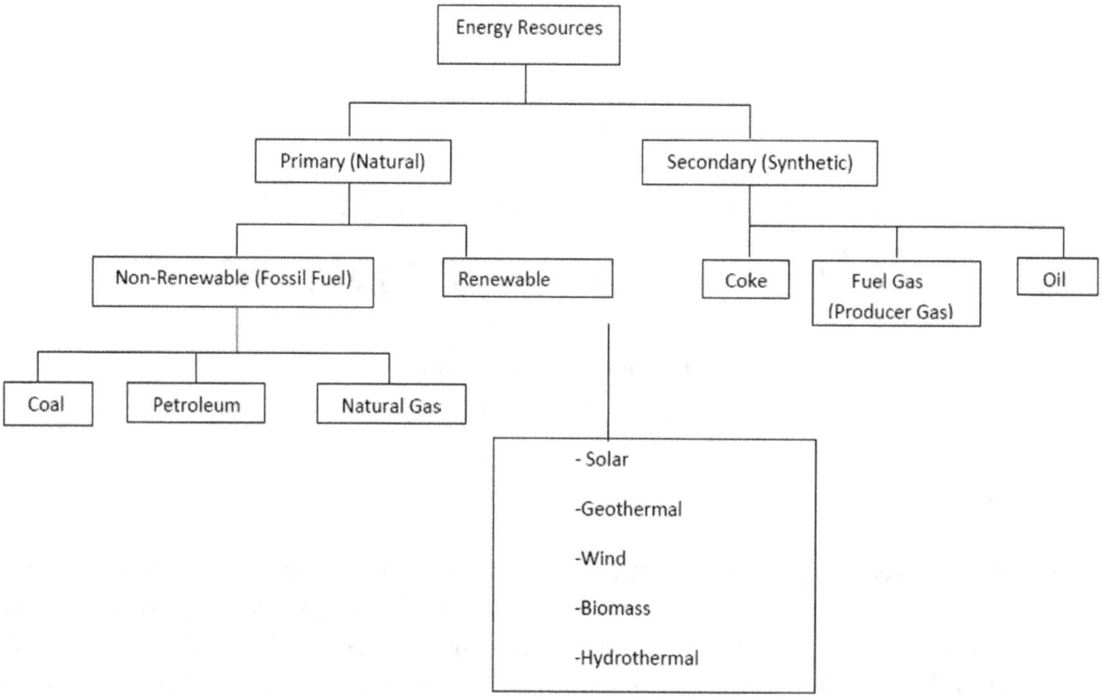

It should be indicated that all fossil fuel reserves come from plants and it takes millions of years to form a reserve below the earth's crust through physico-chemical modifications.

1.1.1 Types of Non-Renewable Sources of Energy

Coal: It is created by the protracted action of geological forces accumulated below the earth's crust on plant and organic matter and is called as "COALIFICATION". Coalification is dependent on both moment and force and it leads to modifications in the acquired plant (Ani & Abubakar, 2015).

Wood ⟶ Peat ⟶ Lignite ⟶ Bituminous ⟶ Anthracite ⟶ Graphite Increase time and magnitude of forces

⟶ ⟶

Almost all physical changes, such as colour, strength, density, and composition; and there is chemical change. It is essential to make chemical changes.

- Oxygen is reduced from 40% in timber to 305% in peat, 20% in lignite, 5% in bituminous and 2% in anthracite carbon.
- For anthracite coal, volatile matter reduces from about 70% for timber to 5% or less.
- Carbon increase from around 30% for timber and peat to 90-95%

Petroleum: It is created from metamorphic procedures comparable to coalification in the earth's crust from the accumulated vegetable and animal matter. It is acquired from crude oil petrol, lubricating oil, fuel oils, etc. (Ani & Abubakar, 2015).

Natural gas: It is used straightforwardly.

1.1.2 Renewable Sources of Energy

- Geothermal: power acquired by tapping under its surface the heat of the earth. To generate electricity, hot subterranean water or steam is used.
- Biogas: manufactured from paper and sugar waste, animal, etc. The item is CH_4.
- Biofuel: plant-derived biodiesel, ethanol, etc.
- Solid biomass: wood fuel, municipal waste biogenic part, certain crops. Biomass mass can be used to generate energy in a variety of ways. Gasification, combustion, fermentation, anaerobic digestion are prevalent techniques. India is very wealthy in biomass(Eureka eLearning, n.d.; Nade et al., 2018)
- Hydrothermal: water energy in the form of KE, difference in temperature
- Solar Energy: Sunlight energy gathered. It can be used in many ways: using photovoltaic cells, 3¤4 generate energy. 3¤4 Use concentrated solar energy to generate electricity. 3¤4 Low efficiency factor for photovoltaic cells.

1.2 Indian Energy Scenario

- India ranks 6th in the globe in total power usage and needs to accelerate the power sector's development to satisfy the country's 8-9 percent financial growth.
- India, wealthy in coal and rich in renewable electricity, has very tiny reserves of hydrocarbons (0.4% of the world's reserves).
- India is a net power importer, with more than 25% of main power needs being met through crude oil and natural gas exports.
- Coal and oil account for 54 percent and 34 percent in energy production, with the remainder contributing to natural gas, hydro and nuclear. In India, the industrial industry consumes 52% of electricity. Primary energy consumption in India is 530 kg of petroleum equivalent / person in 2004 compared to 1240 kg oil equivalent / person in China and 1770 kg oil equivalent / person on the world average.
- With development in the economy, primary energy consumption per individual will increase because energy consumption is an index of economic growth and prosperity in the country (Ren 21, n.d.; Shrivastava et al., n.d.).

1.3 World Energy Consumption

The consumption of world energy is the complete energy used by all human civilization. Typically measured annually, it includes all energy harvested from every source of energy applied to human efforts in every single industrial and technological industry, in every nation. as shown in figure 2.

Figure 2 Energy harnessed from every energy source*(Ren 21, n.d.)*

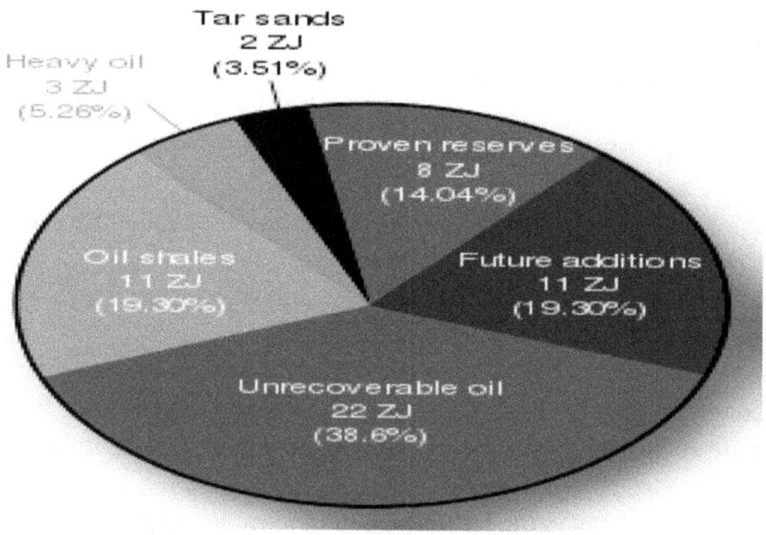

This doesn't include food energy, and it is poorly recorded to what extent direct biomass burning was accounted for. World energy consumption is the energy source metric of civilization and has profound consequences for the socio-economic-political sphere of humanity as shown in figure 3.

Figure 3. Energy Consumption (Ren 21, n.d.)

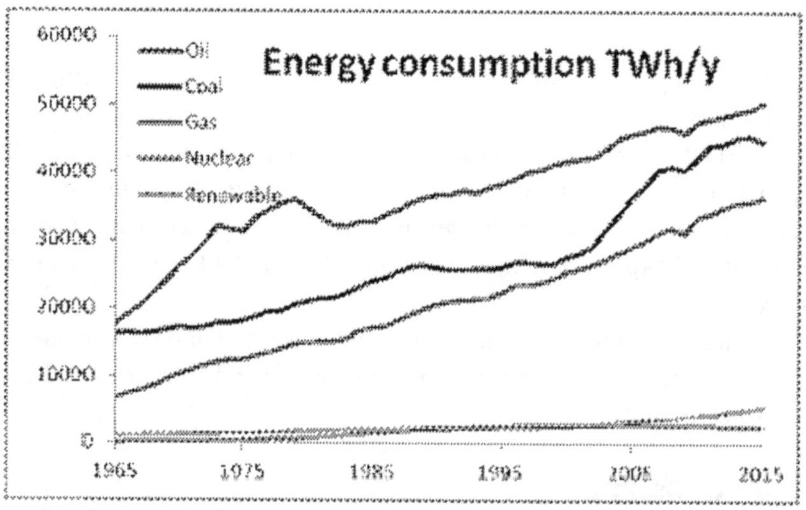

2. ENERGY SUPPLY AND DEMAND

World primary energy supply varies from the world's final energy consumption because much of the energy obtained by humans is lost as other types of energy during the process of its refinement into usable forms of energy and its transportation from its original location of production to customers. For example, it must be refined into gasoline when oil is extracted from the ground so that it can be used in a car and transported to gas stations over long distances where it can be used by consumers. The final consumption of world energy relates to the fraction of the primary energy used by humanity in its final form. (Dey, 2005; Punjab Energy Development Agency, n.d.; Verma, n.d.).

In 2014, the world primary energy supply amounted to 155,481 terawatt-hour (TWh) or 13,541 Mtoe, whereas the world final energy consumption was 109,613 TWh or about 29.5 percent less than the total supply as shown in Table 1. World consumption of final energy involves products such as lubricants, asphalt and petrochemicals that contain chemical energy but are not used as fuel. This non-energy use amounted to 9,404 TWh (809 Mtoe) in 2012.

Table 1. Energy Supply and Consumption (Ministry of New and Renewable Energy, n.d.)

Year	Primary energy supply	Final energy consumption	Electricity generation
1973	71,013 (Mtoe 6,106)	54,335 (Mtoe 4,672)	6,129
1990	102,569	–	11,821
2000	117.687	–	15.395
2010	147,899 (Mtoe 12,717)	100,914 (Mtoe 8,677)	21,431
2011	152,504 (Mtoe 13,113)	103,716 (Mtoe 8,918)	22,126
2012	155,505 (Mtoe 13,371)	104,426 (Mtoe 8,979)	22,668
2013	157,482 (Mtoe 13,541)	108,171 (Mtoe 9,301)	23,322
2014	155,481 (Mtoe 13,369)	109,613 (Mtoe 9,425)	23,816
2015	168,519 (Mtoe 13,647)		

For most types of primary energy resources, the U.S. Energy Information Administration (EIA) regularly publishes a report on world consumption. World energy consumption was estimated at 5,67 some 1020 joules or 157,481 TWh for 2013. In 2008, the total world energy consumption was 143,851 TWh, 133,602 TWh in 2005, 117,687 TWh in 2000, and 102,569 TWh in 1990, according to the IEA. In 2012, roughly 22% of the world's electricity was consumed in North America, 5% in South and Central America, 23% in Europe and Eurasia, 3% in Africa and 40% in the Asia Pacific area. (National Geographic, n.d.).

3. PROBLEM FORMULATION

Fossil fuel is produced by combustion and includes potential energy / chemical energy. Figure 4 demonstrates the following:

Figure 4. Potential energy/chemical energy

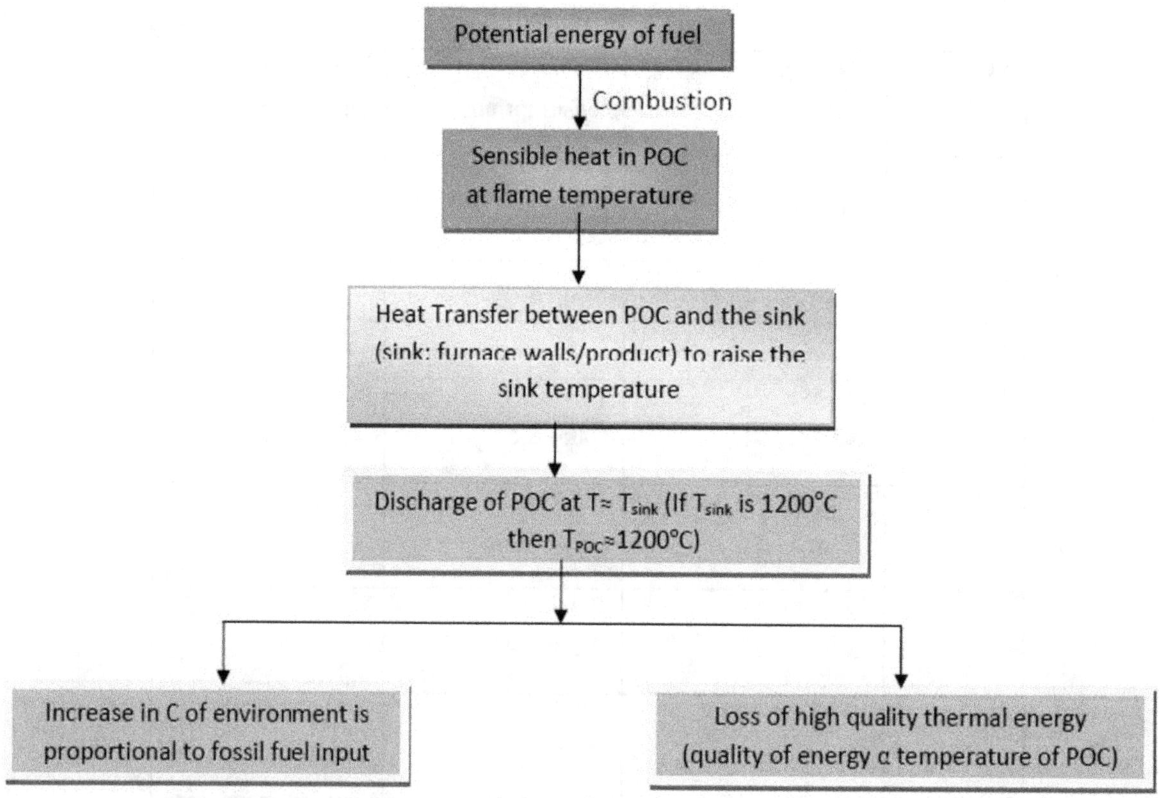

The use of fossil fuel power sources thus refers to sustainability of the atmosphere (enhanced use of fossil fuel increases carbon emissions) and power safety (restricted reserves of fossil fuel).

4. MAJOR PROBLEMS RELATED TO FOSSIL FUEL

4.1 Global Warming

An increase in the average temperature of the Earth's atmosphere and oceans Global temperature on both land and sea increased by 0.6 ± 0.2 °C over the past century Volume of atmospheric carbon dioxide increased from 280 parts per million in 1800 to 367 in 2000, a 31% increase over 200 years(Earth Observatory, n.d.; Srivastava & Misra, 2002).

4.1.1 Our Changing Climate

Since the late 19th century, global mean surface temperatures have increased by 0.5-1.0 ° F.The Northern Hemisphere's snow cover and the Arctic Ocean's floating ice have decreased. Over the previous decade, the global surface temperature has increased by 4-8 inches. Could rise from 1-4.5 ° F (0.6-2.5 ° C) over the next fifty years and from 2.2-10 ° F (1.4-5.8 ° C) over the next decade, as shown in figures 5.

Figure 5. Global Surface Temperature(Ren 21, n.d.)

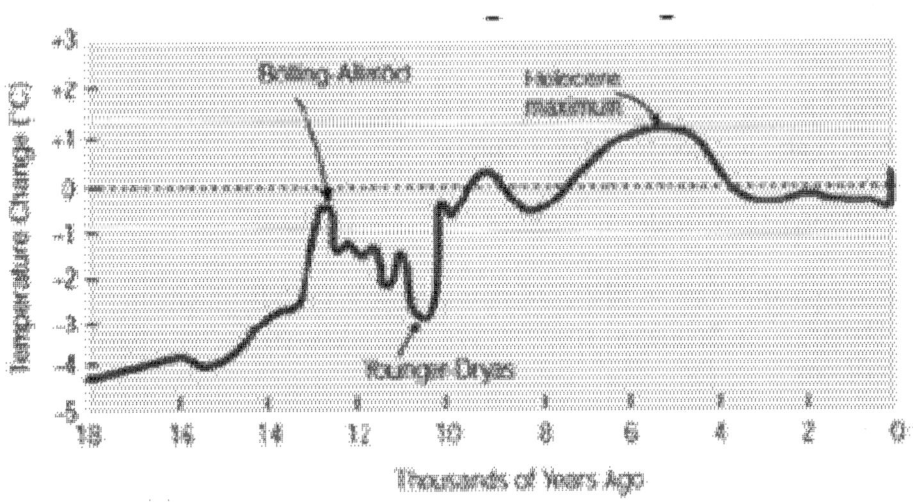

4.1.2 Causes

- Human impacts – Atmospheric greenhouse gasses trap some of the power output, maintaining heat
- Natural impacts – Change in the power production of the sun Volcanoes Water Vapor Clouds

4.1.3 Greenhouse Gases

CO2, methane, oxide nitrous, compounds fluorinated. Since the industrial revolution, atmospheric carbon dioxide levels have increased by 30%, methane has more than doubled, nitrous oxide has increased by 15%. These rises have increased the capacity of the Earth's atmosphere to catch heat.

4.1.4 Greenhouse Gas Emissions

- Fossil fuel combustion, Coal-fired energy plants, automotive exhausts, smokestacks, other waste winds of the human setting add 22 billion tons of carbon dioxide and other greenhouse gasses annually.
- Animal farms, manure, natural gas, rice paddies, landfills, coal and other anthropogenic sources contribute approximately 450 million tons of methane annually.
- Since 1750, atmospheric CO2 and CH4 concentrations have risen by 31% and 149% respectively above pre-industrial levels.

4.1.5 Sources of Green House Gas Emission

- Power Plants
  The combustion of fossil fuels for energy generation accounts for 40% of carbon dioxide emissions.
- Cars
  Twenty percent of carbon dioxide emissions from petrol in car motors and light trucks with a low gas mileage contribute most to global warming.
- Trucks
  Another 13% of carbon dioxide emissions are from vehicles that are mainly used for business reasons
- Airplanes
  Aviation is responsible for 3.5% of worldwide warming, and by 2050 this figure could increase to 15%
- Carbon Dioxide from Buildings: The design of buildings accounts for around 12% of carbon dioxide emissions
- Methane
  Methane is more than 20 times as efficient as CO2 at trapping water in the 2004 environment In the last 100 years, atmospheric methane levels have increased by 145%. Based on sources such as rice paddies, bovine flatulence, fungi in bogs and the manufacturing of fossil fuel. Anaerobic conditions grow in flooded areas and the organic matter in the soil breaks down.
- Nitrous oxide
  Naturally manufactured by oceans and rainforests, man-made nylon and nitric acid manufacturing, fertilizer use in agriculture, catalytic converter vehicles and organic matter burning.

4.2 Deforestation

Deforestation accounts for 25% of all carbon emissions into the environment by burning and cutting approximately 34 million acres of forests annually as shown in figure 6.

Figure 6. Carbon Cycle(Ani & Abubakar, 2015)

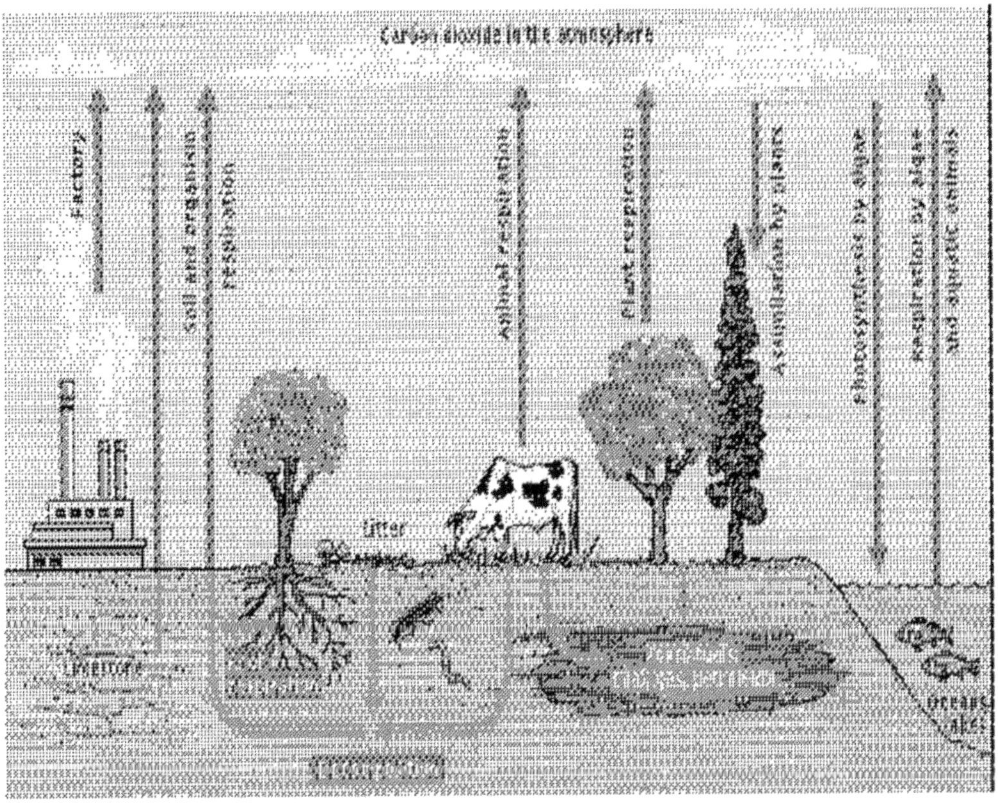

5. RESULTS AND DISCUSSIONS

Negative Effects: Rising Sea Level Change of precipitation and local climate ; acid rain Change of trees and crops produces Expansion of deserts into current rangelands More intense rainstorms Stabilization of ocean currents.

Positive Effects: It can boost plant growth in locations where the limiting factors are CO2 and temperature (preventing photorespiration that can destroy current sugar).

6. SUMMARY OF CHAPTER

Currently, society depends mostly on fossil fuels for electricity (39% natural gas, 24% natural gas, 23% coal, 8% nuclear and 6% more). Fossil fuel accounts for more than 70% of India and other countries ' total power requirements. Future renewable power resources have enormous environmental, political and economic consequences that could alter the world order. Nevertheless, a major role will be played by the geological elements of energy resources.

ACKNOWLEDGMENT

Author is grateful to Chandigarh University for providing the opportunity to complete this chapter successfully.

REFERENCES

Ani & Abubakar. (2015). Feasibility Analysis and Simulation of Integrated Renewable Energy System for Power Generation. *Journal of Energy*.

Dey, D. (2005). Energy and sustainable development in India. *Sustainable Energy Watch, 6*.

Earth Observatory. (n.d.). Retrieved from: http://earthobservatory.nasa.gov

Eureka eLearning. (n.d.). Retrieved from: http://www.eurekaelearning.com

Ministry of New and Renewable Energy. (n.d.). Retrieved from: http://www.mnre.gov.in

Nade, D. P., Potdar, S. S., Pawar, R. P., Mane, S. T., Wategaonkar, S. B., & Janrao, S. C. (2018). Recent state wise statistical review of renewable energy in India. *International Journal for Research in Engineering Application & Management*, 98-106.

National Geographic. (n.d.). Retrieved from: http://www.nationalgeographic.com

Punjab Energy Development Agency. (n.d.). Retrieved from: http://peda.gov.in/main/

Ren 21. (n.d.). *Renewables Now*. Retrieved from: https://www.ren21.net/reports

Shrivastava, S., Saxena, H., & Popat, Y. (n.d.). *Status of solar energy in India*. Academic Press.

Srivastava & Misra. (2002). Climate Change: Impact on Plants. *Plant Physiology, 129*(4), 1421-2.

Verma, N. (n.d.). *Rural Energy Scenario in India: A Case Study of District Sirsa (Haryana)*. Academic Press.

Compilation of References

Fatigue. (2008). In F. C. Campbell (Ed.), *Elements of Metallurgy and Engineering Alloys*. ASM International.

Gomez, J. F., Khodr, H. M., DeOliveira, P. M., Ocque, L., Yusta, J. M., Villasana, R., & Urdaneta, A. J. (2004). Ant colony system algorithm for the planning of primary distribution circuits. *IEEE Transactions on Power Systems*, *19*(2), 996–1004. doi:10.1109/TPWRS.2004.825867

National Geographic. (n.d.). Retrieved from: http://www.nationalgeographic.com

Rezaeiha, A., Montazeri, H., & Blocken, B. (2018). Towards accurate CFD simulations of vertical axis wind turbines at different tip speed ratios and solidities: Guidelines for azimuthal increment, domainsize and convergence. *Energy Conversion and Management*, *156*, 301–316. doi:10.1016/j.enconman.2017.11.026

Watts, P., & Fish, F. E. (2001). The influence of passive, leading edge tubercles on wing performance. *12th International Symposium on Unmanned Untethered Submersible Technology*.

Ask Nature. (n.d.). Available: https://asknature.org/strategy/flippers-provide-lift-reduce-drag/

Bouri, Zeblah, Ghoraf, Hadjeri, & Hamdaoui. (2005). Ant colony optimization to shunt capacitor allocation in radial distribution systems. *Acta Electrotechnica et Informatica*, *5*(1), 4.

Modarres, M., Kaminskiy, M., & Krivtsov, V. (2016). *Reliability Engineering and Risk Analysis: A Practical Guide* (3rd ed.). CRC Press. doi:10.1201/9781315382425

Srivastava & Misra. (2002). Climate Change: Impact on Plants. *Plant Physiology*, *129*(4), 1421-2.

Wang, W. C., Bai, C. J., Lin, Y. Y., & Lin, S. Y. (2015). Computational fluid dynamics analysis of the vertical axis wind turbine blade with tubercle leading edge. *Journal of Renewable and Sustainable Energy*, *7*(3), 033124. doi:10.1063/1.4922192

Aftab, S. M. A., & Ahmad, K. A. (2017). CFD study on NACA 4415 airfoil implementing spherical and sinusoidal Tubercle Leading Edge. *PLoS One*, *12*(8), e0183456. doi:10.1371/journal.pone.0183456 PMID:28850622

Earth Observatory. (n.d.). Retrieved from: http://earthobservatory.nasa.gov

El-Ela, A. A. A., El-Sehiemy, R. A., Kinawy, A.-M., & Mouwafi, M. T. (2016). Optimal capacitor placement in distribution systems for power loss reduction and voltage profile improvement. *IET Generation, Transmission & Distribution*, *10*(5), 1209–1221. doi:10.1049/iet-gtd.2015.0799

Elsayed, E. A. (2012). *Reliability Engineering*. John Wiley & Sons.

Howell, R., Oin, N., Edwards, J., & Durrani, N. (2010). Wind tunnel and numerical study of a small vertical axis wind turbine. *Renewable Energy*, *35*(2), 412–422. doi:10.1016/j.renene.2009.07.025

Ahuja, A., & Pahwa, A. (2005). Using ant colony optimization for loss minimization in distribution networks. In *Proceedings of the 37th Annual North American Power Symposium.* IEEE.] 10.1109/NAPS.2005.1560562

Hahn, G. J., & Meeker, W. Q. (2004). *How to Plan an Accelerated Life Test (E-Book).* ASQ Quality Press.

Kjellin, J., Bulow, F., Eriksson, S., Deglaire, P., Leijon, M., & Bernhoff, H. (2011). Power coefficient measurement on a 12 kW straight bladed vertical axis wind. *Renewable Energy, 36*(11), 3050–3053. doi:10.1016/j.renene.2011.03.031

Miklosovic, Murray, Howle, & Fish. (2004). *Leading-edge tubercles delay stall on humpback whale (Megaptera novaeangliae) flippers.* Academic Press.

Johari, H., Henoch, C., Custodia, D., & Levshin, A. (2009). Effects of leading edge protuberances on hydrofoil performance. *ExHFT-7.*

McPherson, J. (1989). Accelerated testing. In Electronic Materials Handbook Volume 1: Packaging; ASM International Publishing.

Mohamed, M.H., Ali, A.M., & Hafiz, A.A. (2015). CFD analysis for H-rotor Darrieus turbine as a low speed wind energy converter. *Energy Science and Technology, 18*(1), 1-13.

Niknam, T. (2008). A new approach based on ant colony optimization for daily Volt/Var control in distribution networks considering distributed generators. *Energy Conversion and Management, 49*(12), 3417–3424. doi:10.1016/j.enconman.2008.08.015

Annaluru, R., Das, S., & Pahwa, A. (2004). *Multi-level ant colony algorithm for optimal placement of capacitors in distribution systems. In Evolutionary Computation, 2004. CEC2004. Congress on* (Vol. 2). IEEE.

Jones, G., Santer, M., & Papadakis, G. (2018). Control of low Reynolds number flow around an airfoil using periodic surface morphing: A numerical study. *Journal of Fluids and Structures, 76*(January), 95–115. doi:10.1016/j.jfluidstructs.2017.09.009

McPherson, J. (2010). *Reliability Physics and Engineering: Time-to-Failure Modeling.* Springer. doi:10.1007/978-1-4419-6348-2

Seyhan, M., Sarioglu, M., & Akansu, Y. E. (2021, October). Influence of Leading-Edge Tubercle with Amplitude Modulationon NACA 0015 Airfoil. *AIAA Journal, 59*(10), 3965–3978. doi:10.2514/1.J060180

Griffith, A. A. (1921). The phenomena of rupture and flow in solids. *Philos. Trans. R Soc. Lond. A, 221*(582-593), 163–198. doi:10.1098/rsta.1921.0006

Leknys, R. R., Arjomandi, M., Kelso, R. M., & Birzer, C. H. (2018). Thin airfoil load control during post-stall and large pitch angles using leading-edge trips. *Journal of Wind Engineering and Industrial Aerodynamics, 179,* 80–91. doi:10.1016/j.jweia.2018.05.009

Rezaeiha, A., Kalkman, I., & Bert, B. (2017). *Effect of Pitch angle on power performance and aerodynamics of a vertical axis wind turbine.* Academic Press.

Su, C.-T., Chang, C.-F., & Chiou, J.-P. (2005a). Optimal capacitor placement in distribution systems employing ant colony search algorithm. *Electric Power Components and Systems, 33*(8), 931–946. doi:10.1080/15325000590909912

Irwin, G. (1957). Analysis of stresses and strains near the end of a crack traversing a plate. *Journal of Applied Mechanics, 24*(3), 361–364. doi:10.1115/1.4011547

Rostamzadeh, N., Kelso, R.M., Daily, B.B., & Hansen, K.L. (2013). *The effect of Undulating leading-edge modifications on NACA 0021 airfoil characteristics.* Academic Press.

Yan, Y., Avital, E., Williams, J., & Cui, J. (2021). *Aerodynamic performance improvements of a vertical axis wind turbine by leading-edge protuberance.* Academic Press.

Anderson, T. L. (2017). *Fracture Mechanics—Fundamentals and Applications* (3rd ed.). CRC. doi:10.1201/9781315370293

Arunvinthan, S., Pillai, S. N., & Cao, A. (2020). Aerodynamic characteristics of variously modified leading-edge potuberanced (LEP) wind turbine under various turbulent intensities. *Journal of Wind Engineering and Industrial Aerodynamics, 202*, 104188. doi:10.1016/j.jweia.2020.104188

Khaleghinia, J., Roshan, M. Y., Nimvari, M. E., & Salarian, H. (2021, October). Performance improvement of Darrieus wind turbine using different cavity layouts. *Energy Conversion and Management, 246*(15), 114693.

Zhang, Zhao, & Feng. (2008). A Novel Optimization Reconfiguration Algorithm for Power Supply Restoration of Distribution Network. *Power System Technology, 7*, 12.

ASTM E606/E606M. *Standard Test Method for Strain-Controlled Fatigue Testing*; ASTM International: West Conshohocken, PA, USA, 2019.

Carpaneto, E., & Chicco, G. (2004). Ant-colony search-based minimum losses reconfiguration of distribution systems. In *Electrotechnical Conference, 2004. MELECON 2004. Proceedings of the 12th IEEE Mediterranean (Vol. 3)*. IEEE. 10.1109/MELCON.2004.1348215

Saha, U., & Jain, S. (2020). On the influence of blade thickness-to-chord ratio on dynamic stall phenomenon in H-type Darrieus wind rotors. *Energy Conversion and Management, 218.*

Wang, Z., Wang, Y., & Zhuang, M. (2018). Improvement of the aerodynamic performance of vertical axis wind turbines with leading-edge serrations and helical blades using CFD and Taguchi method. *Energy Conversion and Management, 177*, 107-121.

Dorigo, M., & Gambardella, L. M. (1997a). Ant Colonies for the Traveling Salesman Problem. *Bio Systems, 43*(2), 73–81. doi:10.1016/S0303-2647(97)01708-5 PMID:9231906

Global Wind Energy Council. (2021). *Global Wind Report 2021.* GWEC. http://www.indiaenvironmentportal.org.in/files/file/global%20wind%20report%202021.pdf

Gupta, A. K. (2015). Efficient Wind Energy Conversion: Evolution to Modern Design. *Journal of Energy Resources Technology, 137*(5), 051201. doi:10.1115/1.4030109

Ministry of New and Renewable Energy. (n.d.). Retrieved from: http://www.mnre.gov.in

Woo, S., O'Neal, D., Woldemichael, D. E., Atnaw, S. M., & Tulu, M. M. (2021). Improving the Fatigue of Newly Designed Mechanical System Subjected to Repeated Impact Loading. *Metals, 11*(1), 139. doi:10.3390/met11010139

ASTM E399. *Standard Test Method for Linear-Elastic Plane-Strain Fracture Toughness of Metallic Materials*; ASTM International: West Conshohocken, PA, USA, 2020.

Mohamed, M. H., Ghazalla, R. A., & Hafiz, A. A. (2019, December). Synergistic analysis of Darrieus wind turbine using computational fluid dynamics. *Energy, 189*(15), 116214.

Saffar, A., Hooshmand, R., & Khodabakhshian, A. (2011). A new fuzzy optimal reconfiguration of distribution systems for loss reduction and load balancing using ant colony search-based algorithm. *Applied Soft Computing, 11*(5), 4021–4028. doi:10.1016/j.asoc.2011.03.003

Wang, Z., & Zhuang, M. (2017). Leading –edge serrations for performance improvement on a vertical axis wind turbine at low tip-speed-ratio. *Applied Energy, 208*, 1184-119.

ASTM E647. *Standard Test Method for Measurement of Fatigue Crack Growth Rates;* ASTM International: West Conshohocken, PA, USA, 2015.

Guoqing, C. G. W. L. T. (2001). Distribution network reconfiguration for loss reduction using an ant colony optimization method. *Proceedings of the Csu-epsa, 2*.

Lositano, I. C. M., & Danao, L. A. M. (2019). Steady wind performance of a 5 kW three-bladed H-rotor Darrieus Vertical Axis Wind Turbine (VAWT) with cambered tubercle leading edge (TLE) blades. *Energy, 175*, 278–291. doi:10.1016/j.energy.2019.03.033

Sengupta A.R., Biswas, A., & Gupta, R. (2019). Comparison of low wind speed aerodynamics of unsymmetrical blade H-Darrieus rotor-blade camber and curvature signature for performance improvement. *Renewable Energy, 139*, 1412-1427.

ASTM E739-10. *Standard Practice for Statistical Analysis of Linear or Linearized Stress-Life (S-N) and Strain-Life (ε-N) Fatigue Data*; ASTM International: West Conshohocken, PA, USA, 2015.

Chen, J., Yang, H., Yang, M., & Xu, H. (2015). The effect of opening ratio and location on the performance of a novel vertical axis Darrieus turbine. *Energy, 89*(September), 819–834. doi:10.1016/j.energy.2015.05.136

Luo, Y., Wen, F., Wang, S., Hou, R., Wang, S., & Wang, Z. (2021). Numerical study on biomimetic trailing edge in the environment of a high pressure turbine stage. *Aerospace Science and Technology, 115*(August), 106770. doi:10.1016/j.ast.2021.106770

Swarnkar, A., Gupta, N., & Niazi, K. R. (2011). Adapted ant colony optimization for efficient reconfiguration of balanced and unbalanced distribution systems for loss minimization. *Swarm and Evolutionary Computation, 1*(3), 129–137. doi:10.1016/j.swevo.2011.05.004

Braco, R., Prates, P., Costa, J. D. M., & Berto, F. (2018). New methodology of fatigue life evaluation for multiaxially loaded notched components based on two uniaxial strain-controlled tests. *International Journal of Fatigue, 111*, 308–320. doi:10.1016/j.ijfatigue.2018.02.027

Dessoky, A., Bangga, G., Lutz, T., & Kramer, E. (2019, May). Aerodynamic and aeroacoustic performance assessment of H- rotor Darrieus VAWT equipped with wind lens technology. *Energy, 175*(15), 76–97. doi:10.1016/j.energy.2019.03.066

Vuppalapati, S. H. K., & Srivastava, A. (2010). Application of ant colony optimization for reconfiguration of shipboard power system. *International Journal of Engineering Science and Technology, 2*(3), 119–131. doi:10.4314/ijest.v2i3.59181

Chang, C.-F. (2008). Reconfiguration and capacitor placement for loss reduction of distribution systems by ant colony search algorithm. *IEEE Transactions on Power Systems, 23*(4), 1747–1755. doi:10.1109/TPWRS.2008.2002169

Scungio, M., Arpino, F., Focanti, V., Profili, M., & Rotondi, M. (2016, December). Wind tunnel testing of scaled model of a newly developed Darrieus-style vertical axis wind turbine with auxiliary straight blades. *Energy Conversion and Management, 130*(15), 60–70. doi:10.1016/j.enconman.2016.10.033

Weingart, R.G., & Timoshenko, S. P. (2007, Aug.). Father of Engineering Mechanics in the U.S. *Structure Magazine*.

Castelli, M.R., Englaro, A., & Benini, E. (2011). The Darrieus wind turbine: proposal for a new performance predication model based on CFD. *Energy, 36*(8), 4919-4934.

Sánchez, M., Cicero, S., Arroyo, B., & Álvarez, J. A. (2020). Coupling Finite Element Analysis and the Theory of Critical Distances to Estimate Critical Loads in Al6060-T66 Tubular Beams Containing Notches. *Metals, 10*(10), 1395. doi:10.3390/met10101395

Su, C.-T., Chang, C.-F., & Chiou, J.-P. (2005b). Distribution network reconfiguration for loss reduction by ant colony search algorithm. *Electric Power Systems Research*, *75*(2), 190–199. doi:10.1016/j.epsr.2005.03.002

Li, Q., & Xie, L. (2020). Analysis and Optimization of Tooth Surface Contact Stress of Gears with Tooth Profile Deviations, Meshing Errors and Lead Crowning Modifications Based on Finite Element Method and Taguchi Method. *Metals*, *10*(10), 1370. doi:10.3390/met10101370

Shukla, V., & Kaviti, A.K. (2017). *Performance evaluation of profilr modification on straight- bladed vertical axis wind turbine by energy and Spalart Allmaras models*. Academic Press.

Wu, Y.-K., Lee, C.-Y., Liu, L.-C., & Tsai, S.-H. (2010). Study of reconfiguration for the distribution system with distributed generators. *IEEE Transactions on Power Delivery*, *25*(3), 1678–1685. doi:10.1109/TPWRD.2010.2046339

Bangga, G., Dessoky, A., Wu, Z., Rogowski, K., & Hansen, M. O. L. (2020, September). Accuracy and consistency of CFD and engineering models for simulating vertical axis wind turbine loads. *Energy*, *206*(1), 118087. doi:10.1016/j.energy.2020.118087

Carpaneto, E., & Chicco, G. (2008). Distribution system minimum loss reconfiguration in the hyper-cube ant colony optimization framework. *Electric Power Systems Research*, *78*(12), 2037–2045. doi:10.1016/j.epsr.2008.06.009

Hertzberg, R. W., Vinci, R. P., & Hertzberg, J. L. (2020). *Deformation and Fracture Mechanics of Engineering Materials* (6th ed.). John Wiley and Sons Inc.

Giorgetti, S., Pellegrini, G., & Zanforlin, S. (2015). CFD Investigation on the Aerodynamic Interferences between Medium-solidity Darrieus Vertical Axis wind Turbines. *Energy Procedia*, *81*(December), 227–239. doi:10.1016/j.egypro.2015.12.089

Kumar, K. S., & Jayabarathi, T. (2012). Power system reconfiguration and loss minimization for distribution systems using bacterial foraging optimization algorithm. *International Journal of Electrical Power & Energy Systems*, *36*(1), 13–17. doi:10.1016/j.ijepes.2011.10.016

Zupančič, B., Prokop, Y., & Nikonov, A. (2021). FEM analysis of dispersive elastic waves in three-layered composite plates with high contrast properties. *Finite Elements in Analysis and Design*, *193*, 103553. doi:10.1016/j.finel.2021.103553

Abdelaziz, A. Y., Osama, R. A., & El-Khodary, S. M. (2012). Reconfiguration of distribution systems for loss reduction using the hyper-cube ant colony optimisation algorithm. *IET Generation, Transmission & Distribution*, *6*(2), 176–187. doi:10.1049/iet-gtd.2011.0281

IEEE Standard Glossary of Software Engineering Terminology. (2002). *IEEE STD 610.12-1990. Standards Coordinating Committee of the Computer Society of IEEE*. Available online: https://ieeexplore.ieee.org/document/159342

Jain, P. (2016). Wind Energy Engineering (2nd ed.). McGraw-Hill Education.

Ahmad, K. A., Aftab, S. M. A., Razak, N. A., & Rafie, A. S. M. (2016). Mimicking the humpback whale: An aerodynamic perspective. *Progress in Aerospace Sciences*, *84*, 48–69. doi:10.1016/j.paerosci.2016.03.002

Amano, S. (2017). Review of Wind Turbine Research in 21st Century. *Journal of Energy Resources Technology*, *139*(5), 050801. doi:10.1115/1.4037757

CMMI Product Team. (2018). *Capability Maturity Model Integration (CMMI) Version 2.0, Continuous Representation*. Report CMU/SEI-2002-TR-011; Software Engineering Institute.

Dorigo, M., & Gambardella, L. M. (1997b). Ant colony system: A cooperative learning approach to the traveling salesman problem. *IEEE Transactions on Evolutionary Computation*, *1*(1), 53–66. doi:10.1109/4235.585892

Ren 21. (n.d.). *Renewables Now*. Retrieved from: https://www.ren21.net/reports

Kreyszig, E. (2011). *Advanced Engineering Mathematics* (10th ed.). John Wiley and Son.

Song, Y. H., Chou, C. S., & Stonham, T. J. (1999). Combined heat and power economic dispatch by improved ant colony search algorithm. *Electric Power Systems Research*, *52*(2), 115–121. doi:10.1016/S0378-7796(99)00011-5

Grove, A. (1967). Physics and Technology of Semiconductor Device. Wiley International Edition.

Hou, Y-h. (2003). Economic dispatch of power systems based on generalized ant colony optimization method. *Proceedings of the CSEE, 3*, 13.

Laws, P., Saini, J. S., & Kumar, A. (2019). A *Study on OpenFOAM's Overset Mesh Support Using Flow Past NACA 0018 Airfoil*. doi:10.20944/preprints201907.0217.v1

Hou, Y.-H. (2002). Generalized ant colony optimization for economic dispatch of power systems. In *Power System Technology, 2002. Proceedings. PowerCon 2002. International Conference on* (Vol. 1). IEEE.

Minges, M. L. (1989). *Electronic Materials Handbook* (Vol. 1). ASM International.

Karnopp, D. C., Margolis, D. L., & Rosenberg, R. C. (2012). *System Dynamics: Modeling, Simulation, and Control of Mechatronic Systems* (6th ed.). John Wiley & Sons. doi:10.1002/9781118152812

Thanathip. (2004). Economic dispatch by ant colony search algorithm. In *Cybernetics and Intelligent Systems, 2004 IEEE Conference on* (Vol. 1). IEEE.

Pothiya, S., Ngamroo, I., & Kongprawechnon, W. (2010). Ant colony optimization for economic dispatch problem with non-smooth cost functions. *International Journal of Electrical Power & Energy Systems*, *32*(5), 478–487. doi:10.1016/j.ijepes.2009.09.016

Whitesel, H. A. (1957). Capillary two-phase flow Part II. *Refrig. Eng.*, *65*, 35–40.

Abernethy, R. B. (2000). *The New Weibull Handbook*. Reliability Analysis Center.

Huang, S.-J. (2001). Enhancement of hydroelectric generation scheduling using ant colony system based optimization approaches. *IEEE Transactions on Energy Conversion*, *16*(3), 296–301. doi:10.1109/60.937211

Yu, I. K., & Song, Y. H. (2001). A novel short-term generation scheduling technique of thermal units using ant colony search algorithms. *International Journal of Electrical Power & Energy Systems*, *23*(6), 471–479. doi:10.1016/S0142-0615(00)00065-X

Vlachogiannis, J. G., Hatziargyriou, N. D., & Lee, K. Y. (2005). Ant colony system-based algorithm for constrained load flow problem. *IEEE Transactions on Power Systems*, *20*(3), 1241–1249. doi:10.1109/TPWRS.2005.851969

Tippachon, W., & Rerkpreedapong, D. (2009). Multiobjective optimal placement of switches and protective devices in electric power distribution systems using ant colony optimization. *Electric Power Systems Research*, *79*(7), 1171–1178. doi:10.1016/j.epsr.2009.02.006

Falaghi, H., Haghifam, M.-R., & Singh, C. (2009). Ant colony optimization-based method for placement of sectionalizing switches in distribution networks using a fuzzy multiobjective approach. *IEEE Transactions on Power Delivery*, *24*(1), 268–276. doi:10.1109/TPWRD.2008.2005656

Alom, N., & Saha, U. K. (2018). Four Decades of Research into the Augmentation Techniques of Savonius Wind Turbine Rotor. *Journal of Energy Resources Technology*, *140*(5), 050801. doi:10.1115/1.4038785

Ani & Abubakar. (2015). Feasibility Analysis and Simulation of Integrated Renewable Energy System for Power Generation. *Journal of Energy*.

Jun-man, K., & Yi, Z. (2012). Application of an improved ant colony optimization on generalized traveling salesman problem. *Energy Procedia*, *17*, 319–325. doi:10.1016/j.egypro.2012.02.101

Magaziner, I. C., & Patinkin, M. (1989). Cold competition: GE wages the refrigerator war. *Harvard Business Review*, *89*, 114–124.

Wang, L., & Singh, C. (2008). Reliability-constrained optimum placement of reclosers and distributed generators in distribution networks using an ant colony system algorithm. *IEEE Transactions on Systems, Man and Cybernetics. Part C, Applications and Reviews*, *38*(6), 757–764. doi:10.1109/TSMCC.2008.2001573

Vaisakh, K., & Srinivas, L. R. (2011). Evolving ant colony optimization based unit commitment. *Applied Soft Computing*, *11*(2), 2863–2870. doi:10.1016/j.asoc.2010.11.019

Hao, J. (2002). Optimal Unit Commitment Based On Ant Colony Optimization Algorithm. *Power System Technology*, *11*, 6.

Chen, Y. (2008). An Ant Colony Optimization and Particle Swarm Optimization Hybrid Algorithm for Unit Commitment Based on Operate Coding. *Power System Technology*, *6*, 14.

Simon, S. P., Padhy, N. P., & Anand, R. S. (2006). An ant colony system approach for unit commitment problem. *International Journal of Electrical Power & Energy Systems*, *28*(5), 315–323. doi:10.1016/j.ijepes.2005.12.004

Shi, L., Hao, J., Zhou, J., & Xu, G. (2004). Ant colony optimization algorithm with random perturbation behavior to the problem of optimal unit commitment with probabilistic spinning reserve determination. *Electric Power Systems Research*, *69*(2), 295–303. doi:10.1016/j.epsr.2003.10.008

Sum-Im, T., & Ongsakul, W. (2003). Ant colony search algorithm for unit commitment. In *Industrial Technology, 2003 IEEE International Conference on* (Vol. 1). IEEE. 10.1109/ICIT.2003.1290244

Sisworahardjo, N. S., & El-Keib, A. A. (2002). Unit commitment using the ant colony search algorithm. In *Power Engineering 2002 Large Engineering Systems Conference on, LESCOPE 02*. IEEE. 10.1109/LESCPE.2002.1020658

Columbus, C., Chandrasekaran, K., & Simon, S. P. (2012). Nodal ant colony optimization for solving profit based unit commitment problem for GENCOs. *Applied Soft Computing*, *12*(1), 145–160. doi:10.1016/j.asoc.2011.08.057

Chandrasekaran, K., & Simon, P. (2012). Network and reliability constrained unit commitment problem using binary real coded firefly algorithm. *International Journal of Electrical Power & Energy Systems*, *43*(1), 921–832. doi:10.1016/j.ijepes.2012.06.004

Dorigo, M. (2008). Ant Colony Optimization and Swarm Intelligence. In *6th International Conference, ANTS 2008, Brussels, Belgium, September 22-24, 2008, Proceedings (Vol. 5217)*. Springer.

Islam, M., Fartaj, A., & Carriveau, R. (2008). Analysis of the Design Parameters related to a fixed pitch straight bladed vertical axis wind turbine. Wind Engineering, 32. doi:10.1260/030952408786411903

Nade, D. P., Potdar, S. S., Pawar, R. P., Mane, S. T., Wategaonkar, S. B., & Janrao, S. C. (2018). Recent state wise statistical review of renewable energy in India. *International Journal for Research in Engineering Application & Management*, 98-106.

Rosa, J. L., Robin, A., Silva, M. B., Baldan, C. A., & Peres, M. P. (2009). Electrodeposition of copper on titanium wires: Taguchi experimental design approach. *Journal of Materials Processing Technology*, *209*(3), 1181–1188. doi:10.1016/j.jmatprotec.2008.03.021

Seeni, P., Rajendran, P., & Kutty, H. A. (2018). A Critical Review on Tubercles Design for Propellers. *IOP Conf. Series: Materials Science and Engineering, 370.* 10.1088/1757-899X/370/1/012015

Ebrahimi, J., Hosseinian, S. H., & Gharehpetian, G. B. (2011). Unit Commitment Problem Solution Using Shuffled Frog Leaping Algorithm. *IEEE Transactions on Power Systems, 26*(2), 573–581. doi:10.1109/TPWRS.2010.2052639

Jiang, L. L., Maskell, D. L., & Patra, J. C. (2013). A novel ant colony optimization-based maximum power point tracking for photovoltaic systems under partially shaded conditions. *Energy and Building, 58*, 227–236. doi:10.1016/j.enbuild.2012.12.001

Meziane, R., Massim, Y., Zeblah, A., Ghoraf, A., & Rahli, R. (2005). Reliability optimization using ant colony algorithm under performance and cost constraints. *Electric Power Systems Research, 76*(1), 1–8. doi:10.1016/j.epsr.2005.02.008

Samrout, M., Yalaoui, F., Châtelet, E., & Chebbo, N. (2005). New methods to minimize the preventive maintenance cost of series–parallel systems using ant colony optimization. *Reliability Engineering & System Safety, 89*(3), 346–354. doi:10.1016/j.ress.2004.09.005

Sheidaei, F., Shadkam, M., & Zarei, M. (2008). Optimal Distributed Generation allocation in distribution systems employing ant colony to reduce losses. In *Universities Power Engineering Conference, 2008. UPEC 2008. 43rd International.* IEEE.

Niknam, T., Ranjbar, A. M., & Shirani, A. R. (2005). A new approach for distribution state estimation based on ant colony algorithm with regard to distributed generation. *Journal of Intelligent & Fuzzy Systems, 16*(2), 119–131.

Georgilakis, P. S., & Hatziargyriou, N. D. (2013). Optimal distributed generation placement in power distribution networks: Models, methods, and future research. *IEEE Transactions on Power Systems, 28*(3), 3420–3428. doi:10.1109/TPWRS.2012.2237043

Niknam, T. (2005). A new approach based on ant algorithm for Volt/Var control in distribution network considering distributed generation. *Iranian Journal of Science & Technology, Transaction B, 29*(B4), 1–15.

Abdelaziz, A. Y., & Reham, A. (2012). Application of ant colony optimization and harmony search algorithms to reconfiguration of radial distribution networks with distributed generations. *Journal of Bioinformatics and Intelligent Control, 1*(1), 86–94. doi:10.1166/jbic.2012.1007

Falaghi, H., & Haghifam, M.-R. (2007). ACO based algorithm for [72 sources allocation and sizing in distribution systems. In Power Tech, 2007 IEEE. IEEE.

Amato, Bedon, Castelli, & Benini. (2013). Numerical Analysis of the Influence of Tip Devices on the power coefficient of a VAWT. *International Journal of Mechanical, Aerospace, Industrial, Mechatronic and Manufacturing Engineering, 7.*

Chowdhury, S., & Taguchi, S. (2016). *Robust Optimization: World's Best Practices for Developing Winning Vehicles* (1st ed.). John Wiley & Sons Inc. doi:10.1002/9781119212096

Dorigo, M., & Blum, C. (2005). Ant colony optimization theory: A survey. *Theoretical Computer Science, 344*(2), 243–278. doi:10.1016/j.tcs.2005.05.020

Eureka eLearning. (n.d.). Retrieved from: http://www.eurekaelearning.com

Hand, B., Kelly, G., & Cashman, A. (2021). Aerodynamic design and performance parameters of a lift-type vertical axis wind turbine: A comprehensive review. *Renewable and Sustainable Energy Reviews, 139.* doi:10.1016/j.rser.2020.110699

Lin, Z-h. (2003). Generalized ant colony optimization algorithm for reactive power optimization in power systems. *Journal of North China Electric Power University, 2*, 1.

Abou El-Ela, A. A., Kinawy, A. M., El-Sehiemy, R. A., & Mouwafi, M. T. (2011). Optimal reactive power dispatch using ant colony optimization algorithm. *Electrical Engineering, 93*(2), 103–116. doi:10.100700202-011-0196-4

Liu, Z., Wang, X., & Li, C. (2010). *Reactive power optimization in power system based on chaos ant colony algorithm. In CICED 2010 Proceedings*. IEEE.

Mouwafi, M. T., El-Sehiemy, R. A., Abou El-Ela, A. A. & Kinawy, A. M. (2014). A Complete Reactive Power Management Strategy Using Ant Colony Optimization Algorithm. *Emirates Journal for Engineering Research, 19*(3), 19–31.

Gardel, P. (2006). Multiobjective reactive power compensation with an ant colony optimization algorithm. In *AC and DC Power Transmission, 2006. ACDC 2006. The 8thIEE International Conference on*. IET. 10.1049/cp:20060056

Abbasy, A., & Hosseini, S. H. (2007). *Ant colony optimization-based approach to optimal reactive power dispatch: a comparison of various ant systems. In Power Engineering Society Conference and Exposition in Africa, 2007. Power-Africa'07*. IEEE.

Ren, X., Deng, Y., Zhao, C., & Zhao, D. P. (2003). Study on the algorithm for dynamic reactive power optimization of distribution systems. *Zhongguo Dianji Gongcheng Xuebao, 23*(1), 31–36.

Ketabi, A., Alibabaee, A., & Feuillet, R. (2010). Application of the ant colony search algorithm to reactive power pricing in an open electricity market. *International Journal of Electrical Power & Energy Systems, 32*(6), 622–628. doi:10.1016/j.ijepes.2009.11.019

El-Sehiemy, R. A., El Ela, A. A. A., Kinawy, A. M. M., & Mouwafia, M. T. (2012). An ant colony optimization algorithm-based emergency control strategy for voltage collapse mitigation. *International Review of Applied Sciences and Engineering, 3*(2), 147–156. doi:10.1556/irase.3.2012.2.8

Chen & Tang. (2005). Tabu search-ant colony optimization hybrid algorithm based distribution network planning. *Power System Technology, 2*, 5.

Allen, P. (2020). *Design of Experiments for 21st Century Engineers* (1st ed.). Lulu Press.

Dorigo, M., & Stützle, T. (2009). *Ant colony optimization: overview and recent advances. Techreport*. IRIDIA, Universite Libre de Bruxelles.

Elsakka, M. M., Ingham, D. B., Ma, L., & Pourkashanian, M. (2019). CFD analysis of the angle of attack for a vertical axis wind turbine blade. *Energy Conversion and Management, 182*, 154–165. doi:10.1016/j.enconman.2018.12.054

Mishra, N., Gupta, A. S., Dawar, J., Kumar, A., & Mitra, S. (2018). Numerical and Experimental Study on Performance Enhancement of Darrieus Vertical Axis Wind Turbine with Wingtip Devices. *ASME Journal of Energy Resources Technology, 140*(12), 121-201 doi:10.1115/1.4040506

Shrivastava, S., Saxena, H., & Popat, Y. (n.d.). *Status of solar energy in India*. Academic Press.

Dong, Y.-F. (2007). Combination of genetic algorithm and ant colony algorithm for distribution network planning. In *2007 International Conference on Machine Learning and Cybernetics* (Vol. 2). IEEE. 10.1109/ICMLC.2007.4370288

Hu, B., Gu, J., & Wang, Y.-D. (2005). An ant colony optimization based method for power distribution network planning. *Relay, 33*(21), 54–57.

Sun, Shang, & Niu. (2006). Application of Improved Ant Colony Optimization Algorithm in Distribution Network Planning. *Power System Technology, 15*, 21.

Eroğlu, Y., & Seçkiner, S. U. (2012). Design of wind farm layout using ant colony algorithm. *Renewable Energy, 44*, 53–62. doi:10.1016/j.renene.2011.12.013

Karaboga, N. (2009). A new design method based on artificial bee colony algorithm for digital IIR filters. *Journal of the Franklin Institute*, *346*(4), 328–348. doi:10.1016/j.jfranklin.2008.11.003

Gasbaoui, B., & Allaoua, B. (2009). Ant colony optimization applied on combinatorial problem for optimal power flow solution. *Leonardo Journal of Sciences*, *14*, 1–17.

Chang, C. S., Tian, L., & Wen, F. S. (1999). A new approach to fault section estimation in power systems using Ant system. *Electric Power Systems Research*, *49*(1), 63–70. doi:10.1016/S0378-7796(98)00127-8

Teng, J.-H., & Liu, Y.-H. (2003). A novel ACS-based optimum switch relocation method. *IEEE Transactions on Power Systems*, *18*(1), 113–120. doi:10.1109/TPWRS.2002.807038

Chen, Ding, & Zhao. (2006). Ant colony algorithm for solving fault location in distribution networks. *Automation of Electric Power Systems*, *5*, 17.

Foong, W. K., Simpson, A. R., Maier, H. R., & Stolp, S. (2008). Ant colony optimization for power plant maintenance scheduling optimization—A five-station hydropower system. *Annals of Operations Research*, *159*(1), 433–450. doi:10.100710479-007-0277-y

Arun, K. K. M., Strong, S., ElGammal, T., & Amano, R. S. (2015). Development of Novel Self-Healing Polymer Composites for Use in Wind Turbine Blades. *Journal of Energy Resources Technology*.

Dorigo, M., Di Caro, G., & Gambardella, L. M. (1999). Ant algorithms for discrete optimization. *Artificial Life*, *5*(2), 137–172. doi:10.1162/106454699568728 PMID:10633574

Duga, J. J., Fisher, W. H., Buxaum, R. W., Rosenfield, A. R., Buhr, A. R., Honton, E. J., & McMillan, S. C. (1982). *The Economic Effects of Fracture in the United States*. Final Report. NBS Special Publication 647-2. Battelle Laboratories.

Fish, F. E. (2020). *Biomimetics and the Application of the Leading-Edge Tubercles of the Humpback Whale Flipper* (1st ed.). Springer. doi:10.1007/978-3-030-23792-9_1

Punjab Energy Development Agency. (n.d.). Retrieved from: http://peda.gov.in/main/

Wang, B., Liu, D., & Li, X. (2009). An improved ant colony system in optimizing power system PMU placement problem. In *2009 Asia-Pacific Power and Energy Engineering Conference*. IEEE. 10.1109/APPEEC.2009.4918138

Abdelsalam, H. A., Abdelaziz, A. Y., & Mukherjee, V. (2014). Optimal PMU placement in a distribution network considering network reconfiguration. *2014 International Conference on Circuits, Power and Computing Technologies, ICCPCT 2014*, 191–196. 10.1109/ICCPCT.2014.7054997

Abou El-Ela, A. A., Kinawy, A. M., El-Sehiemy, R. A., & Mouwafi, M. T. (2014). Optimal Placement of Phasor Measurement Units with Limited Channels Using Ant Colony Optimization Algorithm. Engineering Research Journal, 37(2), 191-197. doi:10.21608/erjm.2014.66918

Abou El-Ela, A., Kinawy, A., El-Sehiemy, R., & Mouwafi, M. (2014). Ant colony optimizer for phasor measurement units placement. *International Review of Applied Sciences and Engineering*, *5*(2), 127–134. doi:10.1556/irase.5.2014.2.4

Abou El-Ela, A. A., Kinawy, A. M., Mouwafi, M. T., & El-Sehiemy, R. A. (2014). Optimal Placement of Phasor Measurement Units for Power System Observability Using Ant Colony Optimization Algorithm. *16th International Middle- East Power Systems Conference (MEPCON'2014)*, 1-6.

Mouwafi, M. T., El-Sehiemy, R. A., Abou El-Ela, A. A., & Kinawy, A. M. (2016). Optimal placement of phasor measurement units with minimum availability of measuring channels in smart power systems. *Electric Power Systems Research*, *141*, 421–431. doi:10.1016/j.epsr.2016.07.029

da Silva, A. M. (2010). Reliability worth applied to transmission expansion planning based on ant colony system. *International Journal of Electrical Power & Energy Systems, 32*(10), 1077–1084. doi:10.1016/j.ijepes.2010.06.003

Kannan, S., & Mary Raja Slochanal, S. (2005). Application and Comparison of Metaheuristic Techniques to Generation Expansion Planning Problem. *IEEE Transactions on Power Systems, 20*(1), 466–475. doi:10.1109/TPWRS.2004.840451

Rahmani, R., Yusof, R., Seyedmahmoudian, M., & Mekhilef, S. (2013). Hybrid technique of ant colony and particle swarm optimization for short term wind energy forecasting. *Journal of Wind Engineering and Industrial Aerodynamics, 123*, 163–170. doi:10.1016/j.jweia.2013.10.004

Wang, S.-K., Chiou, J.-P., & Liu, C.-W. (2009). Parameters tuning of power system stabilizers using improved ant direction hybrid differential evolution. *International Journal of Electrical Power & Energy Systems, 31*(1), 34–42. doi:10.1016/j.ijepes.2008.10.003

Dey, D. (2005). Energy and sustainable development in India. *Sustainable Energy Watch, 6.*

Dréo, J., & Siarry, P. (2002). A new ant colony algorithm using the heterarchical concept aimed at optimization of multiminima continuous functions. In *International Workshop on Ant Algorithms.* Springer Berlin Heidelberg.]

Matt. (2017). *Self-Healing of Wind Turbine Blades Using Microscale Vascular Vessels. ASME Journal of Energy Resources Technology.*

Toribio, J., Kharin, V., Ayaso, F. J., González, B., Matos, J. C., Vergara, D., & Lorenzoc, M. (2010). Failure analysis of a lifting platform for tree pruning. *Engineering Failure Analysis, 17*(4), 739–747. doi:10.1016/j.engfailanal.2009.08.017

Yadav, R., Mohamed, M. A. R., & Guven, U. (2021). Flow Separation Control using a Bio-Inspired Nose for NACA 4 and 6 Series Airfoils. *Aircraft Engineering and Aerospace Technology, 93*(2), 251–266. doi:10.1108/AEAT-08-2019-0170

Kefayat, M., Lashkar Ara, A., & Nabavi Niaki, S. A. (2015). A hybrid of ant colony optimization and artificial bee colony algorithm for probabilistic optimal placement and sizing of distributed energy resources. *Energy Conversion and Management, 92*, 149–161. doi:10.1016/j.enconman.2014.12.037

Lu, C.-F., Hsu, C.-H., & Juang, C.-F. (2013). Coordinated control of flexible AC transmission system devices using an evolutionary fuzzy lead-lag controller with advanced continuous ant colony optimization. *IEEE Transactions on Power Systems, 28*(1), 385–392. doi:10.1109/TPWRS.2012.2206410

Berbaoui, B. (2010). Optimization of shunt active power filter system fuzzy logic controller based on ant colony algorithm. *Journal of Theoretical and Applied Information Technology, 14*(1), 117–125.

Nakawiro, W., & Erlich, I. (2009). Optimal load shedding for voltage stability enhancement by ant colony optimization. In *Intelligent System Applications to Power Systems, 2009. ISAP'09. 15th International Conference on* (pp. 1-6). IEEE. 10.1109/ISAP.2009.5352886

Akama, M., & Kiuchi, A. (2019). Fatigue Crack Growth under Non-Proportional Mixed Mode Loading in Rail and Wheel Steel Part 1: Sequential Mode I and Mode II Loading. *Applied Sciences (Basel, Switzerland), 9*(14), 2866. doi:10.3390/app9142866

Amano, R., Sunden, B., & Gupta, A. (2017). Special Issue for the Second International Conference on 2016 Next Generation of Wind Energy (ICNGWE). *ASME Journal of Energy Resources Technology.* 10.1115/1.4037711

Mathur, M., Karale, S. B., Priye, S., Jayaraman, V. K., & Kulkarni, B. D. (2000). Ant colony approach to continuous function optimization. *Industrial & Engineering Chemistry Research, 39*(10), 3814–3822. doi:10.1021/ie990700g

Verma, N. (n.d.). *Rural Energy Scenario in India: A Case Study of District Sirsa (Haryana).* Academic Press.

Aamodt, A., & Plaza, E. (1994). Case-based reasoning: Foundational issues, methodological variations, and system approaches. *AI Communications*, *7*(1), 39–59. doi:10.3233/AIC-1994-7104

Aaron, D., & Tsouris, C. (2005). Separation of CO2 from flue gas: A review. *Separation Science and Technology*, *40*(1–3), 321–348. doi:10.1081/SS-200042244

Abedini, M., Moradi, M. H., & Hosseinian, S. M. (2016). Optimal management of microgrids including renewable energy scources using GPSO-GM algorithm. *Renewable Energy*, *90*, 430–439. doi:10.1016/j.renene.2016.01.014

Abu-mouti, F. S. (2011). *Sizing in Distribution Systems via Artificial Bee Colony Algorithm*. Academic Press.

Adefarati, T., Potgieter, S., Bansal, R. C., Naidoo, R., Rizzo, R., & Sanjeevikumar, P. (2019). Optimization of PV-Wind-battery storage microgrid system utilizing a genetic algorithm. *ICCEP 2019 - 7th International Conference on Clean Electrical Power: Renewable Energy Resources Impact*, 633–638. 10.1109/ICCEP.2019.8890204

Adleman, L.M. (1995). On constructing a molecular computer. *DNA Based Computers*, *27*, 1-21.

Adleman, L. (1994). Molecular computation of solutions to combinatorial problems. *Science*, *266*(5187), 1021–1024. doi:10.1126cience.7973651 PMID:7973651

Advantages. (2020). https://www.energy.gov/eere/wind/advantages-and-challenges-wind-energy

Advantages. (2021). https://www.clean-energy-ideas.com/wind/wind-energy/advantages-and-disadvantages-of-wind-energy/

Agency, I. R. E. (2012). *Renewable Energy Technologies: Cost Analysis Series* (Vol. 1). doi:10.1016/B978-0-12-812959-3.00012-5

Aghaei, J., Niknam, T., Azizipanah-Abarghooee, R., & Arroyo, J. M. (2013). Scenario-based dynamic economic emission dispatch considering load and wind power uncertainties. *International Journal of Electrical Power & Energy Systems*, *47*, 351–367. doi:10.1016/j.ijepes.2012.10.069

Ahmadi, S., & Abdi, S. (2016). Application of the Hybrid Big Bang – Big Crunch algorithm for optimal sizing of a stand-alone hybrid PV / wind / battery system. *Solar Energy*, *134*, 366–374. doi:10.1016/j.solener.2016.05.019

Ahmad, J., Imran, M., Khalid, A., Iqbal, W., Ashraf, S. R., Adnan, M., Ali, S. F., & Khokhar, K. S. (2018). Techno economic analysis of a wind-photovoltaic-biomass hybrid renewable energy system for rural electrification: A case study of Kallar Kahar. *Energy*, *148*, 208–234. doi:10.1016/j.energy.2018.01.133

Ahmed, N., Bilal, A., Mahmood, A., Ieee, M., Razzaq, S., Zafar, A., ... Sidhu, S. (2015). Combined emission economic dispatch of power system including solar photo voltaic generation. *Energy Conversion and Management*, *92*, 82–91. doi:10.1016/j.enconman.2014.12.029

Ahn, S.-H., Lee, K.-T., Bhandari, B., Lee, G.-Y., Lee, C. S.-Y., & Song, C.-K. (2012). Formation strategy of renewable energy sources for high mountain off-grid system considering sustainability. *Journal of the Korean Society for Precision Engineering*, *29*(9), 958–963. doi:10.7736/KSPE.2012.29.9.958

Ai, B., Yang, H., Shen, H., & Liao, X. (2003). Computer-Aided Design of PV/Wind Hybrid System. *Renewable Energy*, *28*(10), 1491–1512. doi:10.1016/S0960-1481(03)00011-9

Akanwa, A. O., & Joe-Ikechebelu, N. (2019). The Developing World's Contribution to Global Warming and the Resulting Consequences of Climate Change in These Regions: A Nigerian Case Study. In Global Warming and Climate Change. IntechOpen.

Akcayol, M. A., & Cinar, C. (2005). Artificial neural network based modeling of heated catalytic converter performance. *Applied Thermal Engineering*, *25*(14–15), 2341–2350. doi:10.1016/j.applthermaleng.2004.12.014

Akerkar, R., & Sajja, P. S. (2009). Bio-inspired computing: Constituents and challenges. *International Journal of Bio-inspired Computation*, *1*(3), 135–150. doi:10.1504/IJBIC.2009.023810

Akinci, T. C., Nogay, H. S., Guseinoviene, E., Dikun, J., & Seker, S. (2016). Application of ANN for Short Term Forecasting of Wind Power Density. *15th International Scientific Conference Renewable Energy & Innovative Technologies*, 157-163.

Akın, H. L. (1997). *The body and brain problem in artificial intelligence. Computer and brain*. Pomegranate Publications.

Al Badwawi, R., Abusara, M., & Mallick, T. (2015). A Review of Hybrid Solar PV and Wind Energy System. *Smart Science*, *3*(3), 127–138. doi:10.1080/23080477.2015.11665647

Aldaouab, I., Daniels, M., & Hallinan, K. (2017). Microgrid cost optimization for a mixed-use building. *Proceedings of the IEEE Texas Power and Energy Conference (TPEC)*. 10.1109/TPEC.2017.7868271

Ali, H. H., Al-rub, F. A. A., Shboul, B., & Moumani, H. (2020). *Evaluation of Near-net-zero-energy Building Strategies : A Case Study on Residential Buildings in Jordan*. doi:10.32479/ijeep.10107

Ali, H. M. (2020). Recent advancements in PV cooling and efficiency enhancement integrating phase change materials based systems – A comprehensive review. *Solar Energy*, *197*(June), 163–198. doi:10.1016/j.solener.2019.11.075

Alkadri, S., Ledwos, N., Mirchi, N., Reich, A., Yılmaz, R., Driscoll, M., & Del Maestro, R. F. (2021). Utilizing a multilayer perceptron artificial neural network to assess a virtual reality surgical procedure. *Computers in Biology and Medicine*, *136*, 104770. doi:10.1016/j.compbiomed.2021.104770 PMID:34426170

Al Shamma'a, A. A., & Addoweesh, K. E. (2012). Optimum sizing of hybrid PV/wind/battery/diesel system considering wind turbine parameters using Genetic Algorithm. In *International Conference on Power and Energy*, (pp. 121-126). IEEE. 10.1109/PECon.2012.6450190

Al-Shamma'a, A., & Addoweesh, K. (2014). Techno-economic optimization of hybrid power system using genetic algorithm. *International Journal of Energy Research*, *38*, 1608-1623.

Alshammari, N., & Asumadu, J. (2020). Optimum unit sizing of hybrid renewable energy system utilizing harmony search, Jaya and particle swarm optimization algorithms. *Sustainable Cities and Society*, *60*, 102255. doi:10.1016/j.scs.2020.102255

Amit Arora, S. S. (2015). Techniques for Exploitation of Gas Hydrate (Clathrates) an Untapped Resource of Methane Gas. *Journal of Microbial & Biochemical Technology*, *07*(02), 108–111. doi:10.4172/1948-5948.1000190

Ammari, C., Belatrache, D., Touhami, B., & Makhloufi, S. (2021). *Sizing, optimization, control and energy management of hybrid renewable energy system—A review*. Energy and Built Environment. doi:10.1016/j.enbenv.2021.04.002

Arabali, A., Member, S., & Ghofrani, M. (2014). *Stochastic Performance Assessment and Sizing for a Hybrid Power System of Solar / Wind / Energy Storage*. Academic Press.

Arasi, S. M., & Sasiraja, R. M. (2015). *Optimal location of DG units with exact size for the improvement of voltage stability using SLPSO*. Academic Press.

Arcos-Aviles, D., Pascual, J., Guinjoan, F., Marroyo, L., Sanchis, P., & Marietta, M. P. (2017). Low complexity energy management strategy for grid profile smoothing of a residential grid-connected microgrid using generation and demand forecasting. *Applied Energy*, *205*, 69–84. doi:10.1016/j.apenergy.2017.07.123

Arora, A., Chandrajit, B., & Cameotra, S. S. (2015). Natural Gas Hydrate (Clathrates) as an Untapped Resource of Natural Gas. *Journal of Petroleum & Environmental Biotechnology*, *6*(04).

Arutyunov, V.S., Lisichkin, V. (n.d.). *Energy resources of the 21st century : problems and forecasts . Can renewable energy sources replace fossil fuels?* doi:10.1070/RCR4723

Ashok, S. (2007). *Optimised model for community-based hybrid energy system.* doi:10.1016/j.renene.2006.04.008

Askarzadeh, A. (2013). ScienceDirect A discrete chaotic harmony search-based simulated annealing algorithm for optimum design of PV / wind hybrid system. *Solar Energy*, *97*, 93–101. doi:10.1016/j.solener.2013.08.014

Asvini, M. S., & Amudha, T. (2016). An efficient methodology for reservoir release optimization using plant propagation algorithm. *Procedia Computer Science*, *93*, 1061–1069. doi:10.1016/j.procs.2016.07.310

Atcitty, S., Neely, J., Ingersoll, D., Akhil, A., & Waldrip, K. (2013). *Battery Energy Storage System.* Green Energy Technol. doi:10.1007/978-1-4471-5104-3_9

Augustine, N., Member, S., Suresh, S., Moghe, P., & Sheikh, K. (n.d.). *Economic Dispatch for a Microgrid Considering Renewable Energy Cost Functions*. Academic Press.

Azadeh, A., Babazadeh, R., & Asadzadeh, S. M. (2013). Optimum estimation and forecasting of renewable energy consumption by artificial neural networks. *Renewable & Sustainable Energy Reviews*, *27*, 605–612. doi:10.1016/j.rser.2013.07.007

Azadeh, A., Ghaderi, S. F., & Sohrabkhani, S. (2007). Forecasting electrical consumption by integration of neural network, time series and ANOVA. *Applied Mathematics and Computation*, *186*(2), 1753–1761. doi:10.1016/j.amc.2006.08.094

Azaza, M., & Wallin, F. (2017). Multi objective particle swarm optimization of hybrid micro-grid system: A case study in Sweden. *Energy*, *123*, 108–118. doi:10.1016/j.energy.2017.01.149

Baghban, A., Namvarrechi, S., Phung, L. T. K., Lee, M., Bahadori, A., & Kashiwao, T. (2016). Phase equilibrium modelling of natural gas hydrate formation conditions using LSSVM approach. *Petroleum Science and Technology*, *34*(16), 1431–1438. doi:10.1080/10916466.2016.1202966

Bagul, A., Salameh, Z., & Borowy, B. (1996). Sizing of a Stand-Alone Hybrid Wind-Photovoltaic System using a Three-Event Probability Density Approximation. *Solar Energy*, *56*(4), 323–335. doi:10.1016/0038-092X(95)00116-9

Baimel, D., Tapuchi, S., & Baimel, N. (2016). Smart grid communication technologies. *Journal of Power and Energy Engineering*, *4*(8), 1–8. doi:10.4236/jpee.2016.48001

Bai, W., & Lee, Y. (2016). Modified Optimal Power Flow on Storage Devices and Wind Power Integrated System. *IEEE International Conference on Power and Energy Society General Meeting (PESGM)*. doi: 10.1109/PESGM.2016.7742033

Bakir, H., & Kulaksiz, A. A. (2020). Modelling and voltage control of the solar-wind hybrid micro-grid with optimized STATCOM using GA and BFA. *Engineering Science and Technology, an International Journal*, *23*(3), 576–584. doi:10.1016/j.jestch.2019.07.009

Balat, H. (2005). Wind energy potential in Turkey. *Energy Exploration & Exploitation*, *23*(1), 51-59.

Ballard, A. L. (2002). *Non-ideal hydrate solid solution model for a multi-phase equilibria program, A*. Colorado School of Mines. Arthur Lakes Library.

Banerji, A. (2013). Microgrid: A review. In *2013 IEEE Global Humanitarian Technology Conference: South Asia Satellite (GHTC-SAS)* (pp. 27-35). IEEE.

Bao, Y. (2013). Genetic algorithm based optimal capacity allocation for an independent wind/pv/diesel/battery power generation system. *Journal of Information and Computational Science*, *10*(14), 4581–4592. doi:10.12733/jics20102167

Barbaro, M., & Castro, R. (2020). Design optimisation for a hybrid renewable microgrid: Application to the case of Faial island, Azores archipelago. *Renewable Energy*, *151*, 434–445. doi:10.1016/j.renene.2019.11.034

Barley, C. D., Lew, D. J., & Flowers, L. T. (1997). Sizing wind/photovoltaic hybrids for Bashir, M., Sadeh. J. (2012). *Size optimization of new hybrid stand-alone renewable energy system considering a reliability index.* In *11th international conference on Environment and Electrical Engineering (EEEIC).* Academic Press.

Barton, J. P., & Infield, D. G. (2004). Intermittent. *Renewable Energy*, *19*(2), 441–448.

Bayindir, R., Colak, I., Fulli, G., & Demirtas, K. (2016). Smart grid technologies and applications. *Renewable & Sustainable Energy Reviews*, *66*, 499–516. doi:10.1016/j.rser.2016.08.002

Behzadi, M. S., & Niasati, M. (2015). Comparative performance analysis of a hybrid PV/FC/battery stand-alone system using di_erent power management strategies and sizing approaches. *International Journal of Hydrogen Energy*, *40*(1), 538–548. doi:10.1016/j.ijhydene.2014.10.097

Belfkira, R. (2008). Modelling and optimal sizing of hybrid renewable energy system. *13th Power Electronics and Motion Control Conference, EPE-PEMC 2008*, 1834-1839.

Belfkira, R., Zhang, L., & Barakat, G. (2011). Optimal sizing study of hybrid wind/PV/diesel power generation unit. *Solar Energy*, *85*(1), 100–110. doi:10.1016/j.solener.2010.10.018

Benasla, L., Belmadani, A., & Rahli, M. (2014). Electrical Power and Energy Systems Spiral Optimization Algorithm for solving Combined Economic and Emission Dispatch. *International Journal of Electrical Power & Energy Systems*, *62*, 163–174. doi:10.1016/j.ijepes.2014.04.037

Benatiallah, Kadia, & Dakyob. (2010). Modelling and optimisation of wind energy systems. *JJMIE*, *4*(1).

Benenson, Y., Paz-Elizur, T., Adar, R., Keinan, E., Livneh, Z., & Shapiro, E. (2001). Programmable and autonomous computing machine made of biomolecules. *Nature*, *414*(6862), 430–434. doi:10.1038/35106533 PMID:11719800

Bennell, J., & Sutcliffe, C. (2004). Black–Scholes versus artificial neural networks in pricing FTSE 100 options. *Intelligent Systems in Accounting, Finance & Management*, *12*(4), 243–260.

Berecz, E., & Balla-Achs, M. (1983). *Gas hydrates.* Academic Press.

Bernal, A., José, L., & Rodolfo, D. L. (2009). Multi-objective design and control of hybrid systems minimizing costs and unmet load. *Electric Power Systems Research*, *79*(1), 170–180. doi:10.1016/j.epsr.2008.05.011

Bernal-Agustín, J. L., Dufo-López, R., & Rivas-Ascaso, D. M. (2006). Design of isolated hybrid systems minimizing costs and pollutant emissions. *Renewable Energy*, *31*(14), 2227–2244. doi:10.1016/j.renene.2005.11.002

Beshr, E. (2013). *Comparative study of adding PV/wind energy systems to autonomus micro grid.* Academic Press.

Bevrani, H., François, B., & Ise, T. (2017). *Microgrid dynamics and control.* John Wiley & Sons. doi:10.1002/9781119263739

Bhandari, B., Lee, K. T., Cho, Y. M., Lee, C. S., Song, C. K., & Ahn, S. H. (2013). Hybridization of Multiple Renewable Power Sources for Remote Village Electrification. *Proc. of the International Symposium on Green Manufacturing and Applications.*

Bhandari, B., Lee, K.-T., Lee, C. S., Song, C.-K., Maskey, R. K., & Ahn, S.-H. (2014). A novel off-grid hybrid power system comprised of solar photovoltaic, wind, and hydro energy sources. *Applied Energy*, *133*, 236–242. doi:10.1016/j.apenergy.2014.07.033

Bhandari, R., & Stadler, I. (2011). Electrification using Solar Photovoltaic Systems in Nepal. *Applied Energy*, *88*(2), 458–465. doi:10.1016/j.apenergy.2009.11.029

Bhattacharya, A., & Chattopadhyay, P. K. (2011). Solving economic emission load dispatch problems using hybrid differential evolution. *Applied Soft Computing*, *11*(2), 2526–2537. doi:10.1016/j.asoc.2010.09.008

Bhuiyan, M., & Ali, A. M. (2003). Sizing of a Stand-Alone Photovoltaic Power System at Dhaka. *Renewable Energy*, *28*(6), 929–938. doi:10.1016/S0960-1481(02)00154-4

Bilge, U. (2007). Artificial Intelligence and Expert Systems in Medicine. *Informatics Association of Turkey Congress*, 113–118.

Bishop, C. M. (1995). *Neural networks for pattern recognition*. Oxford University Press.

Blackard, J. A., & Dean, D. J. (1999). Comparative accuracies of artificial neural networks and discriminant analysis in predicting forest cover types from cartographic variables. *Computers and Electronics in Agriculture*, *24*(3), 131–151. doi:10.1016/S0168-1699(99)00046-0

Boonbumroong, U., Pratinthong, N., Thepa, S., Jivacate, C., & Pridasawas, W. (2011). Particle swarm optimization for AC-coupling stand-alone hybrid power systems. *Solar Energy*, *85*(3), 560–569. doi:10.1016/j.solener.2010.12.027

Borges, C. L. T., & Falca, D. M. (2006). *Optimal distributed generation allocation for reliability, losses, and voltage improvement*. doi:10.1016/j.ijepes.2006.02.003

Borhanazad, H., Mekhilef, S., Gounder Ganapathy, V., Modiri-Delshad, M., & Mirtaheri, A. (2014). Optimization of micro-grid system using MOPSO. *Renewable Energy*, *71*, 295–306. doi:10.1016/j.renene.2014.05.006

Borowy, B. S., & Salameh, Z. M. (1996). Methodology for Optimally Sizing the Combination of a Battery Bank and PV Array in a Wind/PV Hybrid System. *IEEE Transactions on Energy Conversion*, *11*(2), 367–375. doi:10.1109/60.507648

Bouchard, C., & Granjean, B. P. A. (1995). A neural network correlation for the variation of viscosity of sucrose aqueous solutions with temperature and concentration. *Lebensmittel-Wissenschaft + Technologie*, *28*(1), 157–159. doi:10.1016/S0023-6438(95)80029-8

Cai, N., Member, S., Thi, N., & Nga, T. (2012). *Economic Dispatch in Microgrids Using Multi-Agent System*. Academic Press.

Calabrese, G. (1947). Generating Reserve Capacity Determined by the Probability Method. *Transactions of the American Institute of Electrical Engineers*, *66*(1), 1439–1450. doi:10.1109/T-AIEE.1947.5059596

Cardoso, G., Brouhard, T., DeForest, N., Wang, D., Heleno, M., & Kotzur, L. (2018). Battery aging in multi-energy microgrid design using mixed integer linear programming. *Applied Energy*, *231*, 1059–1069. doi:10.1016/j.apenergy.2018.09.185

Castillo, E., Guijarro-Berdinas, B., Fontenla-Romero, O., Alonso-Betanzos, A., & Bengio, Y. (2006). A Very Fast Learning Method for Neural Networks Based on Sensitivity Analysis. *Journal of Machine Learning Research*, *7*(7).

Celik, A. N. (2002). *Optimisation and techno-economic analysis of autonomous photovoltaic – wind hybrid energy systems in comparison to single photovoltaic and wind systems*. Academic Press.

Chak, D., & Bengal, W. (1990). *Energy consumption and prospects for renewable energy technologies in an Indian village*. Academic Press.

Chalise, S., Sternhagen, J., Hansen, T. M., & Tonkoski, R. (2016). Energy management of remote microgrids considering battery lifetime. *The Electricity Journal*, *29*(6), 1–10. doi:10.1016/j.tej.2016.07.003

Chang, W. L., Ho, M., & Guo, M. (2004). Molecular Solutions for the Subset-sum Problem on DNA-based Supercomputing. *Bio Systems*, *73*(2), 117–130. doi:10.1016/j.biosystems.2003.11.001 PMID:15013224

Chaouachi, A., Kamel, R. M., Andoulsi, R., & Nagasaka, K. (2012). Multiobjective intelligent energy management for a microgrid. *IEEE Transactions on Industrial Electronics*, *60*(4), 1688–1699. doi:10.1109/TIE.2012.2188873

Chapoy, A., Mohammadi, A. H., & Richon, D. (2007). Predicting the hydrate stability zones of natural gases using artificial neural networks. *Oil & Gas Science and Technology-Revue de l'IFP*, *62*(5), 701–706. doi:10.2516/ogst:2007048

Chauhan, A., & Saini, R. (2014). A review on integrated renewable energy system based power generation for stand-alone applications: Configurations, storage options, sizing methodologies and control. *Renewable and Sustainable Energy Reviews*, *38*, 99-120.

Chedid, R., Karaki, S., & Rifai, A. (2005). A Multi-Objective Design Methodology for Hybrid Renewable Energy Systems. *Proc. of the IEEE on Power Tech*, 1-6. 10.1109/PTC.2005.4524339

Chedid, R., & Rahman, S. (1997). Unit Sizing and Control of Hybrid Wind-Solar Power Systems. *IEEE Transactions on Energy Conversion*, *12*(1), 79–85. doi:10.1109/60.577284

Chellali, F., Khellaf, A., Belouchrani, A., & Recioui, A. (2011). A Contribution to the actualization of the wind map of Algeria. *Renewable & Sustainable Energy Reviews*, *15*(2), 993–1002. doi:10.1016/j.rser.2010.11.025

Chellali, F., Recioui, A., Yaiche, M. R., & Bentarzi, H. (2014). A hybrid wind/solar/diesel stand alone system optimisation for remote areas in Algeria. *International Journal of Renewable Energy Technology*, *5*(1), 12–24. doi:10.1504/IJRET.2014.059658

Chen, G., & Ding, X. (2015). *Optimal economic dispatch with valve loading effect using self-adaptive firefly algorithm*. doi:10.1007/s10489-014-0593-2

Chen, C. L., & Wang, S. C. (1993). Branch-and bound scheduling for thermal generating units. *IEEE Transactions on Energy Conversion*, *8*(2), 184–189. doi:10.1109/60.222703

Chen, S. H., Jakeman, A. J., & Norton, J. P. (2008). Artificial Intelligence techniques: An introduction to their use for modelling environmental systems. *Mathematics and Computers in Simulation*, *78*(2–3), 379–400. doi:10.1016/j.matcom.2008.01.028

Choi, S., & Kim, Y. J. (2021). Artificial neural network models for airport capacity prediction. *Journal of Air Transport Management*, *97*, 102146. doi:10.1016/j.jairtraman.2021.102146

Clerc, M. (2006). Particle swarm optimization. Particle Swarm Optimization. doi:10.1002/9780470612163

Codd, E. F. (1968). *Cellular automata*. Academic Press.

Collett, T. S., Boswell, R., Cochran, J. R., Kumar, P., Lall, M., Mazumdar, A., Ramana, M. V., Ramprasad, T., Riedel, M., Sain, K., Sathe, A. V., & Vishwanath, K. (2014). Geologic implications of gas hydrates in the offshore of India: Results of the National Gas Hydrate Program Expedition 01. In Marine and Petroleum Geology (Vol. 58, Issue PA, pp. 3–28). doi:10.1016/j.marpetgeo.2014.07.021

Connolly, D., Lund, H., Mathiesen, B. V., & Leahy, M. (2010). A review of computer tools for analysing the integration of renewable energy into various energy systems. *Applied Energy*, *87*(4), 1059–1082. doi:10.1016/j.apenergy.2009.09.026

Correa, C. A., Marulanda, G., & Garces, A. (2016). Optimal microgrid management in the Colombian energy market with demand response and energy storage. In *Proceedings of the 2016 IEEE Power and Energy Society General Meeting (PESGM)* (pp. 1-5). IEEE. 10.1109/PESGM.2016.7741905

Coulibaly, P., Anctil, F., & Bobée, B. (2000). Daily reservoir inflow forecasting using artificial neural networks with stopped training approach. *Journal of Hydrology (Amsterdam)*, *230*(3–4), 244–257. doi:10.1016/S0022-1694(00)00214-6

Cristóbal-Monreal, I. R., & Dufo-López, R. (2016). Optimisation of photovoltaic-diesel-battery stand-alone systems minimising system weight. *Energy Conversion and Management*, *119*, 279–288. doi:10.1016/j.enconman.2016.04.050

Da Rosa, A. V., & Ordonez, J. C. (2021). *Fundamentals of renewable energy processes*. Academic Press.

Das, B. K., Al-Abdeli, Y. M., & Kothapalli, G. (2018). Effect of load following strategies, hardware, and thermal load distribution on stand-alone hybrid CCHP systems. *Applied Energy*, *220*, 735–753. doi:10.1016/j.apenergy.2018.03.068

Das, G., De, M., & Mandal, K. K. (2021). Multi-objective optimization of hybrid renewable energy system by using novel autonomic soft computing techniques. *Computers & Electrical Engineering*, *94*, 107350. doi:10.1016/j.compeleceng.2021.107350

Dawoud, S. M., Xiangning, L., Jinwen, S., Okba, M. I., Khalid, M. S., & Waqar, A. (2015)... *Feasibility Study of Isolated PV-Wind Hybrid System in Egypt.*, *1093*, 145–151. doi:10.4028/www.scientific.net/AMR.1092-1093.145

De Santis, E., Rizzi, A., & Sadeghian, A. (2017). *Hierarchical genetic optimization of a fuzzy logic system for energy flows management in microgrids. Appl. Soft Comput. J.* doi:10.1016/j.asoc.2017.05.059

Delgado, C., & Dominguez-Navarro, J. A. (2014). Optimal design of a hybrid renewable energy system. *Proceedings of the Ninth International Conference on Ecological Vehicles and Renewable Energies (EVER).* 10.1109/EVER.2014.6844008

Denai, M. A., Palis, F., & Zeghbib, A. (2004). ANFIS based modelling and control of non-linear systems: a tutorial. *2004 IEEE International Conference on Systems, Man and Cybernetics (IEEE Cat. No. 04CH37583)*, *4*, 3433–3438. 10.1109/ICSMC.2004.1400873

Deshun, W., Yumeng, Z., Qiong, T., Jinhua, X., & Jelei, Y. (2018). Research on planning and configuration of multi-objective energy storage system solved by improved ant colony algorithm. *China International Conference on Electricity Distribution*, 2279–2283. 10.1109/CICED.2018.8592157

Detore, A. W. (1989). *An Introduction to Expert Systems*. Academic Press.

Diaf, S., Belhamel, M., Haddadi, M., & Louche, A. (2008). *Technical and economic assessment of hybrid photovoltaic / wind system with battery storage in Corsica island.* doi:10.1016/j.enpol.2007.10.028

Diaf, S., Notton, G., Belhamel, M., Haddadi, M., & Louche, A. (2008). Design and techno-economical optimization for hybrid PV/wind system under various meteorological conditions. *Applied Energy*, *85*(10), 968–987. doi:10.1016/j.apenergy.2008.02.012

Dincer, I., Rosen, M. A., & Ahmadi, P. (2017). Optimization of Energy Systems. doi:10.1002/9781118894484

Ding, Z., Hou, H., Yu, G., Hu, E., Duan, L., & Zhao, J. (2019). Performance analysis of a wind-solar hybrid power generation system. *Energy Conversion and Management*, *181*(September), 223–234. doi:10.1016/j.enconman.2018.11.080

Ding, X., Guo, Q., Qiannan, T., & Jermsittiparsert, K. (2021). Economic and environmental assessment of multi-energy microgrids under a hybrid optimization technique. *Sustainable Cities and Society*, *65*, 102630. doi:10.1016/j.scs.2020.102630

Dorigo, M. (1992). *Optimization, learning and natural algorithms* (PhD Thesis). Politecnico di Milano, Italy.

Dorigo, M., Birattari, M., & St, T. (2006). *Ant Colony Optimization*. Academic Press.

Dorigo, M., Maniezzo, V., & Colorni, A. (1996). Ant system: Optimization by a colony of cooperating agents. *IEEE Transactions on Systems, Man, and Cybernetics. Part B, Cybernetics*, *26*(1), 29–41. doi:10.1109/3477.484436 PMID:18263004

Drury, E., Denholm, P., Margolis, R., Drury, E., Denholm, P., & Margolis, R. (2011). *The Impact of Different Economic Performance Metrics on the Perceived Value of Solar Photovoltaics*. Academic Press.

Dufo-lo, R. (2009). *Simulation and optimization of stand-alone hybrid renewable energy systems*. doi:10.1016/j.rser.2009.01.010

Dufo-López, R., Bernal-Agustín, J. L., & Contreras, J. (2007). Optimization of control strategies for stand-alone renewable energy systems with hydrogen storage. *Renewable Energy*, *32*(7), 1102–1126. doi:10.1016/j.renene.2006.04.013

Dufo-López, R., Bernal-Agustín, J. L., Yusta-Loyo, J. M., Domínguez-Navarro, J. A., Ramírez-Rosado, I. J., Lujano, J., & Aso, I. (2011). Multi-objective optimization minimizing cost and life cycle emissions of stand-alone PV-wind-diesel systems with batteries storage. *Applied Energy*, *88*(11), 4033–4041. doi:10.1016/j.apenergy.2011.04.019

Ei-Bidairi, K. S., Nguyen, H. D., Jayasinghe, S. D. G., & Mahmoud, T. S. (2018). Multiobjective Intelligent Energy Management Optimization for Grid-Connected Microgrids. *Proceedings of the IEEE International Conference on Environment and Electrical Engineering and IEEE Industrial and Commercial Power Systems Europe (EEEIC / I&CPS Europe)*. 10.1109/EEEIC.2018.8493751

Ekanayaka, J. (2012). *The smart grid: Smart grid technology and applications* (1st ed.). Wiley. doi:10.1002/9781119968696

Elbaz, A., & Guneser, M. T. (1848). Algorithms to Model and Optimize a Stand-Alone Photovoltaic-Diesel-Battery System. *An Application in Rural Libya.*, *3651*, 523–529.

Elbaz, A., & Güneşer, M. T. (2020). Optimal Sizing of a Renewable Energy Hybrid System in Libya Using Integrated Crow and Particle Swarm Algorithms. Advance online publication. doi:10.25046/aj060130

Elgibaly, A. A., & Elkamel, A. M. (1998). A new correlation for predicting hydrate formation conditions for various gas mixtures and inhibitors. *Fluid Phase Equilibria*, *152*(1), 23–42. doi:10.1016/S0378-3812(98)00368-9

El-Keib, A. A., Ma, H., & Hart, J. L. (1994). Environmentally Constrained Economic Dispatch Using the La Grangian Relaxation Method. *IEEE Transactions on Power Systems*, *9*(4), 1723–1729. doi:10.1109/59.331423

El-Khadimi, A., Bchir, L., & Zeroual, A. (2004). Dimensionnement et Optimisation Technico-Economique D'un Système D'Energie Hybride Photovoltaïque-Eolien Avec Système de Stockage. *Revue des Énergies Renouvelables*, *7*, 73–83.

Elsied, M., Oukaour, A., Gualous, H., & Hassan, R. (2015). Energy management and optimization in microgrid system based ongreen energy. *Energy*, *84*, 139–151. doi:10.1016/j.energy.2015.02.108

Emel, G. G., & Taşkın, Ç. (2002). Genetic Algorithms and Their Application Areas. *Journal of Uludağ University Faculty of Economics and Administrative Sciences*, *21*(1), 129–152.

Emir, Ş. (2013). *Comparison of the classification performance of artificial neural networks and support vector machine methods: an application on the prediction of stock market index direction*. Istanbul University.

Energy, G. (2020). *Operation Characteristics of Renewable Energy Sources*. Academic Press.

Engelbrecht, A. P. (2007). *Computational intelligence: an introduction*. John Wiley & Sons. doi:10.1002/9780470512517

Erdinc, O., & Uzunoglu, M. (2012). Optimum design of hybrid renewable energy systems: Overview of different approaches. *Renewable & Sustainable Energy Reviews*, *16*(3), 1412–1425. doi:10.1016/j.rser.2011.11.011

Eshra, A., Shah, S., Song, T., & Reif, J. (2019). Renewable DNA hairpin-based logic circuits. *IEEE Transactions on Nanotechnology*, *18*, 252–259. doi:10.1109/TNANO.2019.2896189

Esmail, M., Golshan, H., & Arefifar, S. A. (2007). *Optimal allocation of distributed generation and reactive sources considering tap positions of voltage regulators as control variables*. Academic Press. doi:10.1002/etep

Esmat, A., & Eibakly, A. M. (2013). *A Novel Energy Management System using Ant Colony Optimization for Micro-grids*. Academic Press.

Eusuff, M., Lansey, K., & Pasha, F. (2007). *Shuffled frog-leaping algorithm : a memetic meta- heuristic for discrete optimization Shuffled frog-leaping algorithm : A memetic meta-heuristic for.* doi:10.1080/03052150500384759

Falaghi, H., Member, S., Haghifam, M., & Member, S. (2007). *ACO Based Algorithm for Distributed Generation Sources Allocation and Sizing in Distribution Systems*. Academic Press.

Fan, Y., Wang, P., Heidari, A. A., Chen, H., HamzaTurabieh, & Mafarja, M. (2022). Random reselection particle swarm optimization for optimal design of solar photovoltaic modules. *Energy*, *239*, 121865. doi:10.1016/j.energy.2021.121865

Farag, A., Al-Baiyat, S., & Cheng, T. C. (1995). Economic load dispatch multi-objective optimization procedures using linear programming techniques. *IEEE Transactions on Power Systems*, *10*(2), 731–738. doi:10.1109/59.387910

Fathima, A. H., & Palanisamy, K. (2015). Optimization in microgrids with hybrid energy systems—A review. *Renewable & Sustainable Energy Reviews*, *45*, 431–446. doi:10.1016/j.rser.2015.01.059

Feynman, R. P. (1960). There's Plenty of Room at the Bottom. *Engineering and Science*, *23*(5), 22–36.

Fs-unep. (2021). https://www.fs-unep-centre.org/wp-content/uploads/2020/06/GTR_2020.pdf

Fu, Y., Xiao, H., Lee, L. H., & Huang, M. (2021). Stochastic optimization using grey wolf optimization with optimal computing budget allocation. *Applied Soft Computing*, *103*, 107154. doi:10.1016/j.asoc.2021.107154

Gamarra, C., & Guerrero, J. M. (2015). Computational optimization techniques applied to microgrids planning: A review. *Renewable & Sustainable Energy Reviews*, *48*, 413–424. doi:10.1016/j.rser.2015.04.025

Ganguly, S., Sahoo, N. C., & Das, D. (n.d.). *Multi-Objective Planning of Electrical Distribution Systems using Particle Swarm Optimization*. Academic Press.

Ganguly, S. (2003). Prediction of VLE data using radial basis function network. *Computers & Chemical Engineering*, *27*(10), 1445–1454. doi:10.1016/S0098-1354(03)00068-1

Garg, S., Shah, S., Bui, H., Song, T., Mokhtar, R., & Reif, J. (2018). Renewable Time-Responsive DNA Circuits. *Small*, *14*(33), 1801470. doi:10.1002mll.201801470 PMID:30022600

Geem, Z. W., & Kim, J. H. (2001). *A New Heuristic Optimization Algorithm : Harmony Search*. Academic Press.

Geleta, D. K., & Manshahia, M. S. (2021). A hybrid of grey wolf optimization and genetic algorithm for optimization of hybrid wind and solar renewable energy system. *Journal of the Operations Research Society of China*, 1–14. doi:10.1007/s40305-021-00341-0

Ghasemian, S., Faridzad, A., Abbaszadeh, P., Taklif, A., Ghasemi, A., & Hafezi, R. (2020). An overview of global energy scenarios by 2040: Identifying the driving forces using cross-impact analysis method. *International Journal of Environmental Science and Technology*, 1–24.

Ghiassi, M., Saidane, H., & Zimbra, D. K. (2005). A dynamic artificial neural network model for forecasting time series events. *International Journal of Forecasting*, *21*(2), 341–362. doi:10.1016/j.ijforecast.2004.10.008

Gibbs, J. W. (1928). *The collected works of J. Willard Gibbs*. Yale Univ. Press.

Gielen, D., Boshell, F., Saygin, D., Bazilian, M. D., Wagner, N., & Gorini, R. (2019). The role of renewable energy in the global energy transformation. *Energy Strategy Reviews*, *24*, 38–50.

Gildardo Gómez, W. D. (2016). *Metodología para la Gestión Óptima de Energía en una Micro red Eléctrica Interconectada* (Unpublished Ph.D. Thesis). Universidad Nacional de Colombia, Medellín, Colombia.

Global. (2021). https://globalwindatlas.info/en/area/Turkey?print=true

Golshan, M. E. H., & Arefifar, S. A. (n.d.). *Distributed generation, reactive sources and network-configuration planning for power and energy-loss reduction*. doi:10.1049/ip-gtd

Gómez, M., López, A., & Jurado, F. (2010). Optimal placement and sizing from standpoint of the investor of Photovoltaics Grid-Connected Systems using Binary Particle Swarm Optimization. *Applied Energy*, *87*(6), 1911–1918. doi:10.1016/j.apenergy.2009.12.021

Grady, S. A., Hussaini, M. Y., & Abdullah, M. M. (2005). Placement of wind turbines using genetic algorithms. *Renewable Energy*, *30*(2), 259–270. doi:10.1016/j.renene.2004.05.007

Green, C., & Tibbetts, C. (1981). Reassociation rate limited displacement of DNA strands by branch migration. *Nucleic Acids Research*, *9*(8), 1905–1918. doi:10.1093/nar/9.8.1905 PMID:6264399

Green, S. J., Lubrich, D., & Turberfield, A. J. (2006). DNA hairpins: Fuel for autonomous DNA devices. *Biophysical Journal*, *91*(8), 2966–2975. doi:10.1529/biophysj.106.084681 PMID:16861269

Gudmundsson, J., & Borrehaug, A. (1996). Frozen hydrate for transport of natural gas. *NGH 96: 2nd International Conference on Natural Gaz Hydrates*, 415–422.

Guo, S., He, Y., Pei, H., & Wu, S. (2020). The multi-objective capacity optimization of wind-photovoltaic-thermal energy storage hybrid power system with electric heater. *Solar Energy*, *195*(August), 138–149. doi:10.1016/j.solener.2019.11.063

Guo, F., Peng, H., & Tang, J. (2016). Genetic algorithm-based parameter selection approach to single image defogging. *Information Processing Letters*, *116*(10), 595–602. doi:10.1016/j.ipl.2016.04.013

Gupta, A., Saini, R. P., & Sharma, M. P. (2006). *Optimised Application of Hybrid Renewable Energy System in Rural Electrification*. Academic Press.

Gupta, G. P., & Jha, S. (2018). Engineering Applications of Artificial Intelligence Integrated clustering and routing protocol for wireless sensor networks using Cuckoo and Harmony Search based metaheuristic techniques. *Engineering Applications of Artificial Intelligence*, *68*(September), 101–109. doi:10.1016/j.engappai.2017.11.003

Gurney, K. (1997). *An introduction to neural networks*. CRC Press. doi:10.4324/9780203451519

Guseinoviene, E., Senulis, A., Seker, S., & Akinci, T. C. (2014, March). Statistical and continuous wavelet analysis of wind speed data in Mardin-Turkey. In *2014 Ninth International Conference on Ecological Vehicles and Renewable Energies (EVER)* (pp. 1-5). IEEE.

Güvenç, U., Sönmez, Y., Duman, S., & Yörükeren, N. (2012). Sharif University of Technology Combined economic and emission dispatch solution using gravitational search algorithm. *Scientia Iranica*, *19*(6), 1754–1762. doi:10.1016/j.scient.2012.02.030

Hakimi, S. M., & Moghaddas-Tafreshi, S. M. (2009). *Optimal sizing of a stand-alone hybrid power system via particle swarm optimization for Kahnouj area in South-East of Iran*. Academic Press.

Halabi, L. M., & Mekhilef, S. (2018). Flexible hybrid renewable energy system design for a typical remote village located in tropical climate. *Journal of Cleaner Production, 177*, 908–924. doi:10.1016/j.jclepro.2017.12.248

Hamadache, I., & Mellal, M. A. (2021). Design optimization of car side safety system by particle swarm optimization and grey wolf optimizer. In M. A. Mellal & G. M. Pecht (Eds.), *Nature-Inspired Computing Paradigms in Systems: Reliability, Availability, Maintainability, Safety and Cost (RAMS+C) and Prognostics and Health Management (PHM)*. Elsevier. doi:10.1016/B978-0-12-823749-6.00006-4

Hameed, A. M. (2012). Optimum sizing of hybrid WT/PV systems via open-space particle swarm optimization. *Second Iranian Conference on Renewable Energy and Distributed Generation (ICREDG)*, 55-60. 10.1109/ICREDG.2012.6190468

Hammerstrom, D. (1993). Working with neural networks. *IEEE Spectrum, 30*(7), 46–53. doi:10.1109/6.222230

Han, Y. (2014). *Microgrid Optimization, Operation, and Control* (Unpublished PhD thesis). Colorado State University.

Hatata, A. Y., Osmana, G., & Aladla, M. M. (2018). An optimization method for sizing a solar/wind/battery hybrid power system based on the artificial immune system. *Sustainable Energy Technologies and Assessments, 27*, 83–93. doi:10.1016/j.seta.2018.03.002

Hatziargyriou, N., Asano, H., Iravani, R., & Marnay, C. (2007). Microgrids: An Overview of Ongoing Research, Development, and Demonstration Projects. *IEEE Power & Energy Magazine, 5*(4), 78–94. doi:10.1109/MPAE.2007.376583

Hayes-Roth, F. (1985). Rule-based systems. *Communications of the ACM, 28*(9), 921–932. doi:10.1145/4284.4286

Haykin, S. (1994). *Neural networks: A comprehensive foundation*. Mc Millan.

Haykin, S., & Network, N. (2004). A comprehensive foundation. *Neural Networks, 2*, 41.

Hazra, S., & Roy, P. K. (2020). Newly developed Swarm Intelligence Algorithms (GOA) applied on real-world non conventional energy based load dispatch Problems. *Advancements of Swarm Intelligence Algorithms for Solving Real-World Problems*, 1-26. doi:10.4018/978-1-7998-9152-9.ch035

Hazra, S., Pal, T., & Roy, P. K. (2019). Renewable Energy Based Economic Emission Load Dispatch Using Grasshopper Optimization Algorithm. *International Journal of Swarm Intelligence Research, 10*(1), 38–57. doi:10.4018/IJSIR.2019010103

Hazra, S., & Roy, P. K. (2015). Economic Load Dispatch Considering Non-smooth Cost Functions Using Predator–Prey Optimization. *Springer Conference on Advances in Intelligent Systems and Computing*, 67–78. doi: 10.1007/978-81-322-2268-2_8

Hazra, S., & Roy, P. K. (2019). Evolutionary Oppositional Moth Flame Optimization for Renewable and Sustainable Wind Energy Based Economic Dispatch. *International Journal of Applied Evolutionary Computation, 10*(4), 65–84. doi:10.4018/IJAEC.2019100104

Hazra, S., & Roy, P. K. (2019). Quasi-oppositional chemical reaction optimization for combined economic emission dispatch in power system considering wind power uncertainties. *Renewable Energy Focus, 31*, 45–62. doi:10.1016/j.ref.2019.10.005

Hazra, S., & Roy, P. K. (2021). Metaheuristic Moth-Flame Optimization Applied on Renewable Wind Energy Incorporating Load Transmit Penetration. *International Journal of Applied Metaheuristic Computing, 12*(1), 185–210. doi:10.4018/IJAMC.2021010110

Hazra, S., Roy, P. K., Hazra, S., & Roy, P. K. (2020). Optimal dispatch using moth-flame optimization for hydro-thermal-wind scheduling problem. *International Transactions on Electrical Energy Systems*, *30*(8), e12460. doi:10.1002/2050-7038.12460

Hazra, S., Roy, P. K., & Sinha, A. (2015). An efficient evolutionary algorithm applied to economic load dispatch problem. *International Conference on Computer, Communication, Control and Information Technology*, 1–6. 10.1109/C3IT.2015.7060129

Heaton, J. (2008). *Introduction to neural networks with Java*. Heaton Research, Inc.

Hebb, D. O. (1949). The organization of behavior: A neuropsycholocigal theory. *A Wiley Book in Clinical Psychology, 62*, 78.

Helal, S. A., Najee, R. J., Hanna, M. O., Shaaban, M. F., Osman, A. H., & Hassan, M. S. (2017). An energy management system for hybrid microgrids in remote communities. *Can. Conf. Electr. Comput. Eng.*

Helioscsp. (2014). *225 MW from concentrated solar power in Jordan*. Author.

Hemeida, A., El-Ahmar, M., El-Sayed, A., Hasanien, H. M., Alkhalaf, S., Esmail, M., & Senjyu, T. (2020). Optimum design of hybrid wind/PV energy system for remote area. *Ain Shams Engineering Journal*, *11*(1), 11–23. doi:10.1016/j.asej.2019.08.005

Hennet, J. C., & Samarakou, M. T. (1986). Optimization of a combined wind and solar power plant. *International Journal of Energy Research*, *10*(2), 181–188. doi:10.1002/er.4440100208

Hesami, S. M., Dehghani, M., Kamali, Z., & Ejraei Bakyani, A. (2017). Developing a simple-to-use predictive model for prediction of hydrate formation temperature. *International Journal of Ambient Energy*, *38*(4), 380–388. doi:10.1080/01430750.2015.1100678

Hetzer, J., Yu, D. C., & Bhattrarai, K. (2008). An economic dispatch model incorporating wind power. *IEEE Transactions on Energy Conversion*, *23*(2), 603–611. doi:10.1109/TEC.2007.914171

Heydari, A., Shayesteh, K., & Kamalzadeh, L. (2006). Prediction of hydrate formation temperature for natural gas using artificial neural network. *Oil and Gas Business*.

Heyrman, B., & Dupré, L. (2013). Efficient Modeling, Control and Optimization of Hybrid Renewable- Conventional Energy Systems. *International Journal of Renewable Energy Research*, *3*(4), 781–788.

Hirbodi, K., Enjavi-Arsanjani, M., & Yaghoubi, M. (2020). Techno-economic assessment and environmental impact of concentrating solar power plants in Iran. *Renewable & Sustainable Energy Reviews*, *120*, 109642. Advance online publication. doi:10.1016/j.rser.2019.109642

Hocaoğlu, F. O., Gerek, Ö. N., & Kurban, M. (2009). A novel hybrid (wind-photovoltaic) system sizing procedure. *Solar Energy*, *83*(11), 2019–2028. doi:10.1016/j.solener.2009.07.010

Holder, G. D., Pradhan, N., Equilibria, P., & Model, T. (1988). *Phase Behaviour In Systems Containing Clathrate Hydrates. A Review Contents Structure of Gas Hydrates*. Academic Press.

Holland, J. H. (1975). Adaptation in Natural and Artificial Systems. University of Michigan Press. doi:10.1137/1018105

Hongxing, Y. (2008). Optimal sizing method for stand-alone hybrid solar–wind system with LPSP technology by using genetic algorithm. *Solar Energy*, *82*(4), 354–367. doi:10.1016/j.solener.2007.08.005

Hopfield, J. J. (1988). Artificial neural networks. *IEEE Circuits and Devices Magazine*, *4*(5), 3–10. doi:10.1109/101.8118

Hossain, M. A., Pota, H. R., Squartini, S., & Abdou, A. F. (2019). *Modified PSO algorithm for real-time energy management in grid-connected microgrids. Renew. Energy.*

Hrayshat, E. S. (2009). Energy for Sustainable Development Techno-economic analysis of autonomous hybrid photovoltaic-diesel-battery system. *ESD*, *13*(3), 143–150. doi:10.1016/j.esd.2009.07.003

İlkiliç, C. (2012). Wind energy and assessment of wind energy potential in Turkey. *Renewable & Sustainable Energy Reviews*, *16*(2), 1165–1173.

Ismail, M. S., Moghavvemi, M., & Mahlia, T. M. I. (2014). Genetic algorithm based optimization on modeling and design of hybrid renewable energy systems. *Energy Conversion and Management*, *85*, 120–130. doi:10.1016/j.enconman.2014.05.064

Ismail, M. S., Moghavvemi, M., & Mahlia, T. M. I. (2014). Genetic algorithm based optimization on modelling and design of hybrid renewable energy systems. *Energy Conversion and Management*, *85*, 120–130.

Jassim, E. I., Abdi, M. A., & Muzychka, Y. (2008). A CFD-based model to locate flow-restriction induced hydrate deposition in pipelines. *Offshore Technology Conference*. 10.4043/19190-MS

Jemaa, A. B. (2013). Optimum Sizing of Hybrid PV/Wind/Battery System Using Fuzzy-Adaptive Genetic Algorithm. *Proceedings of the 3rd International Conference on Systems and Control.*

Jevtic, M., Jovanovic, N., Radosavljevic, J., & Klimenta, D. (2017). *Moth Swarm Algorithm for Solving Combined Economic and Emission Dispatch Problem*. Academic Press.

Jia, K., Chen, Y., Bi, T., Lin, Y., Thomas, D., & Sumner, M. (2017). Historical-Data-Based Energy Management in a Microgrid with a Hybrid Energy Storage System. *IEEE Trans. Ind. Inform.*

Jiang, X., He, C., & Jermsittiparsert, K. (2020). Online Optimal Stationary Reference Frame Controller for Inverter Interfaced Distributed Generation in a Microgrid System. *Energy Reports*, *6*, 134–145.

John, V. T., & Papadopoulos, K. D. G. D. H. (1985). John V. T.; Papadopoulos K. D.; Holder G. D. A Generalized Model for Predicting Equilibrium Conditions for Gas Hydrates. *AIChE Journal*, *31*(2), 252–259. doi:10.1002/aic.690310212

Jouyban, A., Majidi, M.-R., Jabbaribar, F., & Asadpour-Zeynali, K. (2004). Solubility prediction of anthracene in binary and ternary solvents by artificial neural networks (ANNs). *Fluid Phase Equilibria*, *225*, 133–139. doi:10.1016/j.fluid.2004.08.031

Jyoti, R., & Raju, A. B. (2011). *Economic Analysis and Comparison of Proposed HRES for Stand-Alone Applications at Various Places in Karnataka State.*

Kaabeche, A., Belhamel, M., & Ibtiouen, R. (2011). Sizing optimization of grid-independent hybrid photovoltaic/wind power generation system. *Energy*, *36*(2), 1214–1222.

Kabalci, E. (n.d.). *Hybrid renewable energy systems and microgrids*. Academic Press.

Kalantar, M., & Mousavi, G. S. M. (2010). Dynamic behavior of a stand-alone hybrid power generation system of wind turbine, microturbine, solar array and battery storage. *Applied Energy*, *87*(10), 3051–3064. doi:10.1016/j.apenergy.2010.02.019

Kaldellis, J. K., Kapsali, M., Kaldelli, E., & Katsanou, E. (2020). Comparing recent views of public attitude on wind energy, photovoltaic and small hydro applications. *Renewable Energy*, *52*, 197–208. doi:10.1016/j.renene.2012.10.045

Kalogirou, S. A. (2004). *Optimization of solar systems using artificial neural-networks and genetic algorithms*. doi:10.1016/S0306-2619(03)00153-3

Kansal, S., Kumar, V., & Tyagi, B. (2013). Electrical Power and Energy Systems Optimal placement of different type of DG sources in distribution networks. *International Journal of Electrical Power & Energy Systems*, *53*, 752–760. doi:10.1016/j.ijepes.2013.05.040

Karaki, S., Chedid, R., & Ramadan, R. (1999). Probabilistic Performance Assessment of Autonomous Solar-Wind Energy Conversion Systems. *IEEE Transactions on Energy Conversion*, *14*(3), 766–772.

Kari, L. (1997). DNA computing: Arrival of biological mathematics. *The Mathematical Intelligencer*, *19*(2), 9–22. doi:10.1007/BF03024425

Katsigiannis, Y. A., Georgilakis, P. S., Member, S., & Karapidakis, E. S. (2012). *Hybrid Simulated Annealing – Tabu Search Method for Optimal Sizing of Autonomous Power Systems With Renewables*. Academic Press.

Katsigiannis, Y. A., Georgilakis, P. S., & Karapidakis, E. S. (2010). *Genetic algorithm solution to optimal sizing problem of small autonomous hybrid power systems*. Lecture.

Katsigiannis, Y. A., Georgilakis, P. S., & Karapidakis, E. S. (2010). Multiobjective genetic algorithm solution to the optimum economic and environmental performance problem of small autonomous hybrid power systems with renewables. *IET Renewable Power Generation*, *4*(5), 404–419.

Kaul, M., Hill, R. L., & Walthall, C. (2005). Artificial neural networks for corn and soybean yield prediction. *Agricultural Systems*, *85*(1), 1–18. doi:10.1016/j.agsy.2004.07.009

Kellogg, W. (1996). Optimal unit Sizing for a Hybrid Wind/Photovoltaic Generating System. *Electric Power Systems Research*, *39*(1), 35–38.

Kellogg, W. (1998). Generation Unit Sizing and Cost Analysis for Stand-Alone Wind, Photovoltaic, and Hybrid Wind/PV Systems. *IEEE Transactions on Energy Conversion*, *13*(1), 70–75.

Kennedy, J., & Eberhart, R. (1995). Particle swarm optimization. *Neural Networks, 1995. Proceedings, IEEE International Conference On*, *4*, 1942–1948. 10.1109/ICNN.1995.488968

Keyif, E., Hornung, M., & Zhu, W. (2020). Optimal configurations and operations of concentrating solar power plants under new market trends. *Applied Energy*, *270*(May), 115080. doi:10.1016/j.apenergy.2020.115080

Khan, A. A., Naeem, M., Iqbal, M., Qaisar, S., & Anpalagan, A. (2016). A compendium of optimization objectives, constraints, tools and algorithms for energy management in microgrids. *Renewable & Sustainable Energy Reviews*.

Khan, F. A., Pal, N., & Saeed, S. H. (2018). Review of solar photovoltaic and wind hybrid energy systems for sizing strategies optimization techniques and cost analysis methodologies. *Renewable & Sustainable Energy Reviews*, *92*(March), 937–947. doi:10.1016/j.rser.2018.04.107

Khan, M. N., Warrier, P., Peters, C. J., & Koh, C. A. (2016). Review of vapor-liquid equilibria of gas hydrate formers and phase equilibria of hydrates. *Journal of Natural Gas Science and Engineering*, *35*, 1388–1404. doi:10.1016/j.jngse.2016.06.043

Khan, M., & Iqbal, M. (2005). Pre-feasibility study of stand-alone hybrid energy systems for applications in Newfoundland. *Renewable Energy*, *30*, 835–854.

Khan, S. H., Kumari, A., Chandrajit, G. D., & Amit, B. M. (2020). Thermodynamic modeling and correlations of CH 4, C 2 H 6, CO 2, H 2 S, and N 2 hydrates with cage occupancies. *Journal of Petroleum Exploration and Production Technology*, (8), 3689–3709. Advance online publication. doi:10.100713202-020-00998-y

Khare, A., & Rangnekar, S. (2013). Review article A review of particle swarm optimization and its applications in Solar Photovoltaic system. *Applied Soft Computing*, *13*(5), 2997–3006. doi:10.1016/j.asoc.2012.11.033

Khatib, T., Azah, M., & Kamaruzzaman, S. (2012). A software tool for optimal sizing of PV systems in Malaysia. *Modelling and Simulation in Engineering*, 10.

Khatib, T., & Elmenreich, W. (2014). An Improved Method for Sizing Standalone Photovoltaic Systems Using Generalized Regression Neural Network. *International Journal of Photoenergy*, 1–8.

Khatib, T., Mohamed, A., & Sopian, K. (2012). Optimization of a PV / wind micro-grid for rural housing electrification using a hybrid iterative / genetic algorithm : Case study of Kuala Terengganu, Malaysia. *Energy and Building*, *47*, 321–331. doi:10.1016/j.enbuild.2011.12.006

Khatod, D. K., Pant, V., & Sharma, J. (2010). *Analytical approach for well-being assessment*. Academic Press.

Khorasaninejad, E., & Fetanat, A. (2015). Size optimization for hybrid photovoltaic–wind energy system using ant colony optimization for continuous domains based integer programming. *Applied Soft Computing*, *31*, 196–209.

Kingsley, A., Shongwe, T., & Joseph, M. K. (2018). Renewable Energy Integration in Ghana: The Role of Smart Grid Technology. In *International Conference on Advances in Big Data, Computing and Data Communication Systems* (pp. 1-7). IEEE.

Koh, C. A., Sum, A. K., & Sloan, E. D. (2009). Gas hydrates: Unlocking the energy from icy cages. *Journal of Applied Physics*, *106*(6), 061101. Advance online publication. doi:10.1063/1.3216463

Kolawole, S. (2014). Application of Neural Networks For Predicitng The Optimal Sizing Parameters Of Stand-Alone Photovoltaic Systems. *SOP Trans. Appl. Phys.*, 12–16.

Ko, M. J. (2015). Multi-objective optimization design for a hybrid energy system using the genetic algorithm. *Energies*, *8*(4), 2924–2949.

Kornelakis, A., & Marinakis, Y. (2010). Contribution for optimal sizing of grid-connected PV-systems using PSO. *Renewable Energy*, *35*, 1333–1341.

Koutroulis, E. (2006). Methodology for optimal sizing of stand-alone photovoltaic/wind-generator systems using genetic algorithms. *Solar Energy*, *80*(9), 1072–1088.

Koutroulis, E., Dionissia, K. A. P., & Kostas, K. (2006). Methodology for optimal sizing of stand-alone photovoltaic/ wind-generator systems using genetic algorithms. *Solar Energy*, *80*(9), 1072–1088.

Koutroulis, E., & Kolokotsa, D. (2010). Design optimization of desalination systems power-supplied by PV and W/G energy sources. *Desalination*, *258*(1–3), 171–181. doi:10.1016/j.desal.2010.03.018

Kraj, A. (2015). *Intelligent computational infrastructures for optimized autonomous distributed energy generation in remote communities*. Brazil Wind Power.

Kramer, M. A. (1991). Nonlinear principal component analysis using autoassociative neural networks. *AIChE Journal. American Institute of Chemical Engineers*, *37*(2), 233–243. doi:10.1002/aic.690370209

Kröse, B., Krose, B., van der Smagt, P., & Smagt, P. (1993). *An introduction to neural networks*. Academic Press.

Kuang, Y., Zhang, Y., Zhou, B., Li, C., Cao, Y., Li, L., & Zeng, L. (2016). A review of renewable energy utilization in islands. *Renewable & Sustainable Energy Reviews*, *59*, 504–513. doi:10.1016/j.rser.2016.01.014

Kuik, O., Branger, F., & Quirion, P. (2019). Competitive advantage in the renewable energy industry: Evidence from a gravity model. *Renewable Energy*, *131*, 472–481.

Kumari, A., Khan, S. H., Misra, A. K., Majumder, C. B., & Arora, A. (2020). Hydrates of Binary Guest Mixtures: Fugacity Model Development and Experimental Validation. Journal of Non-Equilibrium Thermodynamics, 45(1), 39–58. doi:10.1515/jnet-2019-0062

Kumari, A., Hasan, S., Majumder, K. C. B., Arora, A., & Dixit, G. (2021). Physio - chemical and mineralogical analysis of gas hydrate bearing sediments of Andaman Basin. *Marine Geophysical Researches*, *42*(1), 1–12. doi:10.100711001-020-09423-9

Kumar, K. P., & Saravanan, B. (2019). *Day ahead scheduling of generation and storage in a microgrid considering demand Side management*. J. Energy Storage.

Kumar, M. (2020). *Social, economic, and environmental impacts of renewable energy resources*. Wind Solar Hybrid Renewable Energy System.

Kumar, R., Gupta, R. A., & Bansal, A. K. (2013). Economic analysis and power management of a stand-alone wind/photovoltaic hybrid energy system using biogeography based optimization algorithm. *Swarm and Evolutionary Computation*, *8*, 33–43. doi:10.1016/j.swevo.2012.08.002

Kusakana, K., & Vermaak, H. J. (2013). Hybrid Renewable Power Systems for Mobile Telephony Base Stations in developing Countries. *Renewable Energy*, *51*, 419–425.

Kutsurelis, J. E. (1998). *Forecasting financial markets using neural networks: An analysis of methods and accuracy*. Naval Postgraduate School.

Kvenvolden, K. A. (1999). Potential effects of gas hydrate on human welfare. *Proceedings of the National Academy of Sciences of the United States of America*, *96*(7), 3420–3426. doi:10.1073/pnas.96.7.3420 PMID:10097052

Lagorse, J., Paire, D., & Miraoui, A. (2009). Hybrid stand-alone power supply using PEMFC, PV and battery Modelling and optimization. *2009 International Conference on Clean Electrical Power, ICCEP 2009*, 135–140. 10.1109/ICCEP.2009.5212069

Lakatos, L., Hevessy, G., & Kovács, J. (2011). Advantages and disadvantages of solar energy and wind-power utilization. *World Futures*, *67*(6), 395–408.

Lakshminarasimman, L., & Subramanian, S. (2008). A modified hybrid differential evolution for short-term scheduling of hydrothermal power systems with cascaded reservoirs. *Energy Conversion and Management*, *49*(10), 2513–2521. doi:10.1016/j.enconman.2008.05.021

Lam, A. Y. S., & Li, V. O. K. (2010). Chemical-reaction-inspired metaheuristic for optimization. *IEEE Transactions on Evolutionary Computation*, *14*(3), 381–399. doi:10.1109/TEVC.2009.2033580

Lasseter, R. H. (2007). CERTS Microgrid. *Proceedings of the 2007 IEEE International Conference on System of Systems Engineering*.

Laugier, S., & Richon, D. (2003). Use of artificial neural networks for calculating derived thermodynamic quantities from volumetric property data. *Fluid Phase Equilibria*, *210*(2), 247–255.

Learning, M., & Publishers, K. A. (1988). *Genetic Algorithms and Machine Learning*. Academic Press.

Lee, K. Y., Yome, A. S., & Park, J. H. (1998). Adaptive Hopðeld neural networks for economic load dispatch. *IEEE Transactions on Power Systems*, *13*(2), 519–526. doi:10.1109/59.667377

Leonori, S., Rizzi, A., Paschero, M., & Mascioli, F. M. F. (2018). Microgrid Energy Management by ANFIS Supported by an ESN Based Prediction Algorithm. *Proceedings of the International Joint Conference on Neural Networks (IJCNN).*

Liao, G. (2012). Electrical Power and Energy Systems Solve environmental economic dispatch of Smart MicroGrid containing distributed generation system – Using chaotic quantum genetic algorithm. *International Journal of Electrical Power & Energy Systems*, *43*(1), 779–787. doi:10.1016/j.ijepes.2012.06.040

Liao, X., Zhou, J., Ouyang, S., Zhang, R., & Zhang, Y. (2013). An adaptive chaotic artificial bee colony algorithm for short-term hydrothermal generation scheduling. *International Journal of Electrical Power & Energy Systems*, *53*, 34–42. doi:10.1016/j.ijepes.2013.04.004

Li, H., Eseye, A. T., Zhang, J., & Zheng, D. (2017). *Optimal energy management for industrial microgrids with high-penetration renewables*. Prot. Control Mod. Power Syst.

Li, J., Wei, W., & Xiang, J. (2012). A Simple Sizing Algorithm for Stand-Alone PV/Wind/Battery Hybrid Microgrids. *Energies*, *5*(12), 5307–5323.

Lingfeng, W., & Singh, C. (2007). Compromise between cost and reliability in optimum design of an autonomous hybrid power system using mixed-integer PSO algorithm. *International conference on clean electric power.*

Lipton, R. (1995). DNA solution of hard computational problems. *Science*, *268*(5210), 542–545. doi:10.1126cience.7725098 PMID:7725098

Liu, H., Zhai, R., Fu, J., Wang, Y., & Yang, Y. (2019). Optimization study of thermal-storage PV-CSP integrated system based on GA-PSO algorithm. *Solar Energy, 184*(December), 391–409. doi:10.1016/j.solener.2019.04.017

Liu, N., Yu, X., Wang, C., Wang, J. (2017). Energy Sharing Management for Microgrids with PV Prosumers: A Stackelberg Game Approach. *IEEE Trans. Ind. Inform.*

Liu, L., Zhao, D., Yu, F., Heidari, A. A., Li, C., Ouyang, J., Chen, H., Mafarja, M., Turabieh, H., & Pan, J. (2021). Ant colony optimization with Cauchy and greedy Levy mutations for multilevel COVID 19 X-ray image segmentation. *Computers in Biology and Medicine*, *136*, 104609. doi:10.1016/j.compbiomed.2021.104609 PMID:34293587

Li, Y. Z., Wu, Q. H., Li, M. S., & Zhan, J. P. (2014). Mean-variance model for power system economic dispatch with wind power integrated. *Energy*, *72*, 510–520. doi:10.1016/j.energy.2014.05.073

Lotfi, S., Farid, L. T., & Ghiamy, M. (2013). Optimal design of a hybrid solar-wind-diesel power system for rural electrification using imperialist competitive algorithm. *International Journal of Renewable Energy Research*, *3*(2), 403–411.

Luna, A. C., Meng, L., Diaz, N. L., Graells, M., Vasquez, J. C., Guerrero, J. M. (2018). Online Energy Management Systems for Microgrids: Experimental Validation and Assessment Framework. *IEEE Trans. Power Electron.*

Lü, X., Meng, L., Long, L., & Wang, P. (2021). Comprehensive improvement of camera calibration based on mutation particle swarm optimization. *Measurement*, *110303*. Advance online publication. doi:10.1016/j.measurement.2021.110303

Ma, L., Liu, N., Zhang, J., Tushar, W., Yuen, C. (2016). Energy Management for Joint Operation of CHP and PV Prosumers Inside a Grid-Connected Microgrid: A Game Theoretic Approach. *IEEE Trans. Ind. Inform.*

Ma, Y., Yang, P., Guo, H., & Wu, J. (2012). Power source planning of wind-PV-biogas renewable energy distributed generation system. *Power System Technology*, *9*(624), 1.

Mahesh, A., & Sandhu, K. S. (2015). *Hybrid wind / photovoltaic energy system developments : Critical review and findings.* doi:10.1016/j.rser.2015.08.008

Mahesh, A., & Sandhu, K. S. (2015). Hybrid wind/photovoltaic energy system developments: Critical review and findings. *Renewable and Sustainable Energy Reviews*, *52*, 1135-1147.

Maleki, A., & Askarzadeh, A. (2014). Comparative study of artifcial intelligence techniques for sizing of a hydrogen-based stand-alone photovoltaic/wind hybrid system. *International Journal of Hydrogen Energy*, *39*(19), 9973–9984.

Maleki, A., & Fathollah, P. (2015). Optimal sizing of autonomous hybrid photovoltaic/wind/battery power system with LPSP technology by using evolutionary algorithms. *Solar Energy*, *115*, 471–483.

Maleki, A., Gholipour, M., & Ameri, M. (2016). Electrical Power and Energy Systems Optimal sizing of a grid independent hybrid renewable energy system incorporating resource uncertainty, and load uncertainty. *International Journal of Electrical Power & Energy Systems*, *83*, 514–524. doi:10.1016/j.ijepes.2016.04.008

Maleki, A., & Pourfayaz, F. (2015). Sizing of stand-alone photovoltaic/wind/diesel system with battery and fuel cell storage devices by harmony search algorithm. *Journal of Energy Storage*, *2*, 30–42.

Mandal, B., Roy, P. K., & Mandal, S. (2014). Economic load dispatch using krill herd algorithm. *International Journal of Electrical Power & Energy Systems*, *57*, 1–10. doi:10.1016/j.ijepes.2013.11.016

Markvart, T. (1996). Sizing of hybrid photovoltaic-wind energy systems. *Solar Energy*, *57*(4), 277–281.

Markvart, T., Fragaki, A., & Ross, J. (2006). PV System Sizing using Observed Time Series of Solar Radiation. *Solar Energy*, *80*(1), 46–50.

Marzband, M., Azarinejadian, F., Savaghebi, M., & Guerrero, J. M. (2017). An optimal energy management system for islanded microgrids based on multiperiod artificial bee colony combined with markov chain. *IEEE Systems Journal*.

Max, M. D., Johnson, A. H., & Dillon, W. P. (2005). *Economic geology of natural gas hydrate* (Vol. 9). Springer Science & Business Media.

McCulloch, W. S., & Pitts, W. (1943). A logical calculus of the ideas immanent in nervous activity. *The Bulletin of Mathematical Biophysics*, *5*(4), 115–133. doi:10.1007/BF02478259

Mehrizadeh, M. (2020). Prediction of gas hydrate formation using empirical equations and data-driven models. *Materials Today: Proceedings*. doi:10.1016/j.matpr.2020.06.058

Mellal, M. A., & Williams, E. J. (2019). Optimal replacement strategy of obsolete industrial components under fuzzy data. *Proceedings of the Institution of Mechanical Engineers. Part I: Journal of Systems and Control Engineering*. doi:10.1177/0959651819853483

Mellal, M. A., Adjerid, S., & Benazzouz, D. W. E. J. (2012). Fault detection-isolation and tolerance control for maintenance operators: A bond graph methodology. CIMGLE'2012, Oran, Algeria.

Mellal, M. A., Frik, A., & Boutiche, R. (2021). Reliability optimization of power plant safety system using grey wolf optimizer and shuffled frog-leaping algorithm. In Nature-Inspired Computing Paradigms in Systems: Reliability, Availability, Maintainability, Safety and Cost (RAMS+C) and Prognostics and Health Management (PHM) (pp. 1–13). doi:10.1016/B978-0-12-823749-6.00008-8

Mellal, M. A., Salhi, A., & Márquez, F. P. G. (2021). System availability and cost optimization under failure dependencies by flower pollination and plant propagation algorithms. *Fifteenth International Conference on Management Science and Engineering*, *78*, 469–476. 10.1007/978-3-030-79203-9_36

Mellal, M. A., Adjerid, S., Benazzouz, D., Berrazouane, S., & Williams, E. J. (2013). Obsolescence optimization of electronic and mechatronic components by considering dependability and energy consumption. *Journal of Central South University, 20*(5), 1221–1225. doi:10.100711771-013-1605-9

Mellal, M. A., Al-Dahidi, S., & Williams, E. J. (2020). System reliability optimization with heterogeneous components using hosted cuckoo optimization algorithm. *Reliability Engineering & System Safety, 203*, 107110. doi:10.1016/j.ress.2020.107110

Mellal, M. A., & Pecht, M. (2020). A multi-objective design optimization framework for wind turbines under altitude consideration. *Energy Conversion and Management, 222*, 113212. doi:10.1016/j.enconman.2020.113212

Mellal, M. A., & Salhi, A. (2021a). Multi-objective system design optimization via PPA and a fuzzy method. *International Journal of Fuzzy Systems, 23*(5), 1–9. doi:10.100740815-021-01068-z

Mellal, M. A., & Salhi, A. (2021b, April 30). System reliability-redundancy allocation by the multiobjective plant propagation algorithm. *International Journal of Quality & Reliability Management*. Advance online publication. doi:10.1108/IJQRM-10-2018-0285

Mellal, M. A., & Williams, E. J. (2015). Cuckoo optimization algorithm with penalty function for combined heat and power economic dispatch problem. *Energy, 93*, 1711–1718. doi:10.1016/j.energy.2015.10.006

Mellal, M. A., & Williams, E. J. (2016). Total production time minimization of a multi-pass milling process via cuckoo optimization algorithm. *International Journal of Advanced Manufacturing Technology, 87*(1-4), 747–754. Advance online publication. doi:10.100700170-016-8498-3

Mellal, M. A., & Williams, E. J. (2017). The cuckoo optimization algorithm and its applications. In *Handbook of Neural Computation* (pp. 269–277). Elsevier. doi:10.1016/B978-0-12-811318-9.00014-4

Mellal, M. A., & Williams, E. J. (2018). A survey on ant colony optimization, particle swarm optimization, and cuckoo algorithms. In *Handbook of Research on Emergent Applications of Optimization Algorithms*. IGI Global. doi:10.4018/978-1-5225-2990-3.ch002

Mellal, M. A., & Williams, E. J. (2020). Cuckoo optimization algorithm with penalty function and binary approach for combined heat and power economic dispatch problem. *Energy Reports, 6*, 2720–2723. doi:10.1016/j.egyr.2020.10.004

Mellal, M. A., & Zio, E. (2019a). An adaptive cuckoo optimization algorithm for system design optimization under failure dependencies. *Journal of Risk and Reliability, 233*(6), 1099–1105. doi:10.1177/1748006X19863639

Mellal, M. A., & Zio, E. (2019b). An adaptive particle swarm optimization method for multi-objective system reliability optimization. *Journal of Risk and Reliability, 233*(6), 990–1001. doi:10.1177/1748006X19852814

Mellal, M. A., & Zio, E. (2020). System reliability-redundancy optimization with cold-standby strategy by an enhanced nest cuckoo optimization algorithm. *Reliability Engineering & System Safety, 201*, 106973. doi:10.1016/j.ress.2020.106973

Mellit, A. (2006). Artificial intelligence based-modeling for sizing of a Stand-Alone Photovoltaic Power System: Proposition for a New Model using Neuro-Fuzzy System (ANFIS). *Proceedings of the 3rd International IEEE Conference Intelligent Systems*, 606–611.

Mellit, A. (2007). Sizing of a stand-alone photovoltaic system based on neural networks and genetic algorithms: Application for remote areas. *Istanbul Univ. J. Electr. Electron. Eng., 7*, 459–469.

Mellit, A., Kalogirou, S. A., & Drif, M. (2010). Application of neural networks and genetic algorithms for sizing of photovoltaic systems. *Renewable Energy, 35*(12), 2881–2893. doi:10.1016/j.renene.2010.04.017

Menanteau, P., Finon, D., & Lamy, M. (2010). *Prices versus quantities : choosing policies for promoting the development of renewable energy*. Academic Press.

Mirjalili, S., Mirjalili, S. M., & Lewis, A. (2014). Grey Wolf Optimizer. *Advances in Engineering Software*, *69*, 46–61. doi:10.1016/j.advengsoft.2013.12.007

Mohamed, M. A. (2016). PSO-based smart grid application for sizing and optimization of hybrid renewable energy systems. *PLoS One*, *11*(8).

Mohammadi, A. H., Belandria, V., & Richon, D. (2010). Use of an artificial neural network algorithm to predict hydrate dissociation conditions for hydrogen+ water and hydrogen+ tetra-n-butyl ammonium bromide+ water systems. *Chemical Engineering Science*, *65*(14), 4302–4305.

Mohammadi, A. H., Martínez-López, J. F., & Richon, D. (2010). Determining phase diagrams of tetrahydrofuran+ methane, carbon dioxide or nitrogen clathrate hydrates using an artificial neural network algorithm. *Chemical Engineering Science*, *65*(22), 6059–6063.

Mohammadi, A. H., & Richon, D. (2010). Hydrate phase equilibria for hydrogen+ water and hydrogen+ tetrahydrofuran+ water systems: Predictions of dissociation conditions using an artificial neural network algorithm. *Chemical Engineering Science*, *65*(10), 3352–3355.

Mohammadi, A. H., & Tohidi, B. (2005). A novel predictive technique for estimating the hydrate inhibition effects of single and mixed thermodynamic inhibitors. *Canadian Journal of Chemical Engineering*, *83*(6), 951–961.

Mohammadi, K., Khanmohammadi, S., Khorasanizadeh, H., & Powell, K. (2020). A comprehensive review of solar only and hybrid solar driven multigeneration systems: Classifications, benefits, design and prospective. *Applied Energy*, *268*(March), 114940. Advance online publication. doi:10.1016/j.apenergy.2020.114940

Møller, M. F. (1993). A scaled conjugate gradient algorithm for fast supervised learning. *Neural Networks*, *6*(4), 525–533.

Momoh, J. (2012). Smart Grid Fundamentals of Design and Analysis. Academic Press.

Mondal, A., Misra, S., Patel, L. S., Pal, S. K., & Obaidat, M. S. (2018). DEMANDS: Distributed energy management using noncooperative scheduling in smart grid. *IEEE Systems Journal*.

Moravej, Z., & Akhlaghi, A. (2013). Electrical Power and Energy Systems A novel approach based on cuckoo search for DG allocation in distribution network. *International Journal of Electrical Power & Energy Systems*, *44*(1), 672–679. doi:10.1016/j.ijepes.2012.08.009

Moridis, G. J., Collett, T. S., Boswell, R., Kurihara, M., Reagan, M. T., Koh, C., & Sloan, E. D. (2009). Toward production from gas hydrates: Current status, assessment of resources, and simulation-based evaluation of technology and potential. *SPE Reservoir Evaluation & Engineering*, *12*(5), 745–771. https://doi.org/10.2118/114163-PA

Motaz, A., Namaane, A., & M'sirdi, N. K. (2013). Optimization of hybrid renewable energy systems (HRES) using PSO for cost reduction. *Energy Procedia*, *42*, 318–327.

Motevasel, M., & Seifi, A. R. (2014). Expert energy management of a micro-grid considering wind energy uncertainty. *Energy Conversion and Management*.

Motevasel, M., Seifi, A. R., & Niknam, T. (2013). Multi-objective energy management of CHP (combined heat and power)-based micro-grid. *Energy*, *51*, 123–136. doi:10.1016/j.energy.2012.11.035

Musharavati, F., Khoshnevisan, A., Alirahmi, S. M., Ahmadi, P., & Khanmohammadi, S. (2022). Multi-objective optimization of a biomass gasification to generate electricity and desalinated water using Grey Wolf Optimizer and artificial neural network. *Chemosphere*, *287*, 131980. doi:10.1016/j.chemosphere.2021.131980 PMID:34509018

Naderpour, H., Poursaeidi, O., & Ahmadi, M. (2018). Shear resistance prediction of concrete beams reinforced by FRP bars using artificial neural networks. *Measurement: Journal of the International Measurement Confederation, 126*(June), 299–308. doi:10.1016/j.measurement.2018.05.051

Nadjemi, O., Nacer, T., Hamidat, A., & Salhi, H. (2017). Optimal hybrid PV / wind energy system sizing : Application of cuckoo search algorithm for Algerian dairy farms. *Renewable & Sustainable Energy Reviews*, *70*, 1352–1365. doi:10.1016/j.rser.2016.12.038

Nafeh, A. S. A. (2011). Optimal economical sizing of a PV-wind hybrid energy system using genetic algorithm. *International Journal of Green Energy*, *8*(1), 25–43.

Nafey, A. S., Sharaf, M. A., & García-Rodríguez, L. (2010). Thermo-economic analysis of a combined solar organic Rankine cycle-reverse osmosis desalination process with different energy recovery configurations. *Desalination*, *261*(1–2), 138–147. Advance online publication. doi:10.1016/j.desal.2010.05.017

Najibi, H., Chapoy, A., Haghighi, H., & Tohidi, B. (2009). Experimental determination and prediction of methane hydrate stabilityin alcohols and electrolyte solutions. *Fluid Phase Equilibria*, *275*, 127–131.

Nandi, S. K., & Ghosh, H. R. (2009). A wind-PV-battery hybrid power system at Sitakunda in Bangladesh. *Energy Policy*, *37*(9), 3659–3664.

Nikmehr, N., & Najafi-Ravadanegh, S. (2015). Optimal operation of distributed generations in micro-grids under uncertainties in load and renewable power generation using heuristic algorithm. *IET Renewable Power Generation.*

Nivedha, R. R., Singh, J. G., & Ongsakul, W. (2018). PSO based economic dispatch of a hybrid microgrid system. *Proceedings of the 4th International Conference on Power, Signals, Control and Computation (EPSCICON 2018).*

Nojavan, S., & Jermsittiparsert, K. (2020). Risk-Based Performance of Combined Heat and Power Based Microgrid Using Information Gap Decision Theory. *IEEE Access: Practical Innovations, Open Solutions*, *8*(1), 93123–93132.

Nwulu, N. I., & Xia, X. (2017). Optimal dispatch for a microgrid incorporating renewables and demand response. *Renewable Energy.*

Ogunjuyigbe, A. S. O., Ayodele, T. R., & Akinola, O. A. (2016). Optimal allocation and sizing of of small autonomous power systems with solar and wind energy sources. *IEEE Transactions on Energy Conversion*, *25*(2), 535-545.

Olatomiwa, L., Mekhilef, S., & Ohunakin, O. S. (2016). Hybrid renewable power supply for rural health clinics (RHC) in six geo-political zones of Nigeria. *Sustainable Energy Technologies and Assessments*, *13*, 1–12. doi:10.1016/j.seta.2015.11.001

Onar, O. C., & Khaligh, A. (2015). Energy Sources. In Alternative Energy in Power Electronics (pp. 81-154). Butterworth-Heinemann.

Optimal placement of multi-distributed generation units including different load models using particle swarm optimization. (2011). *Swarm and Evolutionary Computation*, *1*(1), 50–59. doi:10.1016/j.swevo.2011.02.003

Othman, Z., Sulaiman, S. I., Musirin, I., & Mohamad, K. (2015). Bat inspired algorithm for sizing optimization of grid-connected photovoltaic system. *Proceedings of the 2015 SAI Intelligent Systems Conference (IntelliSys)*, 195–200.

Ould Bilal, B., Sambou, V., Ndiaye, P. A., Kébé, C. M. F., & Ndongo, M. (2010). Optimal design of a hybrid solar-wind-battery system using the minimization of the annualized cost system and the minimization of the loss of power supply probability (LPSP). *Renewable Energy*, *35*(10), 2388–2390. doi:10.1016/j.renene.2010.03.004

Ouyang, H., Gao, L., Li, S., Kong, X., Wang, Q., & Zou, D. (2017). Improved Harmony Search Algorithm : LHS. *Applied Soft Computing*, *53*, 133–167. doi:10.1016/j.asoc.2016.12.042

Özyön, S., & Yas, C. (2014). *Electrical Power and Energy Systems Artificial bee colony algorithm with dynamic population size to combined economic and emission dispatch problem.* doi:10.1016/j.ijepes.2013.06.020

Paliwal, P., Patidar, N. P., & Nema, R. K. (2014). Planning of grid integrated distributed generators : A review of technology, objectives and techniques. *Renewable & Sustainable Energy Reviews*, *40*, 557–570. doi:10.1016/j.rser.2014.07.200

Panda, M., & Das, B. (2019). Grey wolf optimizer and its applications: A survey. *Lecture Notes in Electrical Engineering*, *556*, 179–194. doi:10.1007/978-981-13-7091-5_17

Pandi, V. R., Zeineldin, H. H., & Xiao, W. (2013). *Distributed Generation Resources Considering Harmonic and Protection Coordination Limits*. Academic Press.

Paneiro, G., & Rafael, M. (2021). Artificial neural network with a cross-validation approach to blast-induced ground vibration propagation modeling. *Underground Space (China)*, *6*(3), 281–289. doi:10.1016/j.undsp.2020.03.002

Panigrahi, B. K., Pandi, V. R., & Das, S. (2010). Multi-objective fuzzy dominance based bacterial foraging algorithm to solve economic emission dispatch problem. *Energy*, *35*(12), 4761–4770. doi:10.1016/j.energy.2010.09.014

Pan, M., Li, C., Gao, R., Huang, Y., You, H., Gu, T., & Qin, F. (2020). Photovoltaic power forecasting based on a support vector machine with improved ant colony optimization. *Journal of Cleaner Production*, *277*, 123948. doi:10.1016/j.jclepro.2020.123948

Panwar, N. L., Kaushik, S. C., & Kothari, S. (2011). Role of renewable energy sources in environmental protection: A review. *Renewable & Sustainable Energy Reviews*, *15*(3), 1513–1524.

Papari, B., Edrington, C. S., Vu, T. V., & Diaz-Franco, F. (2017). A heuristic method for optimal energy management of DC microgrid. *Proceedings of the IEEE Second International Conference on DC Microgrids (ICDCM).*

Park, Y. S., & Lek, S. (2016). Artificial neural networks: Multilayer perceptron for ecological modeling. In Developments in Environmental Modelling (Vol. 28, pp. 123–140). doi:10.1016/B978-0-444-63623-2.00007-4

Park. (2006). Sequestering carbon dioxide into complex structures of naturally occurring gas hydrates. *Proceedings of the National Academy of Sciences of the United States of America*, *103*(34), 12690–12694. https://doi.org/10.1073/pnas.0602251103

Park, K., Hong, S. Y., Lee, J. W., Kang, K. C., Lee, Y. C., Ha, M.-G., & Lee, J. D. (2011). A new apparatus for seawater desalination by gas hydrate process and removalcharacteristics of dissolved minerals (Na+,Mg2+,Ca2+,K+,B3+). *Desalination*, *274*, 91–96.

Parvez, F., Vasant, P., Kallimani, V., & Watada, J. (2018). A holistic review on optimization strategies for combined economic emission dispatch problem. *Renewable and Sustainable Energy Reviews*, *81*(July), 3006–3020. doi:10.1016/j.rser.2017.06.111

Paul, T. G., Hossain, S. J., Ghosh, S., Mandal, P., Kamalasadan, S. (2018). A Quadratic Programming Based Optimal Power and Battery Dispatch for Grid-Connected Microgrid. *IEEE Trans. Ind. Appl.*

Pedregal, P. (2004). Γ-convergence through Young measures. *SIAM Journal on Mathematical Analysis, 36*(2), 423–440. doi:10.1137/S0036141003425696

Petersen, R., Fredenslund, A., & Rasmussen, P. (1994). Artificial neural networks as a predictive tool for vapor-liquid equilibrium. *Computers & Chemical Engineering, 18*, S63–S67.

Peterson, C., Rgnvaldsson, T., & Lnnbladb, L. (1994). *JETNET 3.0 - A versatile artificial neural network package.* Academic Press.

Petrescu, S., Petre, C., Costea, M., Malancioiu, O., Boriaru, N., Dobrovicescu, A., Feidt, M., & Harman, C. (2010). A methodology of computation, design and optimization of solar Stirling power plant using hydrogen/oxygen fuel cells. *Energy, 35*(2), 729–739. doi:10.1016/j.energy.2009.10.036

Piarehzadeh, H., Khanjanzadeh, A., & Pejmanfer, R. (2012). Comparison of harmony search algorithm and particle swarm optimization for distributed generation allocation to improve steady state voltage stability of distribution networks. *Research Journal of Applied Sciences, Engineering and Technology, 4*(15), 2310–2315.

Pillai, N. V. (2008). *Loss of Load Probability of a Power System.* Munich Personal Repec Archive, Paper No. 6953.

Potukuchi, S., & Wexler, A. S. (1997). Predicting vapor pressures using neural networks. *Atmospheric Environment, 31*(5), 741–753.

Prathyush, M., & Jasmin, E. A. (2018). Fuzzy Logic Based Energy Management System Design for AC Microgrid. *Proceedings of the International Conference on Inventive Communication and Computational Technologies (ICICCT).*

PSD. (2021). http://faculty.etsu.edu/blanton/lab_3_psd.doc

Purwanto, H., Suherman, W. D. A., & Iksan, N. (2021). Renewable energy generation forecasting on smart home micro grid using deep neural network. *International Conference on Artificial Intelligence and Mechatronics Systems.* 10.1109/AIMS52415.2021.9466089

Quarto, M., D'Urso, G., & Giardini, C. (2022). Micro-EDM optimization through particle swarm algorithm and artificial neural network. *Precision Engineering, 73*, 63–70. doi:10.1016/j.precisioneng.2021.08.018

Raal, J. D., & Mühlbauer, A. L. (1994). The Measurement of High Pressure Vapour-Liquid-Equilibria: Part I: Dynamic Methods. *Developments in Chemical Engineering and Mineral Processing, 2*(2-3), 69–87.

Rai, P., Majumdar, G. C., DasGupta, S., & De, S. (2005). Prediction of the viscosity of clarified fruit juice using artificial neural network: A combined effect of concentration and temperature. *Journal of Food Engineering, 68*(4), 527–533.

Rajabioun, R. (2011). Cuckoo optimization algorithm. *Applied Soft Computing, 11*(8), 5508–5518. doi:10.1016/j.asoc.2011.05.008

Rajendra, P. A., & Natarajan, E. (2006). Optimization of integrated photovoltaic–wind power generation systems with battery storage. *Energy, 31*(12), 1943–1954.

Ramesh Babu, N. (2017). *Smart Grid System Modeling and Control.* Apple Academic Press.

Ramoji, K. B. S. K. (2014). Optimal economical sizing of a PV-wind hybrid energy system using genetic algorithm and teaching learning based optimization., *International Journal of Advanced Research in Electrical, Electronics and Instrumentation Engineering, 3*(3).

Ramoji, S. K., Bibhuti, B. R., & Kumar, V. D. (2014). Optimization of hybrid PV/wind energy system using genetic algorithm (GA). *Journal of Engineering Research and Applications, 4*, 29–37.

Rani, D., Jain, S. K., Srivastava, D. K., & Perumal, M. (2013). Genetic Algorithms and Their Applications to Water Resources Systems. In Metaheuristics in Water, Geotechnical and Transport Engineering (pp. 43–78). Elsevier Inc. doi:10.1016/B978-0-12-398296-4.00003-9

Rao, R. S., Ravindra, K., Satish, K., & Narasimham, S. V. L. (2013). *Power Loss Minimization in Distribution System Using Network Recon figuration in the Presence of Distributed Generation.* Academic Press.

Ray, K. S., & Mondal, M. (2011). Similarity-based Fuzzy Reasoning by DNA Computing. *International Journal of Bio-inspired Computation, 3*(2), 112–122. doi:10.1504/IJBIC.2011.039910

Ray, K. S., & Mondal, M. (2011b). Classification of SODAR data by DNA computing. *New Mathematics and Natural Computation, 7*(3), 413–432. doi:10.1142/S1793005711002074

Ray, K. S., & Mondal, M. (2016). Logical inference by DNA strand algebra. *New Mathematics and Natural Computation, 12*(1), 29–44. doi:10.1142/S1793005716500046

Rebai, N., Hadjadj, A., Benmounah, A., Berrouk, A. S., & Boualleg, S. M. (2019). Prediction of natural gas hydrates formation using a combination of thermodynamic and neural network modeling. *Journal of Petroleum Science Engineering, 182*(March), 106270. https://doi.org/10.1016/j.petrol.2019.106270

Recioui, A., & Bentarzi, H. (2021). *Optimizing and Measuring Smart Grid Operation and Control.* IGI Global.

Recioui, A., & Dassa, K. (2017). Design of Standalone Micro-Grid Systems Using Teaching Learning Based Optimization. *Algerian Journal of Signals and Systems, 2*(2), 75–85.

Ripley, B. D. (1996). *Pattern Recognition and Neural Networks.* Cambridge University Press.

Rojas-Zerpa, J. C., & Yusta, J. M. (2015). Application of multicriteria decision methods for electric supply planning in rural and remote areas. In *Renewable and Sustainable Energy Reviews* (Vol. 52, pp. 557–571). Elsevier Ltd., doi:10.1016/j.rser.2015.07.139

Rouholamini, M., & Mohammadian, M. (2016). Heuristic-based power management of a grid-connected hybrid energy system combined with hydrogen storage. *Renewable Energy.*

Roy, P. C., Majumder, A., & Chakraborty, N. (2010). Optimization of a stand-alone Solar PV-Wind-DG Hybrid System for Distributed Power Generation at Sagar Island. *AIP Conference Proceedings, 1298*(1), 260–265.

Roy, P. K. (2013). Teaching learning based optimization for short-term hydrothermal scheduling problem considering valve point effect and prohibited discharge constraint. *International Journal of Electrical Power & Energy Systems, 53*, 10–19. doi:10.1016/j.ijepes.2013.03.024

Roy, P. K., & Hazra, S. (2015). Economic emission dispatch for wind- fossil fuel based power system using chemical reaction optimization. *International Transaction on Electrical Energy Systems, 25*(12), 3248–3274. doi:10.1002/etep.2033

Roy, P. K., Paul, C., & Sultana, S. (2014). Oppositional teaching learning based optimization approach for combined heat and power dispatch. *International Journal of Electrical Power & Energy Systems, 57*, 392–403. doi:10.1016/j.ijepes.2013.12.006

Russell, S. J., & Norvig, P. (2003). Artificial Intelligence: A Modern Approach (2nd ed.). Academic Press.

Sablani, S. S., Baik, O.-D., & Marcotte, M. (2002). Neural networks for predicting thermal conductivity of bakery products. *Journal of Food Engineering, 52*(3), 299–304.

Saffari, M., de Gracia, A., Fernández, C., Belusko, M., Boer, D., & Cabeza, L. F. (2018). Optimized demand side management (DSM) of peak electricity demand by coupling low temperature thermal energy storage (TES) and solar PV. *Applied Energy, 211*, 604–616. doi:10.1016/j.apenergy.2017.11.063

Saini, R. P., & Sharma, M. P. (2010). Integrated renewable energy systems for off grid rural electrification of remote area. *Renewable Energy, 35*(6), 1342–1349. doi:10.1016/j.renene.2009.10.005

Salhi, A., & Fraga, E. S. (2011). Nature-inspired optimisation approaches and the new plant propagation algorithm. *The International Conference on Numerical Analysis and Optimization.*

SAM. (2021, February 18). *CSP Cost Data*. SAM.

Sarkhel, R., Das, N., Saha, A. K., & Nasipuri, M. (2018). Engineering Applications of Artificial Intelligence An improved Harmony Search Algorithm embedded with a novel piecewise opposition based learning algorithm. *Engineering Applications of Artificial Intelligence, 67*(March), 317–330. doi:10.1016/j.engappai.2017.09.020

Sedighizadeh, M., & Sadighi, M. (n.d.). *A Particle Swarm Optimization for Sitting and Sizing of Distributed Generation in Distribution Network to Improve Voltage Profile and Reduce THD and Losses*. Academic Press.

Shaaban, M. F., Member, S., Atwa, Y. M., Member, S., El-saadany, E. F., & Member, S. (2013). *DG Allocation for Bene fi t Maximization in Distribution Networks*. Academic Press.

Shaahid, S. M., & Elhadidy, M. A. (2008). Economic analysis of hybrid photovoltaic–diesel–battery power systems for residential loads in hot regions—A step to clean future. *Renewable & Sustainable Energy Reviews, 12*(2), 488–503.

Shapiro, J. (2001). Genetic Algorithms in Machine Learning BT - Machine Learning and Its Applications: Advanced Lectures. Springer Berlin Heidelberg. https://doi.org/10.1007/3-540-44673-7_7.

Sharaf Eldean, M. A., Rafi, K. M., & Soliman, A. M. (2017). Performance analysis of different working gases for concentrated solar gas engines: Stirling & Brayton. *Energy Conversion and Management, 150*(August), 651–668. doi:10.1016/j.enconman.2017.08.008

Sharma, R., Singhal, D., Ghosh, R., & Dwivedi, A. (1999). Potential applications of artificial neural networks to thermodynamics: Vapor–liquid equilibrium predictions. *Computers & Chemical Engineering, 23*(3), 385–390.

Sharma, S., Bhattacharjee, S., & Bhattacharya, A. (2016). Grey wolf optimisation for optimal sizing of battery energy storage device to minimise operation cost of microgrid. *IET Generation, Transmission & Distribution, 10*(3), 625–637.

Shboul, B., AL-Arfi, I., Michailos, S., Ingham, D., AL-Zoubi, O. H., Ma, L., Hughes, K., & Pourkashanian, M. (2021a). Design and Techno-economic assessment of a new hybrid system of a solar dish Stirling engine instegrated with a horizontal axis wind turbine for microgrid power generation. *Energy Conversion and Management, 245*, 114587. Advance online publication. doi:10.1016/j.enconman.2021.114587

Shboul, B., AL-Arfi, I., Michailos, S., Ingham, D., Ma, L., Hughes, K. J., & Pourkashanian, M. (2021). A new ANN model for hourly solar radiation and wind speed prediction: A case study over the north & south of the Arabian Peninsula. *Sustainable Energy Technologies and Assessments, 46*, 101248. Advance online publication. doi:10.1016/j.seta.2021.101248

Shi, L., Wang, C., Yao, L., Ni, Y., & Bazargan, M. (2012). Optimal power flow solution incorporating wind power. *IEEE Systems Journal, 6*(2), 233–241. doi:10.1109/JSYST.2011.2162896

Siddaiah, R., & Saini, R. P. (2016). A review on planning, con fi gurations, modeling and optimization techniques of hybrid renewable energy systems for off grid applications. *Renewable & Sustainable Energy Reviews, 58*, 376–396. doi:10.1016/j.rser.2015.12.281

Singh, D., Singh, D., & Verma, K. S. (2007). *GA based Optimal Sizing & Placement of Distributed Generation for Loss Minimization*. Academic Press.

Singh, S., Singh, M., & Kaushik, S. C. (2016). A review on optimization techniques for sizing of solar-wind hybrid energy systems. In International Journal of Green Energy (Vol. 13, Issue 15, pp. 1564–1578). Taylor & Francis. doi:10.1080/15435075.2016.1207079

Singh, N., Nyuur, R., & Richmond, B. (2019). Renewable energy development as a driver of economic growth: Evidence from multivariate panel data analysis. *Sustainability*, *11*(8), 2418.

Singh, O., & Verma, A. (2020). Frequency control for stand-alone hydro power plants using ant colony optimization. *IEEE International Conference on Advent Trends in Multidisciplinary Research and Innovation*. 10.1109/ICATMRI51801.2020.9398417

Singh, R. K., & Goswami, S. K. (2009). Optimum Siting and Sizing of Distributed Generations in Radial and Networked Systems. *Electric Power Components and Systems Optimum Siting and Sizing of Distributed Generations in Radial and Networked Systems Optimum Siting and Sizing of Distributed Generations in Radial and Networked Systems*, *5008*(2), 127–145. Advance online publication. doi:10.1080/15325000802388633

Singh, R. K., & Goswami, S. K. (2010). Electrical Power and Energy Systems Optimum allocation of distributed generations based on nodal pricing for profit, loss reduction, and voltage improvement including voltage rise issue. *International Journal of Electrical Power & Energy Systems*, *32*(6), 637–644. doi:10.1016/j.ijepes.2009.11.021

Singh, T. N., Sinha, S., & Singh, V. K. (2007). Prediction of thermal conductivity of rock through physico-mechanical properties. *Building and Environment*, *42*(1), 146–155.

Sinha, S., & Chandel, S. S. (2014). Review of software tools for hybrid renewable energy systems. *Renewable & Sustainable Energy Reviews*, *32*, 192–205. doi:10.1016/j.rser.2014.01.035

Sinha, S., & Chandel, S. S. (2015). Review of recent trends in optimization techniques for solar photovoltaic – wind based hybrid energy systems. *Renewable & Sustainable Energy Reviews*, *50*, 755–769. doi:10.1016/j.rser.2015.05.040

Siva Reddy, S. C. K., & S. K. T. (2012). Exergetic analysis and performance evaluation of parabolic dish Stirling engine solar power plant. *Archives of Thermodynamics*, *33*(4), 23–40. doi:10.1002/er

Sivanandam, S. N., & Deepa, S. N. (2008). *Introduction to Genetic Algorithms*. Springer Berlin Heidelberg.

Sloan. (2003). Fundamental principles and applications of natural gas hydrates. *Nature*, *426*(6964), 353–359.

Sloan, E. D. (1990). Natural gas hydrate phase equilibria and kinetics: Understanding the state-of-the-art. *Revue de l'Institut Français du Pétrole*, *45*(2), 245–266.

Sloan, E. D. Jr, & Koh, C. A. (2007). *Clathrate hydrates of natural gases*. CRC Press.

Solargis. (n.d.). *Solar resource maps of Jordan*. Retrieved July 3, 2020, from https://solargis.com/maps-and-gis-data/download/jordan

Sopian, K. (2008). Optimal operational strategy for hybrid renewable energy system using genetic algorithms. *WSEAS Transactions on Mathematics*, *7*(4), 130–140.

Soroudi, A., Ehsan, M., & Zareipour, H. (2011). A practical eco-environmental distribution network planning model including fuel cells and non-renewable distributed energy resources. *Renewable Energy*, *36*(1), 179–188. doi:10.1016/j.renene.2010.06.019

Soroush, E., Mesbah, M., Shokrollahi, A., Rozyn, J., Lee, M., Kashiwao, T., & Bahadori, A. (2015). Evolving a robust modeling tool for prediction of natural gas hydrate formation conditions _ Elsevier Enhanced Reader.pdf. *Journal of Unconventional Oil and Gas Resources*, *12*, 45–55.

Speer, B. (2015). Role of smart grids in integrating renewable energy. National Renewable Energy Lab. (NREL).

Srivastava, A., & Singh, S. (2020). Implementation of ant colony optimization and particle swarm optimization in economic load dispatch problem using renewable source. *2020 IEEE 17th India Council International Conference*. 10.1109/INDICON49873.2020.9342190

Sriyakul, T., & Jermsittiparsert, K. (2020). Economic scheduling of a smart microgrid utilizing the benefits of plug-in electric vehicles contracts with a comprehensive model of information-gap decision theory. *Journal of Energy Storage*, *32*, 102010.

Stanton, K. N., Giri, J. C., & Bose, A. (2017). *Energy management*. Syst. Control Embed. Syst. Energy Mach.

Starke, A. R., Cardemil, J. M., Escobar, R., & Colle, S. (2018). Multi-objective optimization of hybrid CSP+PV system using genetic algorithm. *Energy*, *147*, 490–503. doi:10.1016/j.energy.2017.12.116

Staub, S., Karaman, E., Kaya, S., Karapınar, H., & Güven, E. (2015). Artificial Neural Network and Agility. *Procedia: Social and Behavioral Sciences*, *195*, 1477–1485. https://doi.org/10.1016/j.sbspro.2015.06.448

Su, S., Lu, C., Chang, R., & Gutiérrez-alcaraz, G. (2011). *Distributed Generation Interconnection Planning : A Wind Power Case Study*. Academic Press.

Suchetha, C., & Ramprabhakar, J. (2018). *Optimization techniques for operation and control of microgrids-Review*. J. Green Eng.

Sukumar, S. (2017). Mix-mode energy management strategy and battery sizing for economic operation of grid-tied microgrid. *Energy*, *118*, 1322–1333.

Sulaiman, M. H., Mustafa, M. W., Azmi, A., Aliman, O., & Rahim, S. R. A. (2012). *Optimal Allocation and Sizing of Distributed Generation in Distribution System via Firefly Algorithm*. Academic Press.

Sulaiman, M., Salhi, A., Selamoglu, B. I., & Kirikchi, O. B. (2014). A plant propagation algorithm for constrained engineering optimisation problems. *Mathematical Problems in Engineering*, *2014*, 1–10. Advance online publication. doi:10.1155/2014/627416

Sulaiman, S. I., Rahman, T. K. A., Musirin, I., Shaari, S., & Sopian, K. (2012). An intelligent method for sizing optimization in grid-connected photovoltaic system. *Solar Energy*, *86*, 2067–2082.

Sumathi, S., & Paneerselvam, S. (2010). *Computational intelligence paradigms : Theory & applications using MATLAB*. CRC Press. doi:10.1201/9781439809037

Su, W., & Wang, J. (2012). Energy Management Systems in Microgrid Operations. *The Electricity Journal*.

Swain, R. K., Barisal, A. K., Hota, P. K., & Chakrabarti, R. (2011). Short-term hydrothermal scheduling using clonal selection algorithm. *International Journal of Electrical Power & Energy Systems*, *33*(3), 647–656. doi:10.1016/j.ijepes.2010.11.016

Tafreshi, S. M. M., Zamani, H. A., Ezzati, S. M., Baghdadi, M., & Vahedi, H. (2010). Optimal unit sizing of distributed energy resources in microgrid using genetic algorithm. *Proceedings - 2010 18th Iranian Conference on Electrical Engineering, ICEE 2010*, 836–841. 10.1109/IRANIANCEE.2010.5506961

Taghikhani, M. A. (2012). *DG Allocation and Sizing in Distribution Network Using Modified Shuffled Frog Leaping Algorithm*. Academic Press.

Taha, M. S., & Mohamed, Y. A. R. I. (2016). Robust MPC-based energy management system of a hybrid energy source for remote communities. *Proceedings of the IEEE Electrical Power and Energy Conference (EPEC)*.

Tanyildizi, H., Özcan, F., Atis, C. D., Karahan, O., & Uncuog, E. (2009). *Advances in Engineering Software Comparison of artificial neural network and fuzzy logic models for prediction of long-term compressive strength of silica fume concrete*. doi:10.1016/j.advengsoft.2009.01.005

Taylor, P., Balamurugan, R., & Subramanian, S. (2007). *Electric Power Components and Systems A Simplified Recursive Approach to Combined Economic Emission Dispatch*. doi:10.1080/15325000701473742

The Guide to Renewable vs. Non-renewable Energy Sources. (n.d.). https://www.gosunpro.com/news/renewable-vs-non-renewable-energy-sources/

The Wind Power. (n.d.). Retrieved May 10, 2020, from https://www.thewindpower.net/windfarms_list_en.php

Thirugnanam, K., Kerk, S.K., Yuen, C., Liu, N., Zhang, M. (2018). Energy Management for Renewable Microgrid in Reducing Diesel Generators Usage with Multiple Types of Battery. *IEEE Trans. Ind. Electron.*

Tina, G., Gagliano, S., & Raiti, S. (2006). Hybrid solar/wind power system probabilistic modelling for long-term performance assessment. *Solar Energy*, *80*(5), 578-588.

Tizhoosh, H. (2005). Opposition-based learning: A new scheme for machine intelligence. *Proceedings of the International Conference on Computational Intelligence for Modelling Control and Automation*, 695–701. 10.1109/CIMCA.2005.1631345

Ton, D. T., Wang, W. M., & Wang, W. T. P. (2011). *Smart Grid R&D by the US Department of Energy to optimize distribution grid operations. In 2011 IEEE Power and Energy Society General Meeting*. IEEE.

Torrecilla, J. S., Otero, L., & Sanz, P. D. (2004). A neural network approach for thermal/pressure food processing. *Journal of Food Engineering*, *62*(1), 89–95.

Trivedi, I. N., Jangir, P., Bhoye, M., & Jangir, N. (2018). An economic load dispatch and multiple environmental dispatch problem solution with microgrids using interior search algorithm. *Neural Computing & Applications*, *30*(7), 2173–2189. doi:10.100700521-016-2795-5

Umam, M. S., Mustafid, M., & Suryono, S. (2021). A hybrid genetic algorithm and tabu search for minimizing makespan in flow shop scheduling problem. *Journal of King Saud University - Computer and Information Sciences*. doi:10.1016/j.jksuci.2021.08.025

Umeozor, E. C., & Trifkovic, M. (2016). Energy management of a microgrid via parametric programming. *IFAC-PapersOnLine*.

Upadhyay, S., & Sharma, M. P. (2014). A review on con fi gurations, control and sizing methodologies of hybrid energy systems. *Renewable & Sustainable Energy Reviews*, *38*, 47–63. doi:10.1016/j.rser.2014.05.057

Vakil-Baghmisheh, M.-T. (2002). Farsi character recognition using artificial neural networks. In *Faculty of Electrical Engineering*. University of Ljubljana.

Venayagamoorthy, G. K., Sharma, R. K., Gautam, P. K., Ahmadi, A. (2016). Dynamic Energy Management System for a Smart Microgrid. *IEEE Trans. Neural Netw. Learn. Syst.*

Wang, L., Member, S., & Singh, C. (2008). *Reliability-Constrained Optimum Placement of Reclosers and Distributed Generators in Distribution Networks Using an Ant Colony System Algorithm*. Academic Press.

Wang, X. Q., Li, X. P., Li, Y. R., & Wu, C. M. (2015). Payback period estimation and parameter optimization of subcritical organic Rankine cycle system for waste heat recovery. *Energy, 88*, 734–745. https://doi.org/10.1016/j.energy.2015.05.095

Waqar, A., Subramaniam, U., Farzana, K., Elavarasan, R. M., Habib, H. U. R., Zahid, M., & Hossain, E. (2020). Analysis of optimal deployment of several DGs in distribution networks using plant propagation algorithm. *IEEE Access: Practical Innovations, Open Solutions, 8*, 175546–175562. doi:10.1109/ACCESS.2020.3025782

Wasilewski, J. (2018). Optimisation of multicarrier microgrid layout using selected metaheuristics. *International Journal of Electrical Power & Energy Systems.*

Watson, J. D., & Crick, F. H. C. (1953). A Structure for Deoxyribose Nucleic Acid. *Nature, 171*(4356), 737–738. doi:10.1038/171737a0 PMID:13054692

Winfree, E., Liu, F., Wenzler, L. A., & Seeman, N. C. (1998). Design and self-assembly of two-dimensional DNA crystals. *Nature, 394*(6693), 539–544. doi:10.1038/28998 PMID:9707114

Wu, L., Park, J., Choi, J., Cha, J., & Lee, K. Y. (2009). A study on wind speed prediction using artificial neural network at Jeju Island in Korea. *Transmission and Distribution Conference and Exposition: Asia and Pacific*. doi:10.1109/TD-ASIA.2009.5356873

Wu, Y., & Chang, S. (2016). *Review of the Optimal Design on a Hybrid Renewable Energy System 2 Case Studies on Small-Scale Hybrid Generation Systems*. Academic Press.

Wuebbles, D. J., & Sanyal, S. (2015). Air quality in a cleaner energy world. *Current Pollution Reports, 1*(2), 117–129.

Wu, J., Zhao, L., Xie, N., Gao, L., Gao, W., Dai, X., & Zhang, J. (2012). Historical weather data supported hybrid renewable energy forecasting using artificial neural network (ANN). *Energy Procedia, 14*, 1035–1040. doi:10.1016/j.egypro.2011.12.1051

Xavier, F. J., Pradeep, A., Premkumar, M., & Kumar, C. (2021). Orthogonal learning-based Gray Wolf Optimizer for identifying the uncertain parameters of various photovoltaic models. *Optik (Stuttgart), 247*, 167973. doi:10.1016/j.ijleo.2021.167973

Xing, X., Meng, H., Xie, L., Li, P., Toledo, S., Zhang, Y., & Guerrero, J. M. (2017). Multi-time-scales energy management for grid-on multi-layer microgrids cluster. *Proceedings of the IEEE Southern Power Electronics Conference (SPEC).*

Xu, C.-G., & Li, X.-S. (2015). Research progress on methane production from natural gas hydrates. *RSC Advances, 5*(67), 54672–54699.

Xu, D., Kang, L., Chang, L., & Cao, B. (2005). *Optimal sizing of standalone hybrid wind / PV power systems using genetic algorithms*. Academic Press.

Xu, X., Hu, W., Cao, D., Huang, Q., Chen, C., & Chen, Z. (2020). Optimized sizing of a standalone PV-wind-hydropower station with pumped-storage installation hybrid energy system. *Renewable Energy, 147*, 1418–1431.

Yang, X. (2010). *Firefly algorithm, stochastic test functions and design optimisation.* Academic Press.

Yang, X., Deb, S., & Behaviour, A. C. B. (2009). *Cuckoo Search via Levy Flights*. Academic Press.

Yang, H., Lu, L., & Burnett, J. (2003). Weather Data and Probability Analysis of Hybrid Photovoltaic-Wind Power Generation Systems in Hong Kong. *Renewable Energy, 28*(11), 1813–1824.

Yang, H., Lu, L., & Zhou, W. (2007). A Novel Optimization Sizing Model for Hybrid Solar-Wind Power Generation System. *Solar Energy, 81*(1), 76–84.

Yang, H., Wei, Z., & Chengzhi, L. (2009). Optimal design and techno-economic analysis of a hybrid solar-wind power generation system. *Applied Energy*, *86*(2), 163–169.

Yang, H., Zhou, W., Lu, L., & Fang, Z. (2008). Optimal sizing method for stand-alone hybrid solar-wind system with LPSP technology by using genetic algorithm. *Solar Energy*, *82*(4), 354–367. https://doi.org/10.1016/j.solener.2007.08.005

Yang, N., Paire, D., Gao, F., & Miraoui, A. (2013). Power management strategies for microgrid—A short review. *Proceedings of the 2013 IEEE Industry Applications Society Annual Meeting*.

Yang, X., Yang, X., Li, Y., Zhang, J., & Kang, Y. (2021). Obstacle avoidance in the improved social force model based on ant colony optimization during pedestrian evacuation. *Physica A*, *583*, 126256. doi:10.1016/j.physa.2021.126256

Yang, Y., Guo, S., Liu, D., Li, R., & Chu, Y. (2018). Operation optimization strategy for wind-concentrated solar power hybrid power generation system. *Energy Conversion and Management*, *160*(January), 243–250. https://doi.org/10.1016/j.enconman.2018.01.040

Yao, F., Dong, Z. Y., Meng, K., Xu, Z., Iu, H. H., & Wong, K. P. (2012). Quantum-inspired particle swarm optimization for power system operations considering wind power uncertainty and carbon tax in Australia 2012. *IEEE Transactions on Industrial Informatics*, *8*(4), 880–888. doi:10.1109/TII.2012.2210431

Yeung, D. S., & Tsang, E. C. C. (1997). A comparative study on Similarity based fuzzy reasoning method. *IEEE Transactions on Systems, Man, and Cybernetics. Part B, Cybernetics*, *27*(2), 216–227. doi:10.1109/3477.558802 PMID:18255859

Younes, M., Khodja, F., & Kherfane, R. L. (2014). Multi-objective economic emission dispatch solution using hybrid FFA (firefly algorithm) and considering wind power penetration. *Energy*, *67*, 1595–11606. doi:10.1016/j.energy.2013.12.043

Yu, B., Yuan, X., & Wang, J. (2007). Short-term hydro–thermal scheduling using particle swarm optimization method. *Energy Conversion and Management*, *48*(7), 1902–1908. doi:10.1016/j.enconman.2007.01.034

Yu, J. J. Q., & Li, V. O. K. (2015). A social spider algorithm for global optimization. *Applied Soft Computing*, *30*, 614–627. doi:10.1016/j.asoc.2015.02.014

Yunus, A., & Çengel, M. A. B. (2015). *Thermodynamic an Engineering Approach* (8th ed.). McGraw-Hill Education.

Yustra, M. A., & Soeprijanto, A. (2012). Optimal distributed generation (DG) allocation for losses reduction using improved particle swarm optimization (IPSO) method. *Journal of Basic and Applied Scientific Research*, *2*(7), 7016–7023.

Zadeh, L. A. (1996). Fuzzy sets. In *Fuzzy sets, fuzzy logic, and fuzzy systems: selected papers by Lotfi A Zadeh* (pp. 394–432). World Scientific.

Zahedi, G., Karami, Z., & Yaghoobi, H. (2009). Prediction of hydrate formation temperature by both statistical models and artificial neural network approaches. *Energy Conversion and Management*, *50*(8), 2052–2059.

Zangeneh, A., Jadid, S., & Rahimi-kian, A. (2009). Promotion strategy of clean technologies in distributed generation expansion planning. *Renewable Energy*, *34*(12), 2765–2773. doi:10.1016/j.renene.2009.06.018

Zayed, M. E., Zhao, J., Li, W., Elsheikh, A. H., Zhao, Z., Khalil, A., & Li, H. (2020). Performance prediction and techno-economic analysis of solar dish/stirling system for electricity generation. *Applied Thermal Engineering*, *164*(April), 114427. https://doi.org/10.1016/j.applthermaleng.2019.114427

Zerpa, L. E., Aman, Z. M., Joshi, S., Rao, I., Sloan, E. D., Koh, C., & Sum, A. (2012). Predicting hydrate blockages in oil, gas and water-dominated systems. *Offshore Technology Conference*.

Zhang, D. Y., & Winfree, E. (2009). Control of DNA strand displacement kinetics using toehold exchange. *Journal of the American Chemical Society*, *131*(47), 17303–17314. doi:10.1021/ja906987s PMID:19894722

Zhao, B. (2006). *A Survey on Application of Swarm Intelligence Computation to Electric Power System*. Academic Press.

Zhou, Y., Zhang, Z., Lin, T. R., & Ma, L. (2013). Maintenance optimisation of a multi-state series-parallel system considering economic dependence and state-dependent inspection intervals. *Reliability Engineering & System Safety*, *111*, 248–259. doi:10.1016/j.ress.2012.10.006

Zhu, Z. (2008). *Computer vision research progress*. Nova Publishers. doi:10.5772/72

About the Contributors

Mohamed Arezki Mellal has a PhD in Mechatronics. He is an Associate Professor (with Accreditation to Supervise Research) at the Department of Mechanical Engineering, Faculty of Engineering Sciences, M'Hamed Bougara University, Algeria and was a Visiting Scholar at CALCE, University of Maryland, College Park, MD, USA. Likewise, he a visiting scholar at various universities. He has published in several journals such as: Reliability Engineering & System Safety, Energy, International Journal of Advanced Manufacturing Technology, ISA Transactions, Journal of Intelligent Manufacturing, Energy Conversion and Management, International Journal of Fuzzy Systems, and conferences. He was also a committee member of over sixty international conferences. He serves as a regular reviewer for several SCI journals. His research interests include mechatronics and soft computing methods for engineering problems.

* * *

Almoataz Y. Abdelaziz (Senior Member, IEEE) received the B.Sc. and M.Sc. degrees in electrical engineering from Ain Shams University, Cairo, Egypt, in 1985 and 1990, respectively, and the Ph.D. degree in electrical engineering according to the channel system between Ain Shams University, Egypt, and Brunel University, U.K., in 1996. He has been a Professor of electrical power engineering with Ain Shams University, since 2007. He was the Vice Dean for Education and Students Affairs in Faculty of Engineering and Technology, Future University in Egypt from 2018-2019. He has authored or coauthored more than 450 refereed journal and conference papers, 35 book chapters, and six edited books with Elsevier, Springer and CRC Press. In addition, he has supervised more than 80 Master's and 35 Ph.D. theses. His research areas include the applications of artificial intelligence, evolutionary and heuristic optimization techniques to power system planning, operation, and control. Dr. Abdelaziz is a senior member in IEEE and member in the Egyptian Sub-Committees of IEC and CIGRE'. He has been awarded many prizes for distinct researches and for international publishing from Ain Shams University and Future University in Egypt. He is the chairman of the IEEE Education Society chapter in Egypt. He is an editor of Electric Power Components and Systems journal, an Editorial Board member, an Editor, an Associate Editor, and an Editorial Advisory Board member for many international journals.

Tahir Cetin Akinci received the B.S. degree in electrical engineering, and the M.Sc. and Ph.D. degrees from Marmara University, Istanbul, Turkey. He is a professor with the Electrical Engineering Department, Istanbul Technical University, Istanbul, Turkey and University of California Riverside

(UCR), WCGEC, Riverside, CA, USA. His research interests include artificial neural networks, deep learning, machine learning, image processing, signal processing, and data analysis.

Amit Arora is currently working as Dean student welfare and Associate Professor of Chemical Engineering at the Shaheed Bhagat Singh State University, Ferozepur, Punjab, India. He has obtained his Ph.D. in Chemical Engineering from the Indian Institute of Technology, Roorkee, India. His area of expertise are gas hydrates, waste to energy, biomass utilization etc.. More precisely he is looking into thermodynamics and kinetic studies of gas hydrates as well as their modelling studies. He has got patent in the field of gas hydrates. He has published more than 50 research papers in various reputed international journals and most of them are related to gas hydrates . He has been awarded many awards for his scientific work from various scientific bodies. He has given many plenary talks in the field of gas hydrates at various international conferences of repute.

Ramazan Caglar received the B.Sc, MSc and PhD diploma in Electrical Engineering from the Istanbul Technical University. His research interests are power system modelling, reliability of electric power system, the effects of deregulation of the electric power system. He is presently an Assoc. Professor. He is active in teaching and research in the control of power systems.

Khaled Dassa is a PhD student at the institute of Electrical and Electronic Engineering, university of Boumerdes, Algeria. He is working on optimization applied to renewable energy system design.

Ragab A. El-Sehiemy, Senior Member IEEE, was born in Menoufia, Egypt, in 1973. He received the B.Sc., Msc and Ph.D. degrees from Menoufia University, Egypt, in 1996, 2005, and 2008, respectively. He is currently Professor with the Electrical Engineering Department, Faculty of Engineering, Kafrelsheikh University, Egypt. He received Prof. Mahmoud Khalifia Award in Power System Engineering from Academy of research and technology in 2016. He is the receipt of El-Shorouk Academy for computers and information in industry from Academy of research and technology in 2021. Prof. El-Sehiemy is one of the top 2% of the effective scientists in 2020 according Elsevier data base. He is an Editor of International Journal of Engineering Research in Africa and an Associate Editor in IEEE Access. Professor El-Sehiemy has more than 150 articles in peer review journals and conferences . His research interests include smart grid and its applications, optimization and AI, and its application to power systems. Renewable energies, power systems protection.

Abdurazaq Elbaz was born in Libya. He received the B.Sc. in electrical engineering from Gharyan University, Gharyan, Libya, in 2009, and the M.Sc. degree from the Department of Electrical and Electronic Engineering, University of Tripoli, Tripoli, Libya, in 2013. He was working in Centre for Solar Energy Research and Studies in Tripoli, Libya. He is currently PhD candidate in the Department of Electrical and electronic Engineering, Karabuk University, Karabuk, Turkey. His research interests are Electrical power system, PV systems. Optimization and Control systems, renewable energy, FACTS devices.

Gokhan Erdemir received the B.Sc., M.Sc., and Ph.D. degrees from Marmara University, Turkey. He was a research scholar with the Department of Electrical and Computer, Michigan State University, East Lansing, MI, USA. He joined the Health Management and Research Center, University of Michigan, Ann Arbor, as a researcher. He is currently an Assistant Professor with the Department of Electrical and Electronics Engineering, Istanbul Sabahattin Zaim University. His research interests include control systems, robotics, and mobile robotics.

Muhammet Guneser was born in Turkey in 1975. He graduated Istanbul Technical University Electronics and Telecommunication Engineering Department in 1999 and PhD. Degree from Karabuk University Electrical and Electronics Engineering Department in 2015. His research interests are Communication on sub-THz Frequencies, Optimization and Control systems, Solar Energy, Energy Management.

Sunanda Hazra was born in 1991 at Indas, Bankura, West Bengal, India. He received a B.E Degree in Electrical Engineering from University Institute of Technology (The University of Burdwan), Burdwan, West Bengal, India in 2012; a M.Tech Degree in Electrical Engineering from West Bengal University of Technology, West Bengal, India in 2014 and a Ph.D from MAKAUT in 2021. Presently, he is working as a Faculty in the department of Electrical & Electronics Engineering, Central Institute of Petrochemicals Engineering & Technology, Haldia, West Bengal, India. His field of research interest includes Economic emission load dispatch, Renewable Energy incorporating multi-objective load dispatch, Solar-Wind-Hydro-Thermal scheduling and evolutionary computing techniques.

Alok Kumar is an alumnus of Shiv Nadar University, Mechanical Engineering, the batch of 2018. Presently working for an ed-tech startup based in Gurugram, Haryana. Aimed to create a significant impact in the field of renewable energy.

Anupama Kumari is a PhD Scholar at Indian Institute of Technology, Roorkee and currently working in the field of Gas Hydrate. Currently, she is working on the project entitled "Development of Thermo Catalytic Technique for Gas Production from Hydrate Bearing Sediments" since December 2017. This project is funded by the "Gas Hydrate Research and Technology Center" and "Oil and Natural Gas Corporation Limited". In this project, her research work is to perform the experiments on the formation and dissociation of gas hydrates in porous sediments. Her PhD thesis topic is "Studies on the Formation and Dissociation of Gas Hydrates in Porous Sediments''. She published two research papers in reputed journals as first author and another two research papers as second author. She has attended five international conferences during her Ph.D. She also worked on the simulation and modeling of gas hydrates formation and dissociation. She performed the thermodynamic modeling of gas hydrate formation to predict the phase equilibrium conditions of gas hydrates. For the thermodynamic modeling; She used the Fugacity model, Artificial Neural Network and Least Square Vector Machine through MATLAB. She also performed the characterization experiments of the gas hydrate bearing sediments of the 'National Gas Hydrate Program' received from GHRTC. For the characterization, she observed the TDS, pH, water content, BET, FESEM, ICPMS, FTIR, XRD and Raman Spectroscopy etc. The area of research during her M.Tech was waste water treatment in which she worked on the phytoremediation of phenol and cyanide through Water Hyacinth and Maize plants.

Praveen Laws has a Doctorate in Computational fluid dynamics. Specialist in performing research in the field of Aerodynamics, Aero-elasticity, VIV & Galloping phenomena, Rotor dynamics, Turbomachinery, Windfarm simulation, Extraction of wind power from moving bluff body and Drone dynamics. Expertise in OpenFOAM, ANSYS Fluent, Matlab, and C++ programming simulations.

Lin Ma is Professor of Fluid Dynamics at the Energy 2050 in the Department of Mechanical Engineering, the University of Sheffield. He has been working for many years on sustainable energy technology and in particular on the modelling of various energy processes and a wide range of industrial fluid flow, heat and mass transfer problems. His recent research areas include carbon capture from power generation and industrial processes, clean coal/biomass/gas combustion, fuel related ash deposition, slagging and fouling, future energy system multi-scale and dynamic simulation, and wind turbine aerodynamics.

Mukund Madhaw is the leading team member of his startup company DHC Trading India Private Limited, Delhi, India, which deals with Ayurvedic medicines and cosmetic items. In this working field, he integrates the essence of India's traditional Ayurveda with western modern management concept, insisting on "taking in everything, growing innovatively and having one's own style". His innovative management philosophies and exploration has provided the global management circle with business. Along with his startup company, he is also contributing his research work in the field of gas hydrates. He studied the equilibrium conditions of gas hydrates in pure water, sea water and in the presence or absence of different inhibitors and promoters. He is currently working on the development of different thermodynamic model for the prediction of phase equilibrium data of gas hydrates.

Chandrajit Majumder is a professor in the department of Chemical Engineering, Indian Institute of Technology, Roorkee, Uttarakhand, India. His area of expertise are gas hydrates, waste water treatment, biofuels, soil remediation etc. He has published for than 300 research papers in various reputed international journals. He has been awarded many awards for his scientific work. He has given many planetary talks at various international conferences.

Santanu Mitra is currently holding an Associate Professor position in the Department of Mechanical Engineering at Shiv Nadar University, and he served as Department Head for six years from 2015 to 2020. He is one of the researchers responsible for the development of strategic energy harvesting techniques and bio-inspired design areas in the country. Dr. Mitra has obtained his Ph.D. from Aerospace Engineering Department, Indian Institute of Technology, Kharagpur and has been actively engaged in teaching and research in bioinspiration & biomimetics, and computational fluid dynamics since 2000.

Mandrita Mondal is Research Associate at Indian Statistical Institute, Kolkata, India. She received Ph.D. in 2017 from Jadavpur University, India. Her research interest includes DNA computing, DNA strand algebra, DNA coding theory and DNA cryptography.

Dennis O'Neal has been the Dean of Engineering and Computer Science at Baylor University since 2012. Prior to joining Baylor University, he served as the Associate Dean for Research for the College of Engineering and as the Head of the Department of Mechanical Engineering at Texas A&M University. He has worked at both Oak Ridge National Laboratories and Sandia National Laboratories. He primary research areas are in refrigeration, heat pumps, and air conditioning systems. He has over 190 refereed publications and two book chapters. His papers have won several awards as best technical papers in ASHRAE. Dr. O'Neal is a fellow of both ASME and ASHRAE.

Punit Prakash is working on CFD and analysis of wind turbine using Ansys and openFoam.

Abdelmadjid Recioui is a Professor at the Institute of Electrical Engineering and Electronics University of Boumerdes, Algeria. He obtained a PhD degree in electrical and electronic engineering option telecommunications from the Institute of Electrical Engineering and Electronics, University of Boumerdes in 2011. He holds also Master (Magister) Electronic System Engineering degree which has been achieved at the Institute of Electrical Engineering and Electronics, University of Boumerdes in 2006. In June 2002, he finished his engineering studies at the institute of Electrical Engineering and Electronics, University of Boumerdes. He is a research Assistant at the laboratory signals and systems from January 2008 to Present in Laboratory: signals and systems, Inst. of Electrical Engineering and electronics, University of Boumerdes. His research interests include Antennas, Wireless Communication Systems, antenna array synthesis and design, capacity enhancement, system optimization, smart antennas, power system protection, power system optimization, power system communications.

Provas Kumar Roy was born in 1973 at Mejia, Bankura, West Bengal, India. He received a B.E Degree in Electrical Engineering from R. E. College, Durgapur, Burdwan, India in 1997; a M.E. Degree in Electrical Machines from Jadavpur University, Kolkata, India in 2001 and a PhD from NIT Durgapur in 2011. Presently, he is working as a Professor in the department of Electrical Engineering, Kalyani Government Engineering College, Kalyani, West Bengal, India. His field of research interest includes economic load dispatch, optimal power flow, FACTS, unit commitment, automatic generation control, power system stabilizer, radial distribution system, state estimation etc. Total citation count of his papers is 3520 (according to Scopus) and his name is included in the list of top 2% of scientists for their research works.

Cihat Seker was born in Turkey, in 1989. He received the B.S. in Electronics and Communication Engineering from the Suleyman Demirel University, Isparta, Turkey, in 2012 and the M.S. degree in Electrical-Electronics Engineering from Karabuk University, Karabuk, Turkey, in 2016. He is currently pursuing the Ph.D. degree in Electrical-Electronics Engineering at Karabuk University, Karabuk, Turkey. His research interest includes wireless communication technologies and design and fabrication of compact microstrip antenna.

Serhat Seker graduated from the Electrical Engineering Department, Istanbul Technical University (ITU). He received the master's degree from the Nuclear Energy Institute, ITU, and the Ph.D. degree from the Electrical Engineering Division, Science and Technology Institute, ITU. He studied the Ph.D. thesis with the Energy Research Centre of the Netherlands (ECN) and worked on signal analysis techniques. He was an Assistant Professor and an Associate Professor with ITU, in 1995 and 1996, respectively. He worked in industrial signal processing with the Maintenance and Reliability Centre, Nuclear Engineering Department, The University of Tennessee, Knoxville, TN, USA, in 1997. He was the Vice Dean with the Electrical and Electronic Engineering Faculty, from 2001 to 2004, and the Department Head of the Electrical Engineering, from 2004 to 2007. He was also the Dean of the Faculty of Electrical and Electronics, from 2013 to 2020.

Seongwoo Woo has a BS and MS in Mechanical Engineering, and he has obtained PhD in Mechanical Engineering from Texas A&M. He majored in energy system such as HVAC and its heat transfer, optimal design and control of refrigerator, reliability design of mechanical components, and failure Analysis of thermal components in marketplace using the non-destructive such as SEM & XRAY. Especially, he developed parametric accelerated life testing (ALT) as new reliability methodology. If there is design fault in the mechanical system that is subjected to repetitive stress, it will fail in its lifetime. Engineer should find the design faults by parametric ALT before product launches. In 1992–1997 he worked in Agency for Defense Development, Chinhae, South Korea, where he has researcher in charge of Development of Naval weapon System. In 2000-2010 he had been working as a Senior Reliability Engineer in Side-by-Side Refrigerator Division, Digital Appliance, SAMSUNG Electronics, where he focused on enhancing the life of refrigerator as using parametric the accelerating life testing. He has published more than 120 research articles, 10 book chapters and edited three books in the field of mechanical engineering. He has published more than 25 papers in reputed journals and has been serving as an editorial board member of Journal of Modern Industry and Manufacturing (JMIM, ISSN 2788-8096).

Aydin Tarik Zengin received the B.S. degree in electrical and electronics engineering from Ege University, Turkey, in 2007, and the M.E. and Ph.D. degrees from the Department of Computer Science and Electrical Engineering, Kumamoto University, Japan, in 2010 and 2013, respectively. He is currently an Assistant Professor at Istanbul Sabahattin Zaim University. His research interests include autonomous systems and control theory.

Index

A

aerodynamic performance 211, 227
ant colony optimization 1-2, 37, 52, 76-77, 143
Artificial neural network (ANN) 1, 4, 74, 102, 168

B

bio-inspiration 213, 226

C

chemical reaction 21-23, 26-27, 33
Concentrated solar power 168, 179
Concentration Ratio 204
Crow Algorithm 154, 156, 158
cuckoo optimization algorithm 1, 3

D

Darrieus Vertical Axis Wind Turbine 211, 225
design defects 241-243, 247, 251, 257-259
dissociation 95-97, 99-101, 105, 108, 132
DNA computing 116-117, 120-121, 123, 126-127, 130, 133-134
DNA logic gates 117, 128
DNA strand displacement 128

E

energy efficiency 124, 133, 169, 199, 242, 246
Energy Management 60, 63, 73, 76-83, 146, 181
Evolutionary Optimization 94

F

Future of ACO 37, 51

G

gas hydrate 95-97, 99-100, 105-109
Genetic Algorithm 74-75, 77-78, 80, 102, 138, 140-142, 166, 168, 204
greenhouse gases 22, 34, 65, 137, 152, 274
grey wolf optimizer 1-2
Grid-connected 72, 76-77, 80-82, 137-138, 153

H

hairpin motif 131-132
HOMER 75, 78, 81, 140, 146
Hybrid Algorithms 138, 145
Hydro-Thermal-Wind (HTW) Scheduling 21

I

inhibitor 95, 97, 99-100

L

leading-edge tubercles 211
Leonard Adleman 117, 121
Levelised Cost of Electricity 166, 168, 184, 204

M

mechanical product 243-245, 247, 259-260
meta-heuristics algorithms 22, 33
Microgrids 60, 63-64, 79, 82-83, 94, 138, 167-168
Multi-objective optimisation 166, 169, 184

N

NACA0018 216, 218, 225-228, 233, 236
NACA2412 225, 228

nature-inspired approach 10

P

parametric ALT 242-247, 250-252, 254-257, 259-260, 265
Pareto front 199
Particle Swarm Algorithm 81
particle swarm optimization 1-2, 76-77, 79, 81, 138, 141-142
Performance analysis 166, 181, 185, 187, 189
plant propagation algorithm 1-2
Power Coefficient 191, 204
Progress of ACO algorithm 37
PV 63-64, 66, 68, 71-80, 82-83, 136, 138, 146-148, 153, 155-156, 158, 168-169, 179, 191

R

Renewable Energy 1, 11-12, 21-23, 29-30, 33, 60-61, 63, 65-74, 76-79, 81-83, 94, 130, 136-139, 146-148, 152-153, 167, 267
Renewable Energy Sources 11-12, 22, 33, 60, 79, 81-83, 94, 137-139, 147-148
Richard J. Lipton 121, 123
Rim Angle 204

S

seasonal wind speed 13-14
Smart Grid 60-64, 66-67, 73, 76, 94
Solar Dish 166, 169, 181, 193, 204
Solar dish Stirling engine 166, 169
spectral analysis 10, 13
Standalone 63, 68, 72, 76-77, 79, 83, 145-146, 158
statistical analysis 10, 13, 33

T

toehold 127-128, 131-132

V

V/Wind Systems 74, 169
Vertical Axis Wind Turbine 211, 225-226

W

Wind speed 10, 13-18, 22, 24-25, 72, 75, 79, 137, 147, 149, 158, 173-174, 176, 184, 189, 191, 196-197, 212, 227
wingtip devices 212

www.ingramcontent.com/pod-product-compliance
Lightning Source LLC
LaVergne TN
LVHW062141300926
849161LV00010B/2594